200 306469 6

£ 50

AF835324

Library use only
I'w ddefnyddio yn y llyfrgell
yn unig

http://library.cardiff.ac.uk
http://llyfrgell.caerdydd.ac.uk

The Pleistocene history of the Middle Thames Valley

PHILIP L. GIBBARD

The Pleistocene history of the Middle Thames Valley

The author is Assistant Director of Research. Subdepartment of Quaternary Research
Botany School, University of Cambridge

The right of the University of Cambridge to print and sell all manner of books was granted by Henry VIII in 1534. The University has printed and published continuously since 1584.

CAMBRIDGE UNIVERSITY PRESS
Cambridge
London New York New Rochelle
Melbourne Sydney

UNIVERSITY COLLEGE
LIBRARY
CARDIFF

Published by the Press Syndicate of the University of Cambridge
The Pitt Building, Trumpington Street, Cambridge CB2 1RP
32 East 57th Street, New York, NY 10022, USA
10 Stamford Road, Oakleigh, Melbourne 3166, Australia

© Cambridge University Press 1985

First published 1985

Printed in Great Britain at
the University Press, Cambridge

Library of Congress catalogue card number: 84–21358

British Library cataloguing in publication data

Gibbard, Philip L.
The Pleistocene history of The Middle Thames Valley
1. Geology–Stratigraphic–Pleistocene
2. Geology–England–Thames, River, Valley
I. Title
551.7'92'09422 QE697

ISBN 0 521 26578 9

23/3

PN

CONTENTS

The sloping number after the title is the page reference

PREFACE

The Thames Valley is rightly regarded as a classic area in British Pleistocene geology, since it includes the largest single spread of terrace deposits in the country. Apart from geological mapping, the only previous large-scale studies of the Middle Thames area were by F. K. Hare and K. R. and C. E. Sealy, both of which were purely morphological mapping projects. There have, of course, also been a number of individual site studies or investigations of small parts of the sequence, but these have led to much misunderstanding and fragmentation of the evidence.

Following my recent studies, Professor R. G. West suggested that it would be desirable, therefore, to present all the evidence in a single volume where it could be properly related and assessed. I have tried to write this book with this aim in mind and it is based on my investigations into the stratigraphy and palaeoenvironments of the terrace deposits of the Thames and certain of its tributary valleys. In contrast to earlier studies, the work reported here depends primarily on investigations of the sedimentary sequences and not on terrace morphology. The latter may then be assessed in the light of these studies. The resulting synthesis provides the first truly geological study of the terrace sequence.

The approach I have adopted throughout the book has been thoroughly interdisciplinary using evidence from all possible sources; these included archaeology, palaeontology and palaeobotany, as well as the lithology and structure of the deposits themselves. This integrated approach allows the fullest interpretation of the Pleistocene sequence, permitting the reconstruction of the river's history from the early Middle Pleistocene up to recent times.

Since the history of the River Thames is perhaps the most widely quoted example of lowland river evolution in Britain, I hope that this detailed study of a crucial area will provide both a practical example of what can be achieved in modern Pleistocene studies and a guide to the concepts of terrace investigations in general. I also hope that it will act as a stimulus to others interested in related topics.

This work represents the culmination of over six years study of the Middle Thames region and indeed of my earlier thesis work in the Vale of St Albans. It was funded by the Natural Environment Research Council, the Leverhulme Trust and the L.S.B. Leakey Foundation to whom I am extremely grateful. There can be little doubt that the work would never have appeared in this form if it were not for the encouragement and assistance of many colleagues. In particular, I would like to thank Miss R. Andrew, Dr K. D. Bennett, Dr D. R. Bridgland, Dr I. D. Bryant, Dr J. A. Catt, Dr D. Collins, Dr G. R. Coope, Ms S. Cranshaw, Mr A. Currant, Dr P. Echlin, Dr A. R. Hall, Mr P. Harding, Dr R. W. Hey, Dr D. T. Holyoak, Dr R. R. Inskeep, Dr M. P. Kerney, Mr J. M. Line, Dr R. C. Preece, Dr. J. E. Robinson, Dr D. Roe, Mr J. Rose, Mr J. Scourse, Mr G. de G. Sieveking, Dr A. J. Stuart, Dr A. J. Sutcliffe, Dr C. Turner, Dr A. G. Wintle and Mr J. J. Wymer. To all these people and others who have contributed their help and support I would like to express my thanks.

I am indebted to all the landowners and contractors who have given me access to their land. The work would not have been possible without the excellent borehole data which was made freely available by the Institute of Geological Sciences, the Eastern and Southeastern Road Construction Units of the Department of Transport, British Gas, the Greater London Council, Surrey County Council, Slough Borough Council, Beaconsfield District Council and the Thames Water Authority.

The manuscript was typed by Ms R. Evans and Ms R. Hockaday, and some of the diagrams were drawn by Ms S. Peglar (Quaternary Research) and Ms L. Rutt and Mr G. Reeve (Birkbeck College, Geography Department drawing office).

As always, I would especially like to thank Professor R. G. West FRS for his constant support, encouragement and inspiration throughout the work.

Finally, I should like to thank Ms Ann Jennison for her tolerance and understanding, and for helping me see it through.

Philip Gibbard
Cambridge, March 1984

Fig. 1. Location map showing the area studied, the localities of important exposures and borehole sections mentioned in the text.

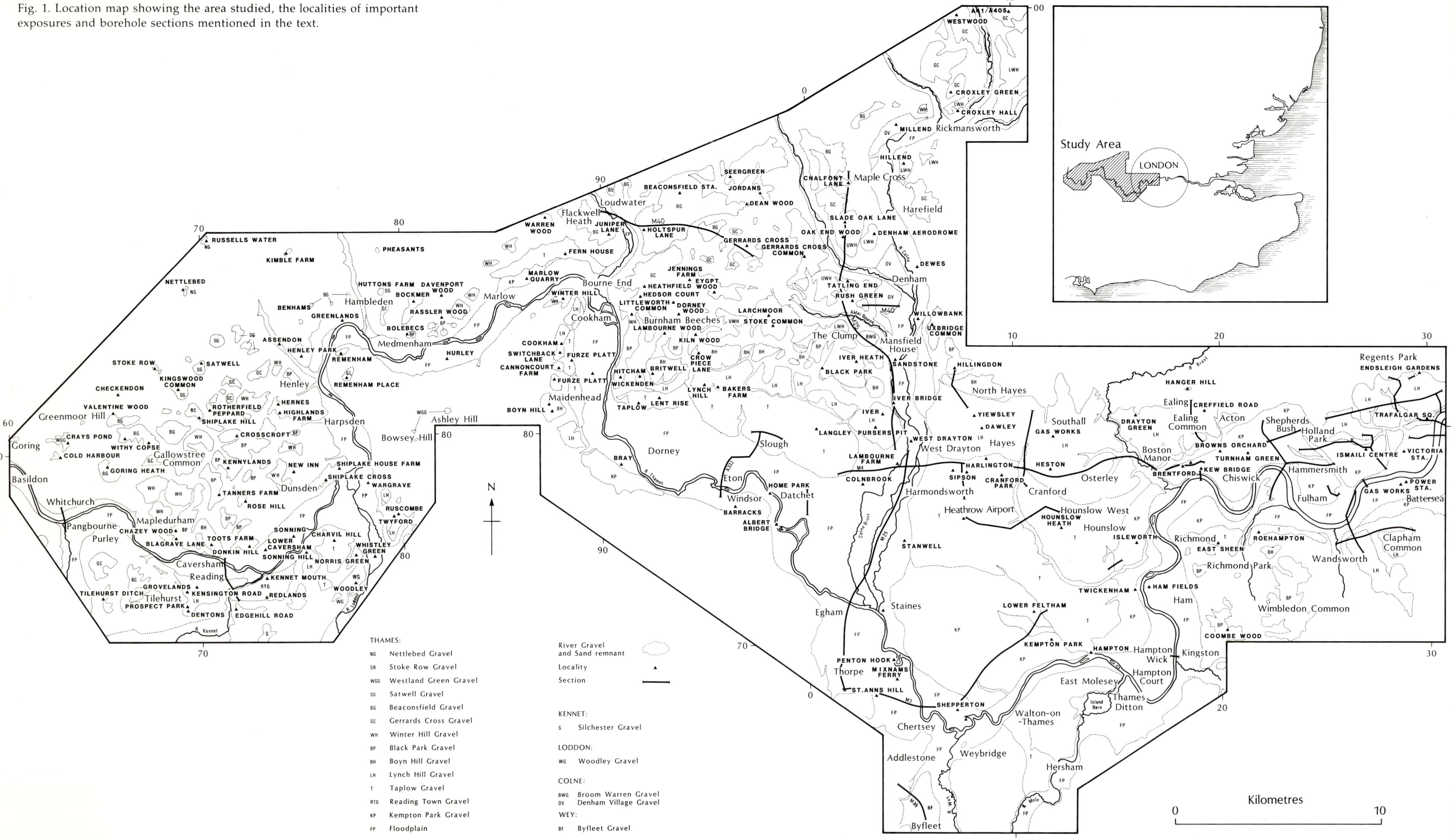

1

Introduction

For over a century the Thames Valley has been recognised as being one of the most important regions for British Pleistocene geology. The valley, together with its extensions into the south Midlands, central Hertfordshire and southern Essex, contains the longest and most complex series of Pleistocene sediments outside East Anglia. In addition, and of critical importance, is its position marginal to an area which was inundated by ice during at least one of the major glacial events of the British Pleistocene. This relationship permits Thames deposits to be linked to glacial sequences of at least two of the Pleistocene glacial stages. The deposits preserved in the London Basin provide a record from the marine transgressions of the Pliocene, through initiation and development of the Thames drainage system, to the diversion of the river during the Middle Pleistocene and its subsequent evolution to the present day.

Within the Thames Valley the central Middle Thames region has provided the classic type area for the Thames terrace sequence. The region, for the purposes of this work, is defined as the area that extends from the Goring Gap in the west – a gorge-like valley, north-west of Reading, where the river cuts through the Chiltern Hills, to Rickmansworth in the north-east and Central London in the east (Figure 1). This area displays a well developed sequence of terrace surfaces and underlying deposits associated with the present and former courses of the river and its tributaries. Since the initial recognition of this terrace sequence, a long series of workers have attempted, using various criteria, to separate and identify the different components of the succession. Apart from the mapping of the terraces by the Geological Survey, the principal important studies have been carried out by archaeologists, for the lower terrace deposits have yielded the largest concentrations of stone artefacts from any area in Britain, and by geomorphologists. Notwithstanding these studies, the purely geological aspects of the Pleistocene deposits have been largely neglected.

This study represents the first comprehensive stratigraphic treatment of the Middle Thames. On the basis of structure, lithology and altitude, it has been possible to separate and define deposits of different stages, thereby providing the essential foundation for identifying and correlating units within the region, and also extending into the Upper and Lower Thames and into the Vale of St Albans.

A descriptive section (Chapter 2) covers the definition of each individual unit of the Middle Thames succession, and is followed by detailed discussions of important themes, including pebble lithology (Chapter 3), sedimentary structures (Chapter 4), the effects of Chalk bedrock solution (Chapter 5), vertebrate fauna (Chapter 6), palaeobotany (Chapter 7) and Palaeolithic archaeology (Chapter 8). These are then synthesised in a palaeogeographic summary (Chapter 9), and the overall stratigraphy correlated with neighbouring regions both up and downstream (Chapter 10).

It is hoped that this study may stimulate other workers to apply similar methods to the Thames itself, as well as to other valley systems, and that the application of formal stratigraphical principles exemplified by the study will encourage further such approaches.

1.1 Topography and geological background

The Middle Thames region occupies the north-western part of the London Basin syncline and is underlain by Chalk and Tertiary rocks (Figure 2). The Chalk outcrops to the north as the Chiltern Hills reaching altitudes of over 250 m OD and forming the basin's northern margin. The Chilterns are covered by varying thicknesses of Clay-with-flints, a polygenetic deposit, which, however, largely consists of weathering products of pre-existing overlying Tertiary rocks (Loveday, 1962). The most extensive Tertiary formations in the area comprise the Reading Beds and London Clay, but in the southern and eastern part the predominantly arenaceous Claygate Beds and sands of the Bagshot Series occur beneath higher ground.

The present valleys of the Thames and its tributaries are deeply incised into the bedrock. Although the river trends generally

Fig. 2. Geological map showing the distribution of Pleistocene sediments in the study area. The map is modified from Geological Survey one-inch sheets 222, 254, 256, 268, 269 and 270.

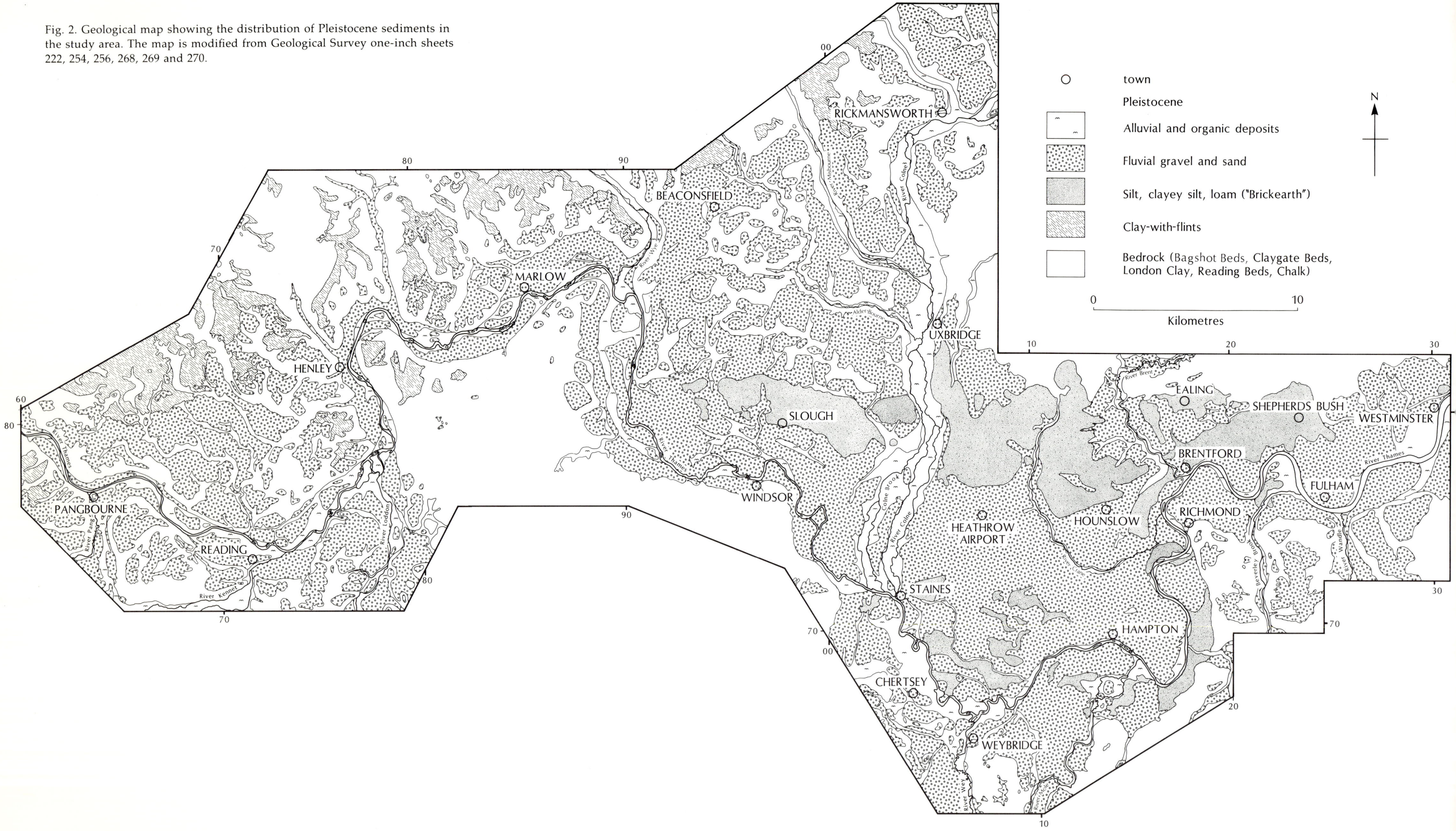

towards the east, its course includes large entrenched meanders which add considerably to its length. The forms of the valley vary substantially with bedrock lithology. Where it lies on Chalk the valley is often narrow, with steep river cliffs, while on the Tertiary clays it is wide with gently sloping sides.

The distribution of Pleistocene sediments varies accordingly. In the Reading–Henley area a series of high-level gravels and sands are distributed along the line of an abandoned valley underlain by Chalk (the so-called Caversham Channel), while the lower level deposits are aligned parallel to the modern valley on the Tertiary rocks. From Sonning downstream as far as Bourne End, the 'Henley loop' country, the river occupies a steep-sided valley along which few gravel and sand outliers occur. In contrast, between Bourne End and Central London the valley is a wide, gently sloping feature, the south-facing slope formed by the widest development of terrace deposits in southern Britain. Throughout the area all the valley sides and interfluve surfaces are greatly dissected by dry valleys. These valleys are incised into surficial and solid rocks alike. The dry valleys grade into the floodplains of flowing streams. Tributary valleys, such as the Pang, Kennet, Loddon, Wye, Colne, Mole, Wey, Brent and Wandle, enter the river at points throughout the area and include gravel and sand spreads similar both in character and position to those of the Thames itself.

Throughout the area there is considerable evidence of valley asymmetry expressed by steeper north-facing and gentler south-facing slopes.

1.2 Previous investigations

The history of research into the Thames Valley system is complex. The evolution of ideas and approaches to investigation of the sequence mirrors the development of British Quaternary studies. The history is, however, too lengthy to present in detail here. Moreover, the Middle and Lower Thames sequence has been the subject of several reviews in recent years, particularly by Hey (1976), Clayton (1977), Green and McGregor (1980) and Jones (1981). It is therefore unnecessary to repeat the reviews in depth and instead a short summary of the major contributions is given. The history and usage of individual unit names is discussed separately at the beginning of each of the unit descriptive sections.

During the nineteenth century a number of workers examined the terrace deposits in the Thames Valley. Whitaker (1864, 1889) was the first to systematically separate terrace accumulations at Maidenhead and, together with Prestwich (1856, 1857, 1890), he laid the foundation of ideas on the evolution of the Thames drainage system. At the end of the century White (1896, 1899, 1904) published a series of papers on the high-level gravel deposits in the western part of the Thames Valley, in which he was able to recognise several terrace stages.

Detailed geological mapping of the Thames Valley was published by Blake (1903), Jukes-Brown and White (1908) and Woodward (1909). Mapping in the Chertsey area led Bromehead (1912) to propose the names Boyn Hill, Taplow and Floodplain to replace the terms 'highest', 'middle' and 'lowest' for the three main terraces then recognised. These names were subsequently used on Geological Survey maps throughout the area by Dewey and Bromehead (1915, 1921), Sherlock and Noble (1922) and Bromehead (1925).

As early as 1905, Salter recognised that the high-level quartz-rich gravels were Thames deposits aligned parallel to the present course as far east as Beaconsfield, but beyond they diverged northwards to Watford. This was also recorded by Sherlock and Noble (1922). In a later paper, Sherlock (1924) recognised the continuation of the high-level gravels into the Vale of St Albans and concluded that they represented the pre-glacial course of the Thames. Sherlock and Noble (1922) also interpreted the Clay-with-flints as an early glacial deposit. This was supported by Barrow (1919) and later termed the 'Chiltern Drift' by Wooldridge (1938).

From his studies of the drainage evolution in the London Basin, Wooldridge (1927, 1938, 1960) and Wooldridge and Linton (1955) divided the Pleistocene deposits into three stages:

- I Summit and high-level drifts, i.e. Red Crag and Pebble Gravels.
- II Glacial and plateau gravels, including the 'Eastern Drift' (Chalky Till) and 'Chiltern Drift', the gravels being subdivided into a series of levels.
- III Valley and terrace drifts.

In the Middle Thames region, Wooldridge (1960) proposed that in stage I times a 'proto-Thames' flowed eastwards across the region into Hertfordshire. By stage II it had been diverted southward by Chiltern ice into the so-called 'Middlesex Depression' (Finchley Valley). During this period quartz-rich gravels were laid down in the Middle Thames to form the 'Higher' and 'Lower' gravel trains. Advance of the Chalky Till ice into the Vale of St Albans and Finchley Valley subsequently caused the final diversion of the river to its present course.

Hare's (1947) detailed geomorphological mapping of the eastern half of the Middle Thames Valley supported Wooldridge's stage II sequence. It also led to the recognition of three new stages; the Harefield, Black Park and Lynch Hill Terraces. Hare's work was later extended upstream by Sealy and Sealy (1956) and into tributary valleys by Thomas (1961).

The abundance of Palaeolithic artefacts in the low-level terrace deposits of Wooldridge's stage III has been well known by archaeologists for over a century. A number of local collectors have investigated the deposits but within the Middle Thames four individuals have made particularly important contributions; Treacher, Brown, Lacaille and most recently Wymer (see Wymer, 1968, for details). Oakley (1936, 1937) worked with Lacaille and Treacher on the geological setting of the Palaeolithic finds, and co-authored an influential review of the younger part of the Thames terrace sequence (King and Oakley, 1936).

Over the years there has been much discussion concerning the age of individual terraces and their climatic implications. Terraces have been variously interpreted as representing cold (Lacaille and Oakley, 1936; Lacaille, 1961b) or warm (Zeuner, 1959) stage accumulations, or even a combination of both. These interpretations were based on the presence of vertebrate finds of 'warm' or 'cold' species together with other fossil evidence as well as on the presence of artefacts within the terrace deposits.

Zeuner (1959) developed a scheme in which he recognised two types of terrace in the Thames system these being separated by their mode of origin: thalassostatic and climatic. The former were thought to be those formed by sea-level variation and were therefore most abundant in the lower reaches of rivers. High sea-level stands during interglacial periods were interpreted as giving rise to terrace accumulation within the valley. Subsequent sea-level lowering in the following cold phase caused downcutting in the valley leaving the terrace remnant on the valley side. Climatic terraces were the most significant in the middle reaches of the river. These were thought to have formed during periods of climatic amelioration when large amounts of detritus were available for deposition, the latter being reworked early in the subsequent cold stage. Downcutting by the river, initiated by sea-level fall, left the deposits isolated as before, to form the terrace.

This scheme has, however, been shown by subsequent investigations to be greatly oversimplified (see discussion in Clayton, 1977; Rose, 1979; Jones, 1981) particularly since the sequences of deposits beneath terrace surfaces were found to be much more complex than can be accommodated in this model. It also places emphasis on the effect of sea-level fluctuation far beyond the estuarine area, a factor which is no longer generally accepted.

In 1965, Hey published his evidence for a 'pre-glacial' Thames flowing from Oxfordshire into Essex via the Goring Gap and the Vale of St Albans based on his studies of the Westland Green Gravels (stage I, Pebble Gravels). This was the first attempt to subdivide the Thames gravels using their lithology and its success encouraged a number of modern investigations. Detailed stratigraphical studies by Gibbard (1974, 1977) demonstrated that the Thames continued to flow through the Vale of St Albans until the advance of the Chalky Till (Anglian) ice. This has subsequently been supported by Green and McGregor (1978). Contrary to Wooldridge's reconstructions, the river never occupied the Finchley Valley which was in fact the valley of the Thames' southbank tributary, the Mole–Wey stream (Gibbard, 1979). The Thames' course in this area was therefore apparently unaffected by an earlier glaciation, modern evidence for which has not been found. The Clay-with-flints (Chiltern Drift) is now generally interpreted as a complex deposit formed partly by the weathering of Tertiary bedrock, but associated particularly at lower levels with other materials intermixed by periglacial slope processes (Loveday, 1962; J. A. Catt, personal communication).

1.3 Terrace stratigraphy

The Pleistocene fluvial deposits in the Middle Thames region are predominantly arranged into altitudinally separable aggradations along valley sides. These deposits are thought to result from a progressive series of incisions into the valley floor bedrock followed by aggradation of alluvial sediment. This alternate series of downcutting and aggradational cycles gives rise to the preservation of a sequence of progressively younger deposits down the valley side, the youngest being those beneath the modern river floodplain. The surfaces preserved on these deposits may present a stairway-like sequence which can be geomorphologically mapped. Such surfaces are conventionally termed *terraces*.

A *terrace* is most often defined as a vestige of a former floodplain of a river (Gilbert, 1877; Leopold, Wolman and Miller, 1964;

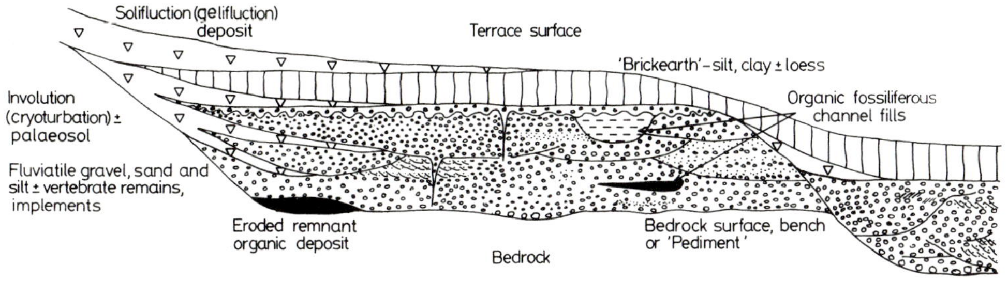

Fig. 3. Transverse section of an idealised terrace in the middle reaches of a river in south-eastern England.

Howard, Fairbridge and Quinn, 1968), although other authors (e.g. Cotton, 1940, p. 27) have suggested that the term should be modified thus 'terraces that border river valleys and mark former levels of flat valley floors such as have resulted from corrasion or valley filling by rivers.' River terraces may or may not occur at more or less constant heights above the present floodplain and usually show a downstream gradient. They are normally separated by slopes or bluffs. The increasing dissection of surfaces and underlying deposits with height up the valley side strongly suggests an accompanying increased antiquity (Howard *et al.*, 1968). Both erosional and depositional terraces are known from river valleys (*sensu* McGee, 1891). Those in the Thames system are of the latter type, the so-called 'cut-and-fill' terraces (Fisk, 1944; Smith, 1949; Quinn, 1957).

Because the subdivision of flights of terraces, particularly in southern Britain, has been morphological, i.e subdivision and downstream correlation has been based exclusively on altitude, this approach has led to the development of morphostratigraphical sequences. Such an approach, however carefully undertaken, is full of pitfalls (for discussion see Johnson, 1944) and largely ignores the underlying deposits.

In the Thames Valley all exposed terrace surfaces have been modified either by later deposition or by periglacial processes. The so-called terrace flat is actually the remains of a much disturbed floodplain surface that originally comprised a complex of channels and bars with amplitudes of as much as 3 m. These features have been flattened by post-depositional solifluction and cryoturbation processes (cf. Jessen and Milthers, 1928). Commonly, valley-side slope wash, loess and soliflucted material may blanket the surface (Johnson, 1944; Howard *et al.*, 1968) giving rise to the commonly observed concave, rather than horizontal, transverse surface profile sloping down into the river valley (Cotton, 1940). Such a form was frequently noted by Hare (1947) in the Middle Thames Valley. In addition, degradation of the surface and its underlying deposits by periglacial slope processes, tributary stream valley erosion, gullying and bedrock solutional subsidence will lead to a progressive deterioration in terrace preservation with time.

In view of these problems of using the terrace surface itself some workers have instead used the so-called terrace *bench* or interface between the fluvial deposits and the underlying bedrock surface for the altitudinal correlation. Such a surface is much less open to later modification and may define the base of the deposits where undisturbed sediments can be demonstrated to overlie it. However, such a surface is very difficult to use in practice and unpredictable for purposes of down-valley correlation. This surface is erosional and as such is seldom horizontal or even gently undulating except on a very gross scale (e.g. Castleden, 1980). In detail, the surface may be highly irregular topographically and consist of scoured channel or trough-like depressions separated by higher less-dissected ridge or whale-backed forms of various scales, but with a potential amplitude of around 2–4 m. Such features were exposed during excavations at Stoke Newington where isolated potholes in the bedrock surface were also observed (Harding and Gibbard, 1984). Further problems including bedrock solutional subsidence producing enclosed hollows often of considerable dimensions and diapirism of clay bedrock producing equally large anomalous highs makes bedrock surface elevation unworkable as the main criterion for correlative purposes, particularly where boreholes are used.

The term *terrace* has been applied to deposits as well as the surface developed on the deposits. In some instances it has also been used to describe the bedrock surface ('bench') beneath the fluvial deposits, by analogy with rock-cut erosional terraces. In order to avoid confusion and to increase precision in stratigraphical usage the Commission of Terraces and Erosion Surfaces of the International Geographical Union has stressed that the term *terrace* should be applied only to the landform and not the material underlying it (Howard *et al.*, 1968). For use in stratigraphy a clear distinction must be made between the terrace surface, which is a morphological feature, and the deposits forming it, which represent a geological sequence (Figure 3). In many previous studies the deposits beneath a terrace surface were assumed to represent a single cycle of aggradation terminating in a floodplain. It was therefore permissible to use the surface as a significant horizon for correlation. However, the complexity of deposits that are found within terrace aggradations necessitates their study by geological stratigraphical techniques if the sequence is to be fully understood (Frye and Leonard, 1954). Moreover, the correlation of terrace deposits using surfaces is itself a questionable approach. Both the 'bench' and the terrace surface are planes of indeterminate history, i.e. they represent points in the sequence where deposition began and where it ceased, respectively. As such they represent surfaces equivalent to hiatuses or bedding planes in a stratified sequence and may be regarded as unconformities. Unconformities almost certainly represent more time than the deposits between them, but in turn can produce no information about that time, unlike the sediments. In short unconformities are gaps, sediments are records. Correlation of deposits using terrace surfaces cannot therefore be regarded as good geological practice.

There is therefore only one truly satisfactory method of correlating deposits, that is to use the information provided by the deposits themselves. Petrographic techniques to determine the differences between units using various parameters such as granulometry, sedimentary structures and the lithology of gravel and sand, for example pebble or granule lithology and heavy-mineral analysis, provide evidence for differentiating between aggradations. Down-valley correlations can then be achieved using standard lithostratigraphical methods (Hedberg, 1976) as recommended by Frye and Leonard (1954), Howard *et al.* (1968) and Gibbard (1981, 1982) among others. The mutual altitudinal relation of deposits of differing lithology can then be ascertained and used to support correlations. This can be further reinforced by relation of deposits to those of known age and affinities, particularly organic deposits that can be correlated either biostratigraphically or chronostratigraphically with sequences elsewhere.

In this way a firm stratigraphical subdivision can be developed which is self-supporting and can be related to the deposits in neighbouring regions whether there is a morphologically expressed terrace sequence or where the sediments are buried beneath a thick cover.

Where samples cannot be obtained for various reasons, such as lack of exposure, downstream gradient of the sedimentary body itself can be used with caution for correlation. However, it must always be remembered that such extensions of gradient are based on the assumption that differential crustal warping has not occurred. Boreholes where available may be used to provide control of sediment body thickness and its relation to other sedimentary units. In correlating fragments of terrace deposits accurate altitudinal control is essential (Johnson, 1944). Because the gradient of terrace aggradations may vary demonstrably from that of the modern stream, it is important that all altitudinal determinations are related to an arbitrary datum (Ordnance Datum or OD in Britain) rather than to the present floodplain or valley floor level (Tricart, 1948).

Nomenclature of terraces and deposits

Because valleys contain a series of terrace levels it is necessary to identify them in some way. Early workers used such terms as lower, middle and upper or first, second and third, etc. (Howard *et al.*, 1968). These systems depend on terraces remaining in the same order throughout the length of the valley. Following geological surveying of the Windsor area, Bromehead (1912) commented 'consecutive numbering has frequently been used, but since the 100-Feet Terrace is spoken alike as the first, second or third . . . the interpretation of the literature is attended with much difficulty. For these reasons the adoption of local names for each area is preferable . . . pits, being well-known and identifiable, even if future work should reveal higher or intermediate terraces.' Howard (1959) supported this view concluding that 'numerical systems of terrace nomenclature lack merit and [that] names should be applied as in stratigraphical nomenclature.' The binomial system of nomenclature was adopted by the Geological Survey in the Thames Valley following Bromehead's observations, and names such as Taplow Terrace and Boyn Hill Terrace became established. It is all the more curious therefore that the Institute of Geological Sciences has returned to the numerical system in recent years.

During the present study the numerical system was thought inappropriate and not sufficiently stratigraphically rigorous for use in complex terrace sequences such as those found in the Thames drainage system. In the study presented here a two-fold system has been developed. The geomorphological mapping of terrace surfaces recognised from type localities, as recommended by Howard (1959) and presented by Hare (1947), Sealy and Sealy (1956) and Allen (1978), is retained as the basic morphostratigraphical scheme. A separate parallel lithostratigraphical subdivision of sedimentary aggradations is proposed, implemented using the techniques summarised above. Where possible the morphostratigraphical scheme as well as biostratigraphical and chronostratigraphical data are related to this subdivision. The content of vertebrate remains, palaeobotany and Palaeolithic artefact assemblages of each aggradation are also discussed.

2

Lithostratigraphy

The Pleistocene deposits of the Middle Thames Valley have been studied in detail using exposures and borehole sections, the location of which are shown in Figure 1. The deposits have been subdivided following the International Stratigraphic Code (Hedberg, 1976) in which aggradations are defined and delimited as far as possible as formal units from type sections. These units are considered to be of member status because this is the most appropriate hierarchical level and is compatible with neighbouring areas e.g. Vale of St Albans (Gibbard 1974, 1977). The Thames members may be grouped into a larger unit for convenience. The term Middle Thames Valley Gravel Formation is proposed for this purpose.

In order to obtain as detailed a sequence as possible the deposits have been investigated using several interdisciplinary methods.

In the field a careful examination of the entire area was undertaken to locate and study exposures. All exposures were cleaned, and logged with all significant features being recorded. All sections were then levelled to Ordnance Datum. Palaeocurrent measurements were undertaken wherever possible and fabric determinations made whenever appropriate. Where appropriate, sections were sampled for pebble lithological counts, grain-size analysis and heavy-mineral analysis. Where organic deposits were encountered they were also sampled either in section or by borehole. Duplicate samples were then subdivided for various palaeontological analyses. If possible samples for absolute radiometric and thermoluminescence techniques were also collected. Where no exposure was available trial holes were dug or augered and larger diameter boreholes were put down to investigate the nature of the sediments.

The methods for individual general techniques used were as follows:

1. Pebble lithological counts were made on samples of pebbles in the 33–8 (-5 to -3 phi) range. Before counting all samples were washed and sieved to remove superfluous material. Pebbles were then sorted into lithological groups, clasts of questionable lithology being broken. The numbers of pebbles in each group were then counted and the percentages of each of the component lithologies determined. The groups recognised follow those previously mentioned by the author. (Gibbard, 1974, 1979, 1982). When sampling, care was taken to avoid the zone of periglacial disturbance immediately beneath the surface, in order to obtain unmodified gravel (cf. Gibbard, 1982).
2. Palaeocurrent analyses were measured on foreset beds in sand and pebbly sand following the recommendations of Potter and Pettijohn (1963). The results are expressed as vector means.
3. Pebble long axis fabric studies were performed using standard techniques summarised by Gibbard (1978).
4. Grain-size analyses were undertaken using the hydrometer and sieve techniques following the ASTM (1964) standard procedure using 50 g of air-dried sediment matrix of less than 2 mm (-1 phi) diameter. Results are shown as cumulative percentage curves.
5. Scanning electron microscopy of quartz sand grains of 200 μm diameter was undertaken following the recommendations of Krinsley and Doornkamp (1973).
6. Borehole data were collected from various sources and interpreted where necessary by the author. Wherever possible suspect records have been checked either by comparison with neighbouring sections, or other records.

The borehole data have been assembled from several sources. In the descriptive sections the source of individual borehole logs is not stated, for the sake of brevity. However, all borehole data have been obtained from the Institute of Geological Sciences borehole records library (IGS), unless otherwise stated. Other sources of borehole logs include the Eastern and Southeastern Road Construction Units (ERCU, SERCU), the Greater London Council (GLC), Slough Borough Council (SBC), Beaconsfield District Council (BDC), Surrey County Council Highways Department (SCC), British Gas (BG) and the Thames Water Authority (TWA). The records for all the holes are also available from the author.

Fig. 4. Standard sediment symbols used in all geological sections in this work.

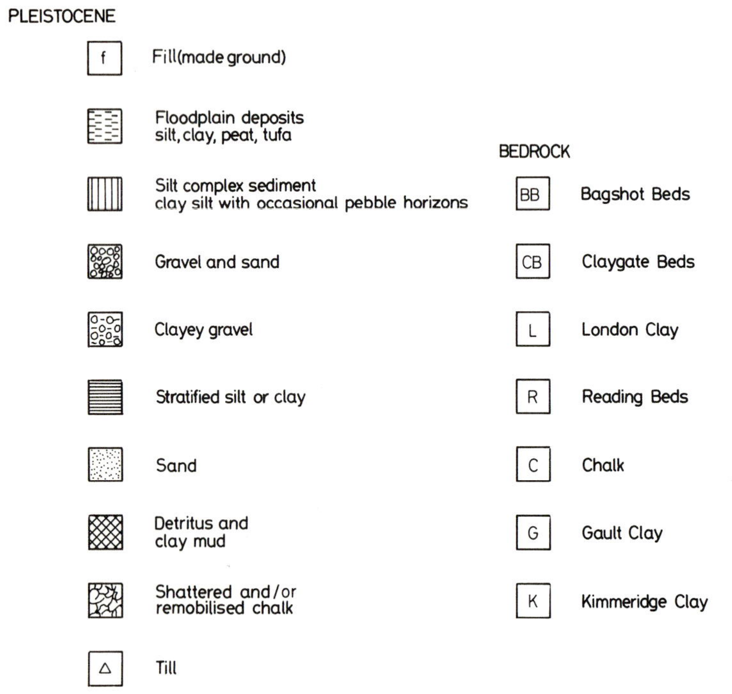

Fig. 5. Long profile section of the high-level gravel spreads in the Goring–Henley area showing the correlations discussed in the text.

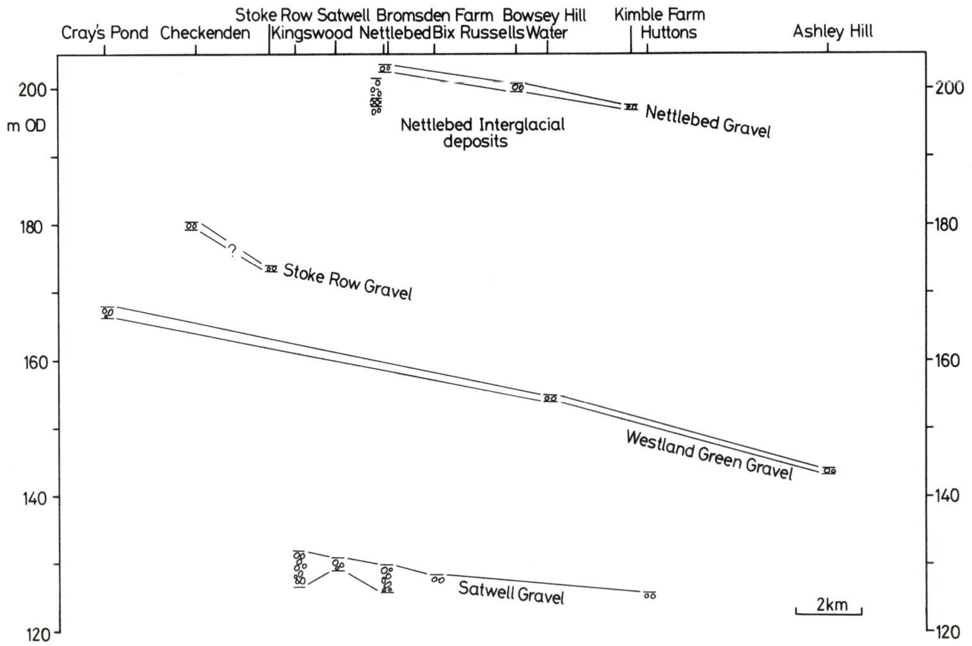

Sections drawn from these borehole records have been interpreted in the light of field experience and checking by the author. The localities of sections illustrated are shown in Figure 1. All sections have been drawn in uniform style using the sediment symbols shown in Figure 4 to aid comparison.

In addition to the borehole sections long profile summary geological sections have been constructed for each of the major aggradational units defined. These sections are based on all the data sources available and discussed in the text.

The members defined using these methods are described below in stratigraphical order and their distribution is shown in Figure 1. Tributary valley equivalents are discussed immediately following the appropriate Thames member to which they relate. The interpretations of the various types of evidence are presented in subsequent chapters.

2.1 Nettlebed Gravel

On the western side of Windmill Hill, Nettlebed (SU702872), up to 1.5 m of gravel with a sandy matrix rests on Reading Beds. The gravel reaches a maximum of 206 m OD, which is 5 m below the gravel-free London Clay hilltop.

The gravel, previously described by White (1896), Jukes-Brown and White (1908) and Horton (1977) lacks bedding structures and may not be entirely *in situ*. However, it has probably not moved far, to judge from the lack of gravel on the summit. The gravel comprises a high proportion of rounded 'Tertiary-type' flint (32–39%), together with angular flint (42–53%), vein quartz (10–15%) and quartzite (4%). The abundance of rounded flint compared with its frequency in lower altitude deposits clearly separates this deposit lithologically from the latter. The presence of quartz and other far-travelled lithologies indicates that the link with the Midlands existed at the time of deposition. White (1896) suggested that all the material could have originated from the Reading Beds, but this seems rather unlikely in view of the abundance of quartz. The source of the rounded flint is, however, the underlying Tertiary strata.

Scanning electron microscopy studies (SEM) of quartz sand grains from the gravels showed that the sand was predominantly derived from the underlying Tertiary rocks. This interpretation is based on the presence of secondary diagenetic quartz and a few V-shaped notches of pre-diagenetic marine history in seven out of fifteen grains examined. The remaining grains showed no V-notches at all which indicates that they were transported by fluvial processes.

The identification of possible laterally equivalent remnants of this gravel unit is extremely difficult because of the highly degraded nature of the deposit. The Henley geological map (sheet 254) shows several 'Pebble Gravel' outcrops of which Nettlebed is one. All these outcrops have been investigated and possible equivalents are shown in Figure 5. Correlation of these patches with that at Nettlebed is necessarily tentative and is based on height and similarity of lithological assemblage. On this basis the deposits at Russell's Water (SU714889) and Kimble Farm (Kilridge Wood) (SU749888) may be correlatives of the Nettlebed Gravel. Satisfactory samples could be obtained from neither of these localities. The former is based on a shallow ditch section although 3.5 m of gravel was observed here by White (1908), and the sample size was rather small from the latter (see Appendix). Other nearby high-level gravel patches such as Greenmoor Hill (SU647814: 182 m OD) comprise a high proportion of white coated angular flint and few quartz pebbles and therefore appear not to be equivalents of this unit.

As a part of the widespread Pebble Gravels, the origin of these deposits has long been the subject of controversy. Jukes-Brown and White (1908, pp. 79–80) believed that they were derived from lower Eocene pebble beds which had been 'fitfully dispersed and regrouped by progressive movement down the Chalk dip-slope.' Wooldridge (1927) believed that the Nettlebed Gravel was equivalent to the Lenham Beds and had been laid down on a wave-cut platform of the Pliocene Diestian Sea. He concluded that the deposit was higher and thus older than the Hertfordshire Pebble Gravels which he recognised as fluviatile. Nevertheless, he stated that they 'had a certain resemblance' to the Hertfordshire spread on compositional grounds (Wooldridge, 1927). Subsequently, he modified his views and invoked an early Pleistocene Calabrian age (Wooldridge, 1960). The marine origin of the Pebble Gravels was only partially supported by Hey, Krinsley and Hyde (1971) based on SEM studies of sand grains, although much of their material indicated fluvial deposition. It is certainly possible, as in the Nettlebed samples, that the sand grains showing a 'marine' morphology could have been reworked.

In spite of the rather unsatisfactory nature of the three sample localities mentioned here and the distribution of the 'Pebble Gravel' in general, it is clear that plotting the outcrops (Figure 5) suggests that a gradient of approximately 90 cm km^{-1} is present.

This gradient is comparable to, if slightly steeper than, that of the lower units (see below). It implies that the Nettlebed Gravels were laid down, as suggested by Horton (1977, p. 18), by the River Thames flowing from Oxfordshire north-eastwards across the Middle Thames region.

2.2 Nettlebed interglacial deposits

During resurveying of the area by the IGS, temporary excavations were made at Priests' Hill, a small hillock lying just below Windmill Hill on the north side of the village (SU701872: 203 m OD). The objective was to re-examine the series of pebbly sands and silts recorded overlying the Reading Beds in abandoned clay pits. It was shown that these deposits fill a channel cut into the Reading Beds and that, unexpectedly, the major silt bed was highly organic (Horton, 1977).

A resistivity survey and further boreholes carried out by I. Bryant, C. Turner and the author have shown that the channel trends NNE–SSW. The lowest sediments comprise a white to orange brown pebbly sand at least 1 m thick. This is overlain by bedded dark brown organic sandy and clayey silts with occasional pebbles, which attain a thickness of at least 3 m in the centre of the channel. This bed is overlain by at least 1.6 m of grey brown clayey gravel to pebbly clay. This is finally succeeded by 1.3 m of brown to grey mottled sandy clayey gravel that reaches up to the ground surface.

The lower part of the sequence at least may be of fluvial origin. The pollen assemblages contained within the organic bed indicate that it accumulated during the Pre-temperate, Early-temperate and the beginning of the Late-temperate biozones of an early Middle Pleistocene interglacial stage (Chapter 7).

It is important that the relationship of the Nettlebed Gravel to this deposit be determined in order to establish the relative age of the former. The lithology of the gravel (Nettlebed D) overlying the organic deposits is very similar to that of the Nettlebed Gravel on Windmill Hill (Nettlebed B, C, E) (see Appendix). The gravel overlying the channel fill, however, has a very clay-rich matrix and in places it is a pebbly clay closely resembling the Reading Beds rather than a gravel. Although there is little difference in altitude between the base of the Windmill Hill spread and the base of that capping the channel on Priests' Hill (4.7 m), the mode of occurrence of the deposit suggests that it may have originated by solifluction of Nettlebed Gravel mixed with Reading Beds clay on top of the interglacial sediments. Alternatively, it is possible that the two gravel units were once continuous, so that the organic deposits pre-date the Windmill Hill gravel. On balance the former explanation would seem more plausible in view of the altitudinal difference, the lithology of the gravels overlying the organic bed and the local bedrock lithology which would produce slopes very susceptible to solifluction. The resolution of this problem must await the results of further investigations which are under way at present.

However, the Nettlebed interglacial deposits form an extremely important marker horizon in the high-level sequence of the Middle Thames region.

2.3 Stoke Row Gravel

South of Stoke Row village, old overgrown workings in the Reading Beds expose a thin gravel deposit. The gravel reaches a maximum height of 174 m OD (SU686834) where it is 60 cm thick, but thickens markedly northwards. According to White (1896) it was about 1 m thick on average, but he also observed the same gravel filling a pipe-like hollow about 4.5 m deep in the Reading Beds near the village.

Sections seen by the author in small trenches showed no current-bedding structure, but White (1896) described the deposit as a 'gravel mixed with red, ochreous sand, becoming coarser towards the bottom of the pit.' In a second excavation he also recorded indistinct bedding.

The gravel differs significantly from that found at Nettlebed (Section 2.1) in its increased content of far-travelled materials, notably vein quartz, red quartzite, and rare pebbles of sandstone, grit and Palaeozoic chert (see Appendix). White (1896) also recorded sarsens from this deposit. A pebble count underlines the striking difference in composition and includes 30.5% angular flint and 50% vein quartz. A very similar count was obtained by Hey (1965, table 1) and led him to correlate these gravels with his quartz-rich Westland Green Gravels. However, this correlation could perhaps be questioned for two reasons. Firstly, Hey's count shows slightly differing proportions of lithologies than those present at his other localities. Secondly, the gradient necessary to link the Stoke Row outcrop downstream with Ashley Hill (SU824812) would seem to be rather steep (over 100 cm km^{-1}) in comparison with that linking Hey's (1965, figure 3) other localities downstream (50 cm km^{-1}). While it is conceivable that the gradient might steepen near the Goring Gap region, this does not happen in later terrace aggradations and therefore must be explained. Two explanations are possible. It has been suggested by Hey (1965, pp. 416–17) that the Westland Green Gravels may represent outwash from glacial ice, an interpretation reinvoked by Shotton *et al.* (1980) to explain the possibility of an early glaciation in the south Midlands. A steepened gradient could be explained by proximity to the ice front as has been noted in modern environments (Fahnestock, 1963). However, it is equally possible that the increased gradient is incorrect and an intermediate stage with a similar pebble lithological assemblage to that of the Westland Green stage might be represented. This is supported by the recognition of a remnant at the corresponding lower elevation for correlation with the Westland Green stage at Cray's Pond (see below). For these reasons it is preferable to exclude the Stoke Row Gravel from the Westland Green member for the present.

Equivalent gravel remnants that can be correlated with the Stoke Row Gravel are absent with the possible exception of that at Checkenden (SU669827: 182 m OD) where a highly degraded outlier of gravel of similar composition is found. A satisfactory sample of this material could not be obtained since the gravel has a very clayey matrix and is partially intermixed with the Clay-with-flints. This deposit was also mentioned by White (1896) who thought that it could be equated with that at Stoke Row (Figure 5). Throughout the area there are few other potential localities that might be equivalent to this member, although McGregor and Green (1978) did record sites that may represent downstream correlates.

2.4 Westland Green Gravel

The occurrence of a division of the 'Pebble Gravels' richer in vein quartz and quartzite than the higher main spread was first noted by Salter (1896) and White (1896), and later mentioned by Wooldridge (1927). A re-examination of these quartzose gravels led Hey (1965) to assign them to a separate lithostratigraphical member, the Westland Green Gravels (type section TL422215). He recognised this subdivision in Pebble Gravel outcrops from Goring to Essex; their main area of development being south Hertfordshire and eastern Essex. He subsequently also extended the unit into East Anglia (Hey, 1980). The gradient, sedimentary structures and SEM studies of sand grains (Hey *et al.*, 1971) have shown that these gravels were laid down by the early River Thames.

In the area considered here, Hey (1965) assigned four localities to his unit. They were Hodgemoor Wood (SU967939), Bowsey Hill (SU807802), Ashley Hill (SU824812) and Stoke Row (SU686835). The last-mentioned site is thought by the author to be too high and of differing composition, it has therefore been excluded from this member (see Section 2.3). Hey's other observations generally have been confirmed during the present study.

Lithologically, the gravel contains high frequencies of far-travelled rock types notably vein quartz, quartzite, sandstone, grit and Palaeozoic chert (Hey, 1965; McGregor and Green, 1978). The origin of the gravel lithologies present in this stage has been discussed by Hey (1965, pp. 414–17) and by Green and McGregor (1978) (see discussion below). All authors agree that the materials represent a mixture of locally derived material and debris transported from outside the London Basin.

Assuming that the exclusion of the Stoke Row deposits is correct, where are the lateral equivalents of the Westland Green Gravels? Extension of the Westland Green member surface gradient (Hey, 1965; figure 3) eastwards from Ashley Hill precisely intersects with the well developed terrace-like surface at Cray's Pond (SU637805: 168 m OD) and Cold Harbour (SU635800) (Figure 5). Blake (1903) recorded Pebble Gravel overlying Reading Beds at both localities. Sections in gravel over 1.5 m deep at the former place have yielded a pebble assemblage (see Appendix) comparable with that of the Westland Green. The gravels here are therefore tentatively correlated with this unit and if this is correct the Stoke Row Gravel must represent a separate aggradational event.

2.5 Satwell Gravel

Overgrown sections in a terrace-like outlier at 131 m OD at Satwell (type section SU706839) revealed 2 m of strongly iron-stained, quartz-rich stratified sand and gravel. A pebble count sample from these deposits is given in the Appendix. The composition appears to be distinct from those of higher and lower units based particularly on the proportions of vein quartz (42%) and angular flint (34%).

Two further gravel patches at comparable altitudes (Figure 5) may also represent remnants of this unit. At Kingswood Common (SU694825) a distinct terrace-like surface occurs at 132 m OD and

Fig. 6. Section along the M40 Motorway from near Gerrards Cross to Loudwater, constructed using borehole records (source: ERCU).

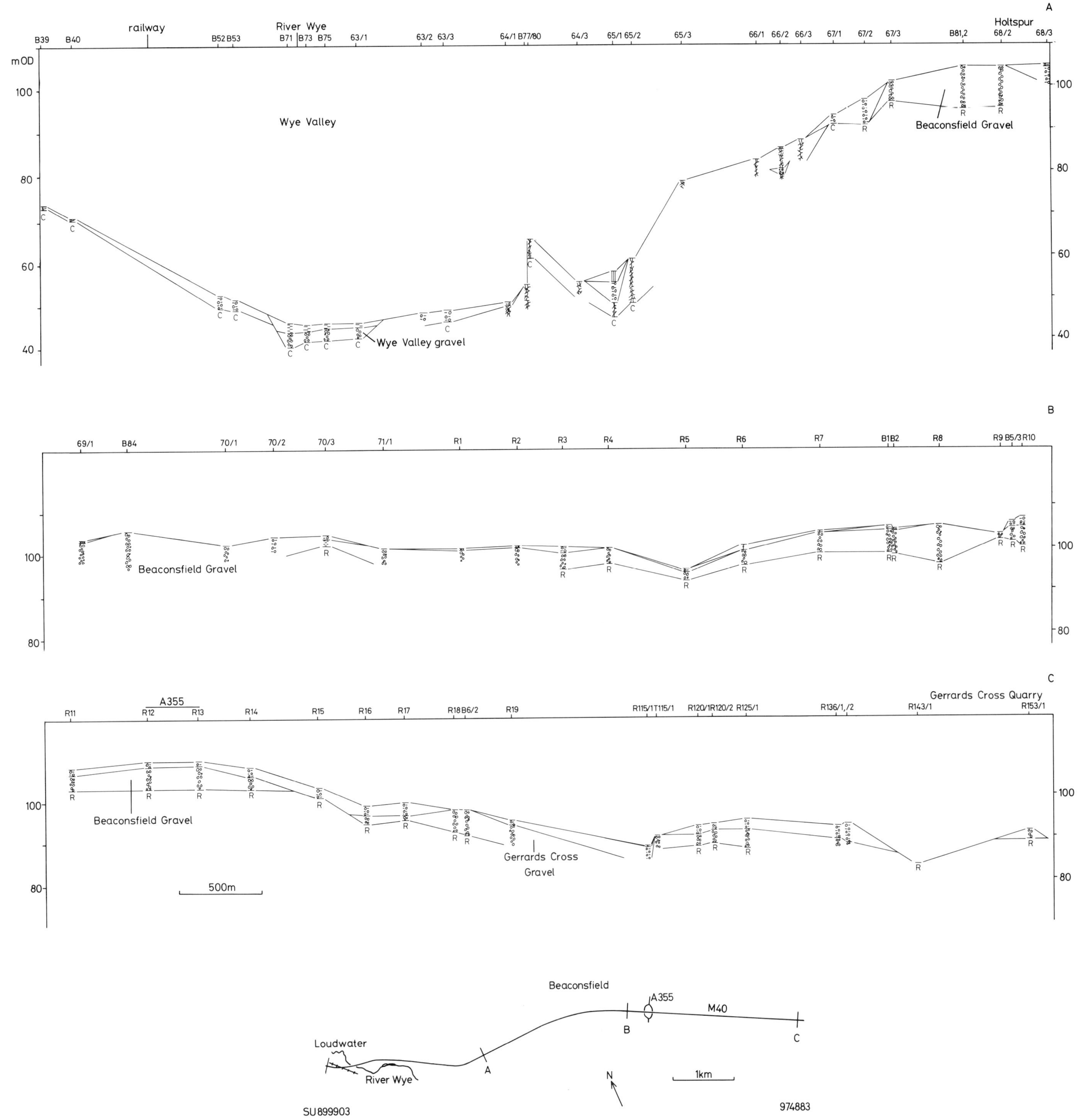

is underlain by up to 6 m of quartzose gravels according to White (1896). Secondly, a borehole at Bromsden Farm, near Highmoor (SU716847: 130 m OD) recorded 4.1 m of gravel and sand resting on Chalk. No exposure could, however, be found at either place during the present study to confirm correlation of these localities.

Downstream three small quartz-rich gravel remnants are present at Bix (SU729846, 733845, 725849: 128–129 m OD) and they would fall altitudinally in this unit. Finally, the gravel spread at Hutton's Farm, Hambleden (SU794870: 126 m OD) comprises involuted, quartz-rich, iron-stained gravel visible in shallow ditch sections.

Plotting of these gravel remnants indicates that they all lie on a gradient of 60 cm km^{-1} which is closely similar to that of the lower gravel members (Figure 5). The deposits are therefore thought to represent an aggradational phase in the Thames sequence, pre-dating the Beaconsfield Gravel and post-dating the Westland Green Gravel, on the basis of altitudinal distribution.

2.6 Beaconsfield Gravel

Originally termed the Higher gravel train by Wooldridge (1938, pp. 639–40), this unit was recognised by him from the area north of Reading, to Marlow and as far west as Chalfont St Giles (Figure 1). Detailed geomorphological mapping in the Bourne End – Colne Valley region led Hare (1947, p. 308) to conclude that 'nothing systematic emerges from a morphological survey of the surface.' However, a re-examination of sections and borehole records by the author indicates that this unit can be recognised as a distinct terrace-like aggradation throughout the Middle Thames Valley. Since the unit is consistent in height, lithological composition and stratigraphical position, it has been assigned the formal lithostratigraphical name: Beaconsfield Gravel (Gibbard, 1978, p. 16).

The type section from which these gravels were named is at Beaconsfield Station (SU940912: 112 m OD). Here up to 6 m of interbedded gravel and sand were temporarily exposed on the north side of the railway line. The fine to medium gravels are sandy, matrix-supported and strongly iron-stained. The sands are tabular cross-bedded and have yielded palaeocurrent directions.

The deposits underlie an irregular and much degraded terrace surface at 108–115 m on which much of Beaconsfield is built. The

Fig. 7. Long profile section showing the deposits correlated with the Beaconsfield Gravel (above) and those correlated with the Gerrards Cross Gravel (below).

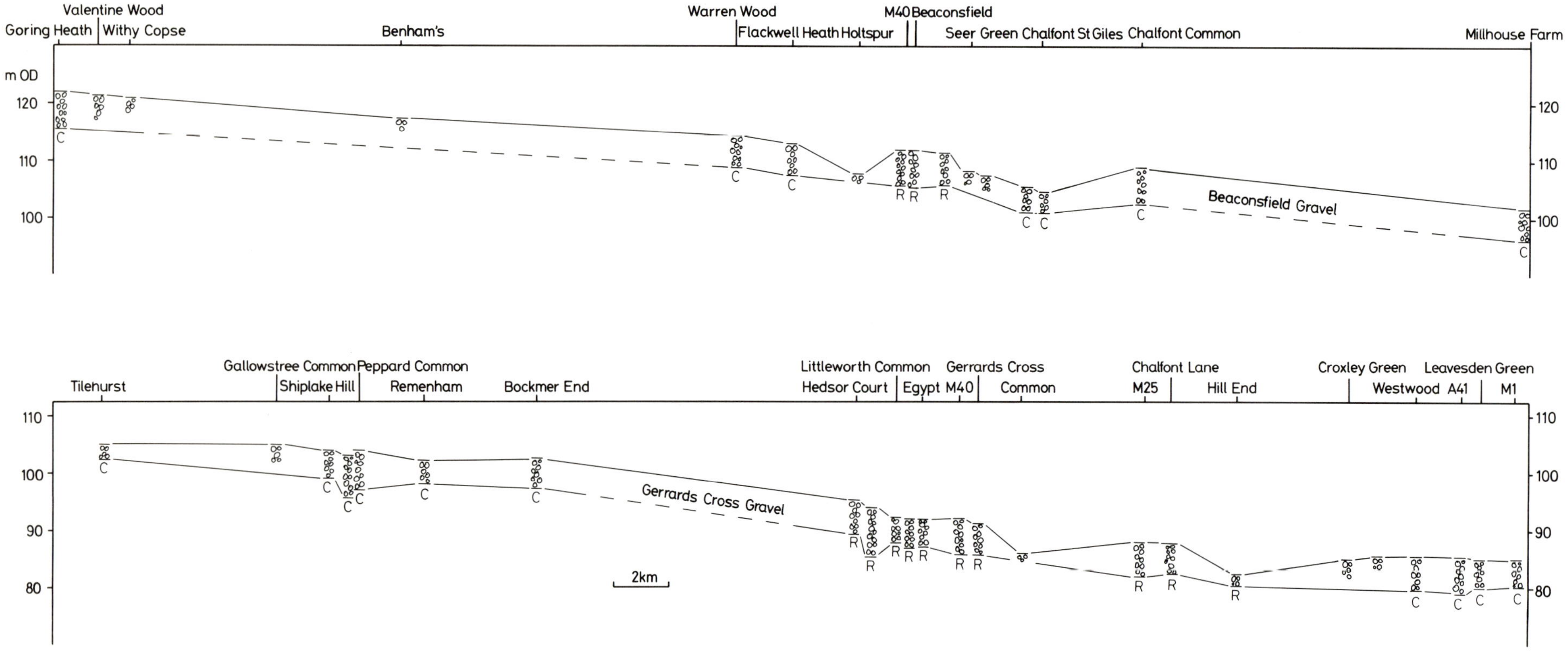

surface is heavily dissected, but can be traced to the south-west and south (Figure 1). An exposure in Holtspur Lane (SU921894: 108 m OD) showed that current-bedded sandy gravels of comparable composition occur beneath the surface here. Boreholes in this surface, sunk for the M40 Motorway, recorded a maximum of 10 m of gravels and sand resting on bedrock, although the average thickness appears to be about 7 m (Figure 6). A number of other small exposures of sandy bedded gravel have been seen around Beaconsfield, and at several localities the sediments are distorted by collapse possibly resulting from bedrock solution.

The deposits and the terrace developed on them can be traced to the north-east of Beaconsfield to the area around Seer Green and Chalfont St Giles, as noted by Hare (1947, p. 308). Several small exposures at Deans Wood (SU971906), Seer Green (SU963920) and Jordans (SU977910) show gravels, the upper 1–1.5 m of which is disturbed by periglacial cryoturbation and ice wedge casts. The deposits can be correlated with those at Beaconsfield by their height (Figure 7) and lithological content. A number of boreholes indicated shallow gravel thicknesses in this area, the maximum being 6 m at Butler's Cross (SU977926). North of these places, Hare (1947, p. 308) recorded a marked bounding bluff at the northern extremity of the surface.

Between the Misbourne and the Colne valleys, a dissected irregular surface occurs at 106–108 m on the interfluve at Newlands Park (TQ008938) and Horn Hill (TQ014927). No exposure was found in these deposits, however a borehole at the latter (TQ012926) recorded 9 m of gravel and sand resting on Reading Beds.

Further to the north-east, on the interfluve between the Colne and Gade valleys, the gravels could not be found. However, heavily degraded, isolated patches of gravel at accordant heights can be observed as far east as the M1 Motorway (Gibbard, 1978), for example at Millhouse Farm (TL114037: 102 m OD). A number of gravel localities in this western part of the Vale of St Albans were also assigned to the Higher gravel train on the basis of their altitude and pebble lithology by Green and McGregor (1978).

West of the Wye Valley, Sealy and Sealy (1956, pp. 372–3) recorded several high-level gravel patches. The nearest of these to Beaconsfield is Flackwell Heath. Wooldridge (1938) suggested that the deposits underlying this place may form part of his Higher gravel train. Sealy and Sealy (1956, p. 372) remarked about the flat, terrace-like surface here at 110–113 m OD. A section observed in Juniper Lane (SU905894) exposed 2–2.5 m of orange yellow interbedded gravel and tabular cross-bedded sand. This gravel is markedly richer in flint (see Appendix) than those on the western side of the hill and, as suggested by Sealy and Sealy (1956, p. 373), may have been deposited by the proto-Wye stream near its confluence with the Thames.

In the Henley–Marlow area, few other large outcrops of this unit are present. At Warren Wood (SU873900: 115 m OD) a temporary excavation showed that the hill is underlain by stratified quartzose gravels and sands up to 5 m thick. These again can be correlated by height and lithological composition with those at Beaconsfield (Figure 7). A number of small gravel outliers at accordant heights were also recorded by Sealy and Sealy (1956, pp. 372–3), but were not exposed. These include those at Sedgmoor Farm (SU896892), north Wilfrid Wood (SU899888), Ridge Wood (SU780864) and Benham's (SU760862).

West of Henley, three further outliers appear to be correlatives of the Beaconsfield Gravel, since they again occur at accordant heights (Figure 7) and are of similar composition (see Appendix). Sections in the well developed terrace surface at Valentine Wood (SU663808: 124 m OD) and Withy Copse (SU677804: 121 m OD) revealed quartz-rich gravel and sand. Pebble counts from both localities show lithologies comparable to those of this member. At nearby Goring Heath (SU635793: 126 m OD) a pipetrench section exposed 3 m of orange red gravel and sand, the upper 2 m much disturbed by periglacial cryoturbation. A nearby borehole (SU635784) recorded 7.5 m of quartzose gravel and sand resting on Chalk.

The series of pebble counts (see Appendix) show that the Beaconsfield Gravel comprises 34–51% angular flint, 8–12% rounded flint, 23–43% vein quartz and 8–17% quartzite. Low percentages (1–2%) of Lower Greensand chert in the eastern part of the area only, igneous and metamorphic rocks, Palaeozoic chert and other minor erratics are also present. The high variability in composition found in these counts may, in part, reflect the generally poor quality of available sections, many of which had to be sampled at 2 m below the present surface. The counts from Flackwell Heath (see Appendix) are markedly different from the main series since they contain over 85% flint with little or no quartz component, as noted above. With the exception of this assemblage, the unit comprises large amounts of far-travelled material derived from north of the Goring Gap as also found in the Westland Green Gravels (Section 2.4). The continued occurrence of this assemblage indicates that the river was still draining a similar catchment to that of the preceding stage.

Similar pebble assemblages were obtained by Green and McGregor (1978) from the eastern part of the area. They noted the lower frequency of rounded flint in comparison to that present in assemblages from the higher, earlier gravels (see discussion below). The gravels of this stage can therefore be separated from the older units on this basis and using the generally lower frequency of vein quartz.

Green and McGregor (1978) also reported the occurrence of pebbly clay 'colluvial' lenses associated with the stratified gravels. Lenses of massive pebbly clay 20–30 cm thick interstratified in sediments were observed during this study at some sections. Their mode of occurrence and contrastingly local lithology suggests that they originated in a similar way to those discussed below (section 2.7).

The present work indicates that the deposits were laid down by a high-energy bed-load dominated river as shown by the sedimentary structures. It flowed towards the east-north-east as shown by the palaeocurrent measurements and the gradient of the unit (Figure 7). The mean gradient obtained by plotting all the outcrops is 45 cm km^{-1}. The recognition of the Beaconsfield Gravel as a discrete unit resting on bedrock and extending from Goring to St Albans (Figure 62a) indicates that the River Thames occupied this course during this aggradational phase. Contrary to Wooldridge's (1938, 1960) ideas, the Thames can be followed into the Vale of St Albans at this stage, as suggested by both Green and McGregor (1978) and Gibbard (1978).

2.7 Gerrards Cross Gravel (Harefield Terrace)

The gravels of this member were originally recognised as a continuous unit by Wooldridge (1938, p. 640). However, because of the degraded nature of the surface, and the possibility that there might be more than one stage represented, he did not recognise a terrace surface, but instead assigned the sediments to a broad grouping called the Lower gravel train. In his reappraisal of the detailed morphology of the Bourne End – Colne Valley district, Hare (1947) renamed this unit the Harefield Terrace after its type locality at Harefield, near Uxbridge (Figure 1). Extension of Hare's work into the western part of the Middle Thames Valley by Sealy and Sealy (1956) showed that the terrace surface could also be recognised as far upstream as Kidmore End. Downstream, based on the assumption that the river was flowing through Wooldridge's (1938, 1960) 'Middlesex Depression', Hare (1947) did not trace the terrace further to the north-east. Gibbard (1974,

1977), however, showed that gravels and sands at Leavesden Green and elsewhere in the western end of the Vale of St Albans, are lateral equivalents of those in the Middle Thames area. This conclusion has recently been supported by Green and McGregor (1978).

On the basis of sections at Leavesden Green (TL105004) Gibbard (1977) proposed a lithostratigraphical member unit, the Leavesden Green Gravel for the fluvial gravels and sands underlying the terrace in this area. However, since the type section is unsatisfactory in that it was only temporarily exposed and it occurs in the Vale of St Albans, it was thought necessary to redefine the unit from a more permanent open section in the Middle Thames region. It is therefore proposed that a new member, the Gerrards Cross Gravel, be used to include all the deposits of the unit discussed here. The type section is that at Gerrards Cross quarry (SU968887), where the deposits rest on bedrock.

The deposits beneath the surface at 92 m OD at Gerrards Cross are up to 6 m thick and rest on an irregularly channelled surface of Reading Beds. The sediments may be divided into two groups: stratified, sorted gravels and sands, and frequently massive, pebbly clays. The first group predominates and comprises fine to medium, matrix-supported gravels with occasional larger clasts. The gravels occur in sub-horizontal units, often showing alternating coarser and finer interbeds 10–30 cm in thickness. Elsewhere massive lenses of gravel, occasionally with an imbricate structure, occur. Pebble count samples from this gravel are given in the Appendix. Cross-cutting channel-fill sequences up to 1.5 m thick occur throughout the sections at various levels and contain beds of gravel and tabular cross-bedded pebbly sand. Sand present as 'chute-type' channel fills 20–30 cm thick, and also as large more persistent beds, is more common in the basal 2 m of the sequence. This may reflect reworking of sand from the underlying bedrock. However, above, the gravel predominates. A number of palaeocurrent measurements from this section indicate flow towards the north-east. Fine sediments are generally rather rare, but light brown clayey silt bands representing shallow infilled channels 5–35 cm thick have been seen. The massive pebbly clays interdigitate with the stratified sediments and on the western side of the pit this material immediately underlies the present ground surface. The origin and structure of this material will be discussed below. At several places in the quarry large infilled funnel-shaped hollows contained brown pebbly clay, as well as disturbed gravel and sand. The hollows, up to 10 m across, often extend through the entire thickness of the bedded deposits and into the bedrock beneath.

At Westwood Quarry (TQ075994: 88 m OD) a very similar sequence of deposits 8 m thick is exposed. The form of the hollows suggests that they originated by bedrock solutional collapse. They have also been described by McGregor and Green (1983). Here again the basal part of the sequence is rich in sand present as tabular cross-bedded channel fills within poorly stratified often rather coarse gravel. The upper part of the sequence is again dominated by subhorizontally stratified gravel. Interbedded brown pebbly clay is again present and in one place was over 1 m thick (see below). As at Gerrards Cross the modern surface is underlain by similar material 2.5 m thick, that was heavily cryoturbated to a depth of almost 2 m and strongly reddened by iron-staining. Throughout the pit the sediments were disturbed by much small-scale normal step faulting. The faulting increased in intensity downwards and indicates collapse by release from beneath. Examination of the Chalk bedrock surface showed that it was highly irregular, varying in level by 2–3 m in places. At one point a small graben-like structure 80 cm across, consisting of almost vertical, converging normal faults with downthrows of 50 cm, was present. Gravel that had collapsed from above filled the structure, while an overlying gravel bed with an erosional base resting on that beneath was undisturbed. It appears therefore that this and probably the other collapse features result from syndepositional bedrock collapse (see below).

A number of other exposures of gravels and sands in the appropriate altitudinal range with structures resembling those described above occur and have been correlated with this unit by pebble lithology and palaeocurrents.

At Chalfont Lane (TQ018917) and Hill End (TQ048923) pits in degraded Harefield Terrace surfaces exposed stratified gravels and sands with interbedded pebbly clay units exactly like those already described. The deposits at both sites show collapse structures, producing vertical bedding in places, but undisturbed areas are also present. At Hill End, in at least two places, lenses of 'rubble' chalk, comprising angular chalk fragments and flints in a putty-like matrix were found interstratified with the gravels. At both localities the Chalk formed a series of pinnacle-like structures up to 2.5 m high at the base. At some places, a narrow bed of Tertiary-type blue grey rounded flint pebbles in a brown clay matrix was visible between the Chalk and the overlying gravels. Since the basal Tertiary deposits occur in the immediate vicinity and even underlie part of the gravel sequence, it seems reasonable to suggest that this material represents the remains of a veneer of Tertiary beds resting on the Chalk.

A large temporary exposure 2.5 km north-east of Westwood at the intersection of the A405 and A41 roads north of Watford (TQ097997) showed about 6 m of strongly iron-stained gravel and sand resting on a very irregular bedrock Chalk surface and overlain by 2 m of brown pebbly clay (Figure 7).

The deposits continue south of Westwood quarry and have been temporarily exposed to a depth of 2 m in Whippendell Wood (TQ076984). The dissected surface can be traced to Croxley Green (TQ070962) where a second temporary excavation revealed 2.5 m of stratified gravel of comparable composition to those at Westwood (see Appendix). A number of boreholes penetrate the deposits in this area. A section along the A405 (M25) road, north of Chandlers Cross passes next to Westwood and the boreholes record this unit. A well drilling at TQ081983 (82 m OD) proved what is almost certainly this unit, 6 m in thickness.

The deposits do not seem to occur north of Rickmansworth. On the eastern side of the Colne, the Hill End sequence is not matched by exposures at Harefield, although on the basis of Hare's (1947) mapping the deposits probably occur here. West of the Colne, a number of boreholes for the M25 Motorway penetrated gravel continuous altitudinally with that at Chalfont Lane quarry. These recorded over 5 m of gravels resting on Reading Beds and strongly dissected by later dry valleys. Trench sections in nearby Slade Oak Lane (TQ017898) and adjacent M25 Motorway cuttings confirmed the occurrence of quartzose stratified gravels beneath the strongly degraded surface in this area.

From Chalfont Common, the deposits can be followed to Gerrards Cross where quartz-rich gravels occur beneath the common itself (SU999884: 86 m OD). This terrace remnant was considered too low for inclusion in the main Harefield unit by Hare (1947, p. 310), but the surface here is probably degraded.

The M40 Motorway forms the southern boundary of the Gerrards Cross quarry (the type section for this unit) and boreholes for the motorway recorded up to 7 m of the gravels and sands overlain by pebbly clay, seen in the quarry faces (Figure 6). A 2 m deep ditch section at Burtley Wood (SU957890: 92 m OD) confirmed that the gravels continue on the south side of the motorway. The surface is much dissected (Figure 6), but the relationship of the unit to the higher Beaconsfield Gravel, described above, is clearly illustrated. The base of the older member occurs at least 10–14 m above that of the Gerrards Cross Gravel member.

Further to the south-west, numerous shallow exposures record the occurrence of quartz-rich gravel beneath the Harefield Terrace surface. They include those at Egypt Woods (SU954867: 92 m OD), Heathfield (SU921868: 97 m OD), Jenkins Farm Wood (SU944877: 91 m OD) and Littleworth Common (SU932860: 91 m OD). All these sites are correlated with this unit using pebble composition and altitude (Figure 6). The sample from Broad Lane, Wooburn (SU920881: 99 m OD) is probably from Wye Valley gravel to judge from its high content of flint and the Hedsor Court sample (SU919864: 95 m OD) seems to be analogous to the gravels from the Flackwell Heath area in Beaconsfield Gravel. A lower gravel remnant near Cliveden (SU914856: 89 m OD) referred to a minor, slightly lower part of this terrace by Hare (1947, p. 310) and correlated with their Rassler Terrace by Sealy and Sealy (1956, p. 374) was examined. The gravel here is very thin, not exceeding 2.6 m according to a borehole (SU914855) and ditch sections nearby. The patch seems to represent either an eroded remnant of the Gerrards Cross Gravel or possibly the northern margin of the lower, Winter Hill Gravel (see Section 2.8). Apart from exposures, a number of boreholes recorded quartzose gravel and sand at accordant heights (Figure 7) for correlation with this unit, south-west of Gerrards Cross. The gravel thickness varies from 4–6 m on average, but greater depths are recorded from where the deposits overlie Chalk and/ or Tertiary bedrock. Such a variation, where not related to slope processes, probably results from the bedrock collapse seen in exposures.

Between Bourne End and Medmenham no trace of this unit has been found. However, at Bockmer End although no exposure was available, a borehole (SU810863: 102.5 m OD) in Sealy and Sealy's Harefield Terrace recorded 5.3 m of quartzose gravel that occurred at the expected altitude. Further upstream at Remenham Place (SU778830: 102 m OD) an old gravel pit section exposed 4 m of gravel and sand, the basal part of which was horizontally stratified, resting on Chalk, while in Henley Park (another Sealy and Sealy locality) a ditch at SU755839 showed 1.25 m of disturbed quartzose gravel with some vertical bedding overlain again by pebbly clay. The base of the gravels was not seen here, but the form of the bedding suggested disturbance possibly by bedrock collapse. Boreholes confirm these depth observations at both Remenham and Henley Park and pebble counts together with their height suggest correlation with the Gerrards Cross unit (see Appendix).

Upstream the gravels are aligned along the Caversham Channel course but are not well exposed. They have been correlated using pebble lithology and height where possible. On the northern side of the channel, the deposits occur beneath Rotherfield Peppard Common (SU708818 and 709816: 104 m OD). Here they reach a thickness of 7 m and rest on Chalk. At nearby Peppard Hill a borehole (SU703815) recorded 8 m of similar gravels. A pit near this locality at Shiplake Hill (*c.*SU706811: 104 m OD), described by White (1896, 1908), exposed 5 m of stratified gravel and sand. Further west the fragment at Gallowstree Common (SU685799) can be correlated with this unit where again degraded remnants occur in trench sections beneath a marked terrace surface at 105 m OD. Blake (1903) also recorded 2 m of gravel in a quarry at this place. On the south side of the channel a minor patch at Dunsden (SU741777) noted by Sealy and Sealy (1956) falls altitudinally into this unit.

Only one outlier of these gravels occurs south of the Thames in the Reading area. At Tilehurst, gravel underlying the plateau surface at 100–105 m OD was mapped as Rassler by Sealy and Sealy (1956). However, as can be seen from Figure 7, it falls into line with this unit. The composition of these gravels was originally described by Blake (1903) and later by Walder (1967). A small ditch (SU652733) exposed gravels of comparable composition to those described by these authors and to those of this unit showing them to be of Thames and not Kennet origin. The latter contain almost only flint (Blake, 1903; Walder, 1967; Gibbard, 1982). Up to 3.2 m of the Tilehurst gravel overlying London Clay were recorded in a boring at 104 m OD at Beal's Farm (SU652733). Blake (1903, p. 69) mentioned that these deposits contain interstratified clay bands in places.

The pebble composition of the Gerrards Cross Gravel is shown in the Appendix. It comprises 38–61% angular flint, 3–19% rounded flint, 18–30% vein quartz and 7–15% quartzite. The gravels also contain small amounts of Palaeozoic chert, sandstone, Greensand chert (in the eastern part of the area) and minor amounts of igneous and metamorphic rocks. As discussed above, the bulk of the material is reworked from local Chalk and Tertiary rocks. The notably lower content of rounded flint compared to higher units may result from smaller areas of Tertiary pebble beds available for erosion at this level and also from reworking of older gravels. The abundance of far-travelled material, as in the Beaconsfield Gravel, suggests a continued link with a source distant from the Goring Gap. Green and McGregor (1978) have recorded comparable pebble assemblages from gravels equivalent to these and in particular noted the occurrence of characteristic igneous rocks. These rocks include rhyolitic tuffs and other acidic igneous rocks of possible Welsh origin. Although they are present in higher members, these rocks reach their acme in this unit, comprising up to 0.6% of the total pebble assemblage (Green, Hey and McGregor 1980). Noting this occurrence, McGregor and Green (1978) followed Hey (1965) in suggesting possible glacial transport into some part of the Thames catchment from north Wales.

In order to test this hypothesis a SEM study of (200 μm) quartz sand grain morphology was undertaken using samples from Gerrards Cross and Westwood quarries. Only one possible glacial surface-textured sand grain was found out of a total of 30 grains examined, the majority showing either aeolian or marine characters probably inherited from their original source. However, since the morphology of quartz grains from periglacial environments is not well known, it is possible that the 'glacial' grain may have been formed under conditions of freeze–thaw rather than glacial crushing. The case for lowland glaciation in the vicinity at this time therefore remains unproven.

Brown pebbly clay beds are frequently found in sections in the Gerrards Cross Gravel both interstratified and overlying the fluvial gravels and sands. They vary considerably in thickness from a few centimetres to over 2 m, occur as lenses, wedges or tongues of sediment, and have irregular erosional bases that often show incorporation of gravel and sand from beneath. Overlying fluvial deposits always show marked erosional contact with these pebbly clay units. No gradation between the two facies has ever been found. The pebble assemblage generally comprises over 90% flint predominantly of angular frost-shattered type, although vein quartz, quartzite and chalk pebbles occur. An example of pebble compositions from these beds is given in the Appendix. The sediments frequently also contain clasts of light grey, red or brown silty clay. The matrix of the beds is variable. It is normally massive, but the sediment is poorly admixed; sand stringers or silt laminae may be present. The latter rise down the dip of the deposit and are often highly distorted and deformed around pebbles, but show flow-type banding.

Pebble fabric measurements have been determined from several undisturbed beds and two examples are shown in Figure 8. They show a strong concentration of clasts in one quadrant of the diagrams with minor transverse maxima. Such fabrics typify those from flow tills and other mass-flow deposits (Lundqvist, 1949; Galloway, 1961; Watson, 1969; Boulton, 1971; French, 1971).

Particle-size analyses from four pebbly clay beds (Figure 9)

Fig. 8. Two-dimensional pebble long axis fabric measurements from beds of brown pebbly clay at Chalfont Lane and Hill End quarries using standard techniques (Gibbard, 1974).

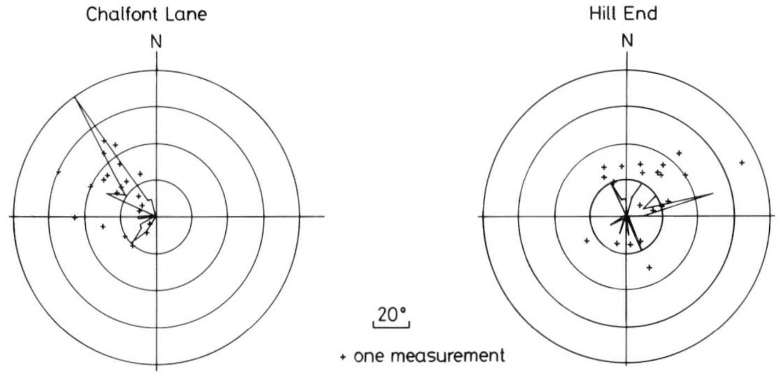

Fig. 9. Cumulative frequency curves of grain size of four pebbly clay beds interstratified with fluvial gravel and sand correlated with the Gerrards Cross Gravel.

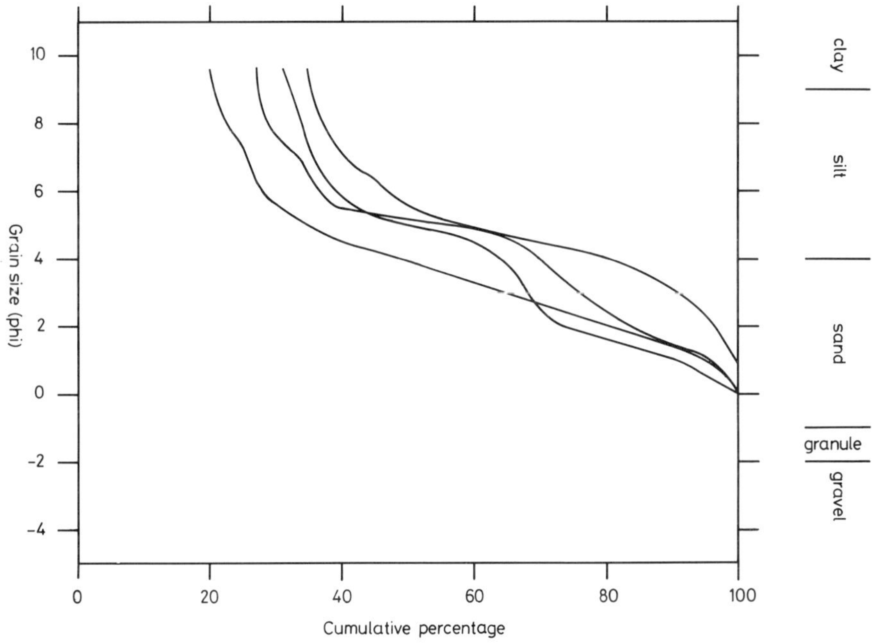

show a wide variation in distribution. All the deposits are very poorly sorted and have bimodal (and in one case trimodal) distributions as can be seen from the cumulative percentage curves. The high clay content of the samples prevents the calculation of statistical grain-size parameters.

Taking all these factors into account it seems likely that the pebbly clay bands were deposited by mass flow rather than fluvial processes. The markedly local pebble composition, together with the abundance of frost-shattered pebbles indicates a local source subjected to cold climate weathering. The mixture of these pebbles with a clay matrix and non-durable clay clasts strongly suggests active incorporation and intermixing of weathered clays in the immediate vicinity. The source for all these materials is probably the Tertiary beds and Clay-with-flints available locally throughout the area. Such bedrock would provide unstable slopes under periglacial conditions and downslope flow by solifluction onto the neighbouring river deposits would be expected (cf. Galloway, 1961).

2.8 Winter Hill Gravel (Winter Hill Terrace)

The term 'Winter Hill Terrace' was first used by Saner and Wooldridge (1929) and later Ross (1931) to describe the surface developed on gravels covering the Chalk on high ground south of the Thames near Bourne End. Both Saner and Wooldridge (1929) and Wooldridge (1938) discussed the extent of the feature and mentioned its wide distribution from near Goring to the Colne. Hare (1947) refined Wooldridge's work and demonstrated that the Winter Hill could be subdivided into two major units: a lower Black Park Terrace (Section 2.9) and a higher Winter Hill Terrace (*sensu stricto*). He also showed that the latter could be divided into lower and upper facets in the Bourne End – Colne Valley area. Sealy and Sealy (1956) extended Hare's work westwards to the Goring Gap and on the basis of morphological mapping suggested that the 'lower' Winter Hill Terrace floored the Caversham Channel, north of Reading. This has been shown to be incorrect by Clarke and Dixon (1981).

Both Wooldridge (1938, 1960) and Hare (1947) thought that the Thames flowed through the so-called 'Middlesex Depression' across north-west London during the Winter Hill stage and therefore did not trace the terrace northwards. However, Gibbard (1974, 1977) showed that the Westmill Gravel member underlying Chalky Till in the Vale of St Albans was the equivalent of the gravel underlying the Winter Hill Terrace (*s.s.*) of Hare (1947, p. 310). The river therefore continued to flow through the Vale of St Albans during this stage. This conclusion was supported by Green and McGregor (1978).

The type locality for the Winter Hill Gravel is a small old pit at SU880863, where 3.5 m of gravel were exposed near Rondels on Winter Hill (Figure 1). An adjacent borehole (SU879863: 75.5 m OD) recorded 5.9 m of gravel. The difference in thickness may result either from downslope slippage or more probably from irregularities in the Chalk bedrock surface possibly caused by collapse as described above (Section 2.7). A second exposure at SU883866 showed stratified, matrix-supported, sandy gravel, 2 m thick, resting on Chalk. The surface here lies at 77 m OD and is degraded, no true terrace surface being preserved, as noted by Hare (1947, p. 311).

The terrace and underlying deposits are better represented east of the river. East of Cliveden, a degraded area of gravel altitudinally continuous with this unit is present. Near the Canadian Hospital (SU916847), a borehole recorded 3.4 m of gravel and sand overlying Reading Beds. Ditch sections confirm its stratification. About 500 m to the north, at Bourne End Road, a trench (cf. Section 2.7) showed cryoturbated gravel of possible Winter Hill type, based on the pebble lithology (see Appendix and below). It would appear, however, to be rather high for inclusion in this unit and could therefore be redeposited older gravel or possibly a River Wye tributary facet.

In the Dorney Wood – Burnham Beeches area, the unit is well developed and can be correlated with Winter Hill by height and pebble lithology. The wide terrace surfaces present were mapped by Hare (1947). At Dorney Wood itself (SU937856: 76 m OD) a large pit exposed stratified silty gravel 5 m deep.

A number of shallow sections showed cryoturbated gravel in neighbouring Dorney Wood (SU944858), while a borehole on Crabtree Heath (80 m OD) found 4.6 m of gravel containing a pebbly clay band 20 cm thick. Several shallow ditches revealed unbedded gravel beneath Burnham Beeches at about 74–75 m OD. Near Wingroves (SU946846) sandy stratified gravel was temporarily exposed and a narrow lens of tabular cross-bedded sand gave a palaeocurrent direction towards the south-east. Similar exposures occur in Burnham Beeches (SU953849) and East Burnham Common (SU953844). Four boreholes on East Burnham Common (74–75 m OD) recorded gravel and sand 3.0–6.2 m thick overlying Reading Beds (Figure 10). It is in the East Burnham – Farnham Common area that Hare (1947, figure 13) mapped the

Fig. 10. Long profile section showing the deposits correlated with the Winter Hill Gravel (above) and those correlated with the Black Park Gravel (below).

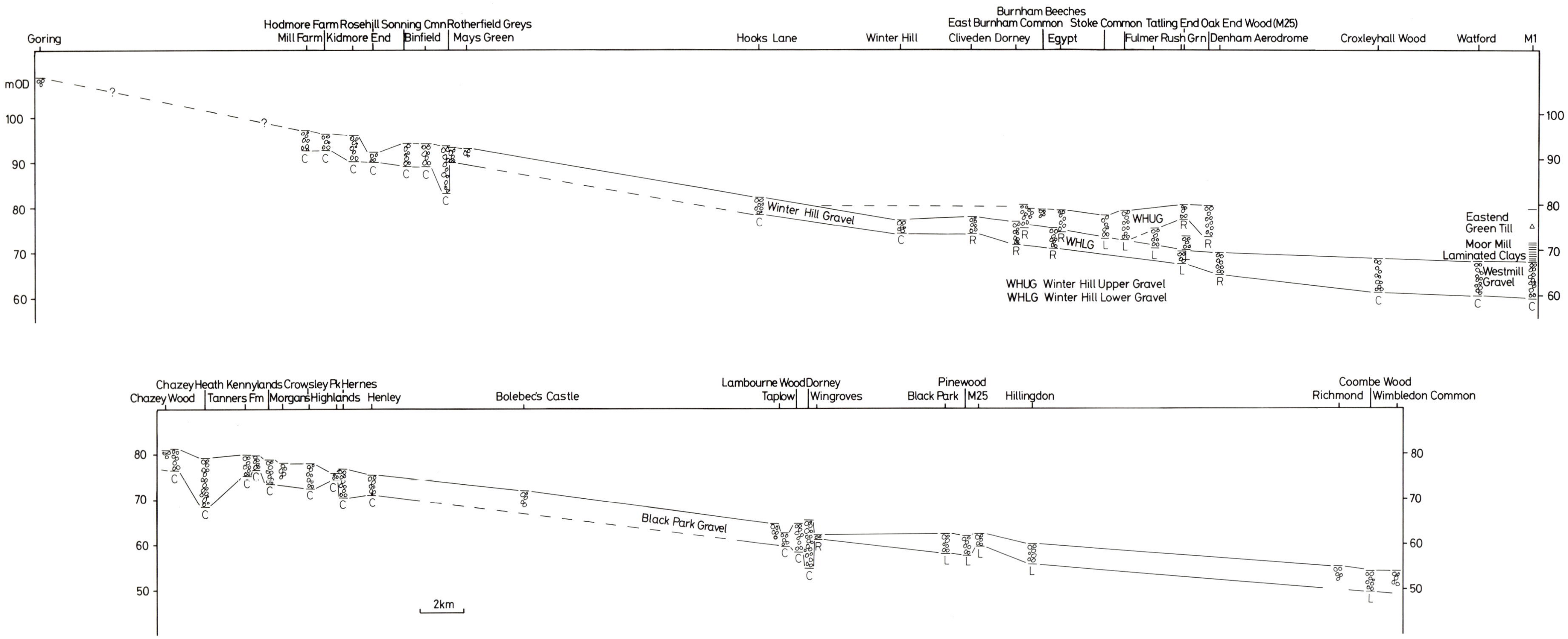

division of the terrace into lower and upper facets. These terrace surfaces are easily distinguished in the field, although small trench sections suggested that the underlying gravels were continuous beneath and between both surfaces. Exposures in the upper subunit surface were absent except for a ditch at Egypt (SU957853: 76 m OD) where 1 m of cryoturbated gravels occurred, but no satisfactory sample could be obtained. However, two boreholes (Figure 10) at 78 and 79 m OD recorded respectively 3.7 and 4.8 m of gravel and sand overlying bedrock here.

The most well developed remnant of the upper subunit surface occurs at Stoke Common of which Hare (1947, p. 311) remarked 'there is no more convincing terrace in the London Basin' (Figure 10). The surface is almost flat at 78–79 m OD. Exposures were rare here, but a small, temporary excavation (SU983857) revealed 2.5 m of gravel and sand cryoturbated in the upper 1.5 m, and showing large-scale cross-stratification below. These gravels can be correlated lithologically with those at Winter Hill. In some places, a thin veneer of light grey clay rested on the gravels. Sherlock and Noble (1922) described a section at Larchmoor (*c.* SU982852) where 3 m of bedded gravel and sand were visible. Several boreholes penetrated the deposits; for example at SU982858 (78 m OD), 0.9 m of brown pebbly clay overlay 5.6 m of gravel and sand. The sequence is therefore quite uniform across the area.

Further east the surface is degraded and fragments of this unit beneath Fulmer Common (TQ003847) may be post-depositionally disturbed to judge from the surface irregularity. This area was mapped as a fragment of the terrace's lower facet by Hare (1947). Boreholes here, for example TQ013848, recorded up to 6.0 m of gravel overlying London Clay at the accordant height for correlation with this unit.

North of the Alder Bourne Valley, equivalent deposits again occur. A large London Clay working at Rush Green (TQ021859: 70 m OD) exposed 3–4.5 m of coarse to medium gravel filling cross-cutting channels, with coarse horizontally stratified, bar-like, matrix-supported gravel between. These deposits alternated with current-bedded often pebbly sands 10–15 cm thick, giving a palaeocurrent direction towards 063° (average of nine readings). In the upper 1.5 m the sediments became more sandy and less coarse. Overlying the gravels and sands with a sharply defined base was 50 cm of indistinctly stratified brown pebbly clay. The stratification consisted of narrow alternating coarser (silt) and finer (clay) laminae a few millimetres in thickness. Above, the clay was massive and contained many pebbles possibly intermixed by cryoturbation (cf. Figure 30).

It was possible to trace the sediments continuously northwards from the 'lower' to the 'upper' subunit in exposures for the M25 Motorway, and a section near Tatling End (TQ012873: 81 m OD) confirmed the occurrence of lithologically equivalent *in-situ* gravel, 3.2 m thick, beneath Hare's (1947) 'upper' surface. This rather low figure suggests that these deposits may have been partially degraded, when compared to thicknesses found further west. Boreholes for the M25 Motorway penetrated the gravels at Tatling End (Figure 11), but because of gravel removal did not demonstrate the relationship of the two subunits.

On the Misbourne–Colne interfluve the deposits recur. Boreholes for the M25 Motorway continue northwards through Oak End Wood (Figure 11). The section shows Tertiary bedrock overlain by up to 7 m of gravel and sand. The relation of the 'upper' subunit to the higher degraded Gerrards Cross Gravel can also be seen. The base of the latter member is a maximum of 7.5 m above that of the Winter Hill Upper Gravel here. Small pits in Oak End Wood at approximately 80 m OD showed cryoturbated gravel that can be equated with this unit altitudinally and lithologically. However, at TQ018892 (Motorway pit P1160), the fine granular gravel was overlain by 1 m of unbedded brown to grey silty clay, also noted by Sherlock and Noble (1922, p. 41). An isolated borehole in Nockhill Wood (TQ023893: 79 m OD) recorded pebbly sand and clay 2.7 m thick resting on 3.5 m of gravels. At nearby Denham Aerodrome, Hare (1947) mapped the 'lower' Winter Hill Terrace surface at 72–73 m OD, and here a ditch section (TQ034811) showed 1.5 m of gravel of composition comparable with this unit. One current-bedding measurement here suggested a northward palaeocurrent flow for these gravels. Up to 5 m of gravel and sand were present in a borehole at TQ034887, and confirm that the deposits occur here (Figure 11).

This unit is not found again in the Denham area, but can be traced to Rickmansworth. At Croxleyhall Wood (Long Valley Wood: TQ075948) between the Chess and the Gade streams, sections occur in old gravel workings that were originally described by Smith and Dewey (1915) and Sherlock and Noble (1922, p. 44). Parts of these sections produced abundant flint implements and on the basis of their descriptions are not equivalent to this unit. However, examination of the area around Croxley Green Station (Figure 1) suggests that a surface referable to Hare's (1947) 'lower' Winter Hill Terrace is present (cf. Whitaker, 1889). Re-examination of the old pit sections shows that the stratified gravels contain sedimentary structures similar to those described from Rush Green and palaeocurrent measure-

Fig. 11. Section along the M25 Motorway from Thorney to Maple Cross (Horn Hill), constructed from borehole records (source: ERCU).

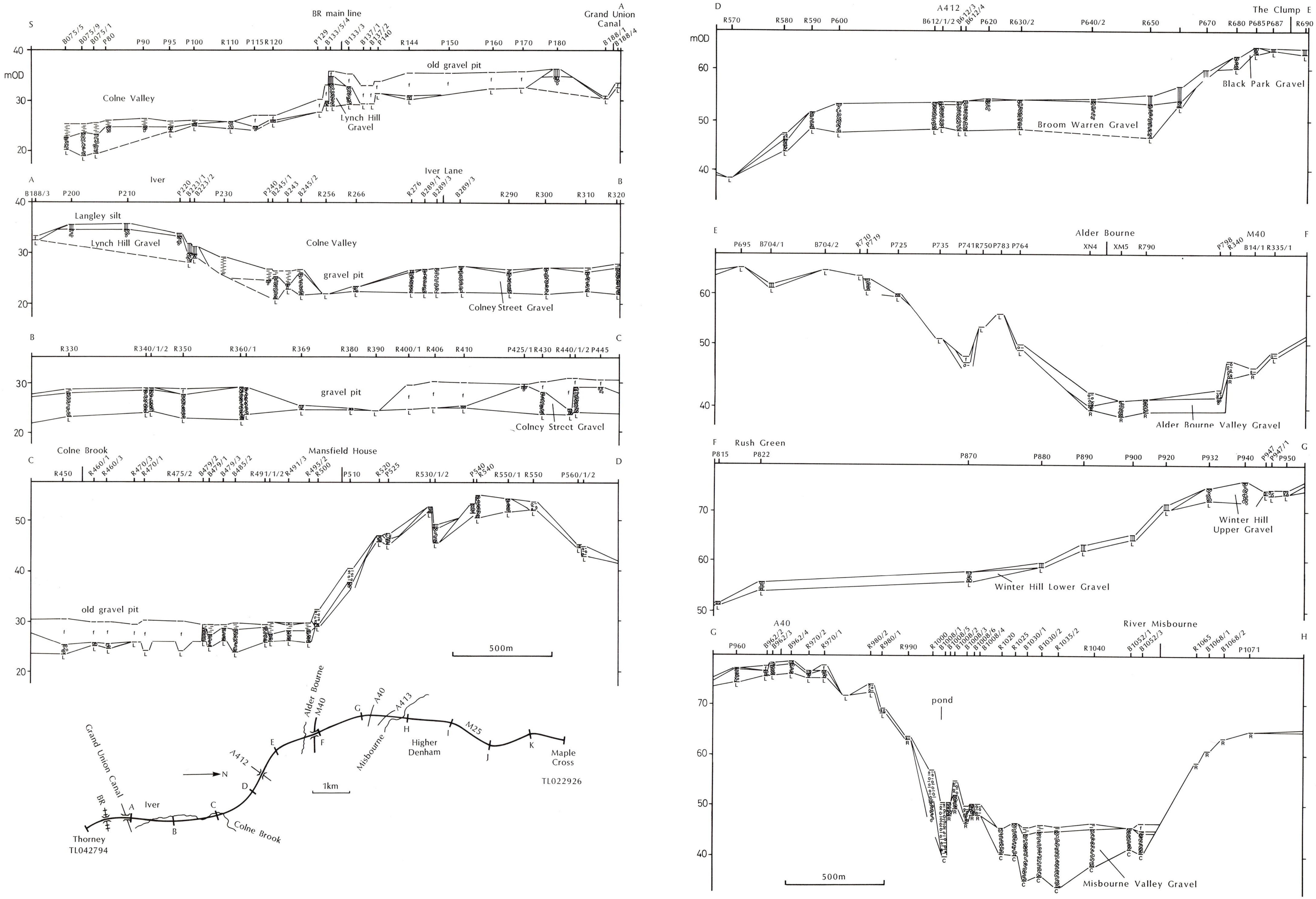

Fig. 11. continued

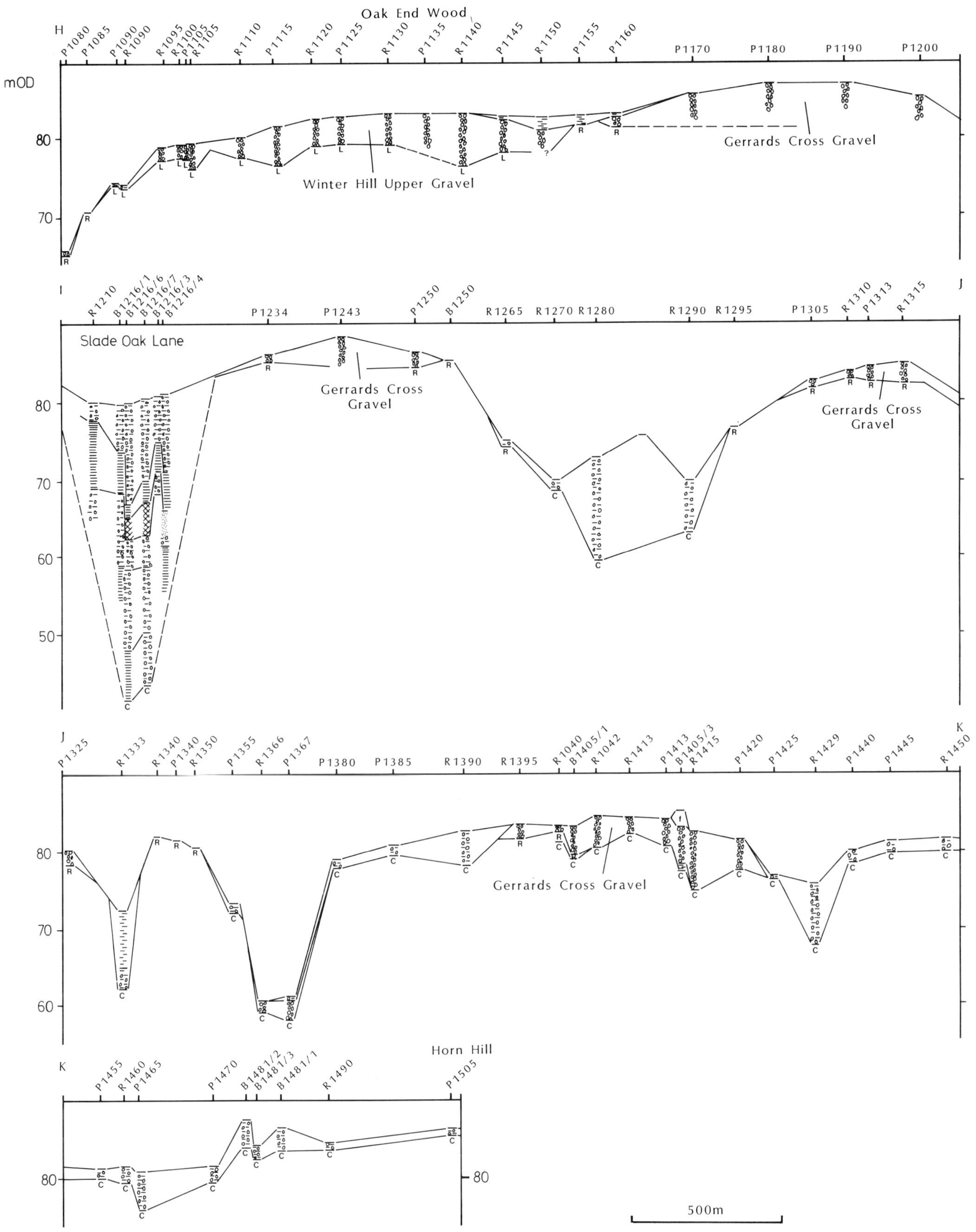

ments indicate flow towards the east-north-east, i.e. opposite to the Gade or the Colne. A pebble count from the gravel shows that it is referable to this unit with which it is correlated. The section is overlain by 1 m of cryoturbated clayey gravel of probable solifluction origin. The relationship of these deposits to those described by Smith and Dewey (1915) must remain somewhat obscure because of a lack of modern sections. It is, however, probable that the implement-rich deposits they described occurred as lower, younger terrace deposits of the Gade or Colne overlain by 'brickearth' and banked against the higher level Winter Hill Gravel (cf. Smith and Dewey 1915, figure 3) as suspected by Sherlock and Noble (1922, p. 39) (see Section 2.13 below). Because of the similarity of Colne gravels to those of the Thames (discussed below) the junction between the units may have been overlooked when it was exposed.

Upstream from Winter Hill, deposits at the accordant height for correlation with this unit occur on the northern valley side west of Marlow at Hooks Corner (SU831852: 82 m OD), the southern part of Rassler Wood (SU822854) and, according to Sealy and Sealy (1956), also at Forty Green (SU836866). Boreholes at the first two places confirm the occurrence of 3.8 m of gravel and sand (Figure 10). A small temporary section at SU825855 in Rassler Wood exposed gravel of a comparable composition to that at Winter Hill (see Appendix).

Rassler Wood was taken as the type locality for a terrace surface intermediate between the Harefield and Winter Hill by Sealy and Sealy (1956, p. 374). A number of small excavations in the northern part of Rassler Wood and neighbouring Davenport Wood (SU822862) indicate that the deposits underlying the surface above the 90 m level are not stratified fluvial deposits, but unbedded chalk-rich gravel and chalky sand (for pebble count see the Appendix) probably formed by mass movement down the valley side. The deposits closely resemble those found by White (1904), Whitaker (1889) and the present author beneath the valley-side fan-like landforms between Henley and Medmenham (cf. Section 2.18). No evidence for a separate Rassler terrace has been found during this study and therefore the name has been abandoned.

Between Rassler Wood and Henley no remnants of this unit occur. However, west of Henley boreholes at Cowfields Farm (SU734820: 94 m OD) and near Rotherfield Greys (SU736820) recorded 3 and 12 m of gravels, respectively, at the expected height resting on Chalk. This great variation in depth may result from bedrock collapse, since on the neighbouring interfluve in Crosscroft a small chalk pit (SU721803: 92 m OD) exposed stratified sands and gravels of similar composition to those of this unit, overlying brown pebbly clay, both of which had collapsed into a vertical Chalk pipe 1.5 m deep. At nearby Blountscourt Farm (SU719804: 94 m OD) 5.2 m of gravel were present.

On the basis of altitude, the deposits mapped by Sealy and Sealy (1956, p. 381) at Kidmore End (SU698793: 93 m OD) and at Hodmore Farm (SU683783: 96 m OD) as Rassler fall into line with the Winter Hill Gravel unit. A borehole near the latter site recorded 3.3 m of gravel overlying the Chalk, while at nearby Mill Farm (SU677780: 97 m OD), 4.6 m of gravel is present. The same unit extends southward and underlies a well developed terrace surface at Mapledurham, (e.g. at SU685769: 97 m OD). The gravel composition suggests correlation with this unit. The furthest east this unit can be identified is at Goring (SU614809: 110 m OD) where a ditch exposed 2 m of quartzose gravel (see Appendix) at the correct height for tentative correlation.

On the south side of the Caversham Channel, the gravels occur beneath clear terrace surfaces mapped by the Sealys (1956) as Rassler, at Mays Green (SU743797) and at Binfield Heath (SU739788) at 93 m OD. An exposure at the New Inn, Binfield Heath (SU747785), displayed gravel with a composition similar to those of this unit (see Appendix). An earlier exposure near here was described by Treacher (1896) and Blake (1903). An adjacent borehole (SU742788: 93.5 m OD) proved 5.2 m of gravel overlain by 2.3 m of pebbly clay at this place. At Rosehill Kiln (SU723722) an old pit originally described by Shrubsole (1890), now redeveloped for housing, exposed 2 m of stratified gravel containing a 1 m thick bed of bedded pebbly sand beneath a degraded surface at 92 m OD. A pebble count from the section confirms correlation with this unit.

The pebble counts from the Winter Hill Gravel are shown in the Appendix. They comprise 59–80% angular flint, 5–10% rounded flint, 7–21% vein quartz and 3–13% quartzite. Samples from east of Henley contain small amounts of Greensand chert. West of Henley, this material was not found, although it was reported from Binfield Heath by Treacher (1896, and in White, 1896). The counts also contain small, consistent amounts of igneous and metamorphic rocks, Palaeozoic chert, brown sandstone, grit and other minor exotic rocks.

Variations in pebble lithology seem to arise by proximity to the valley side and down-valley changes (see Chapter 3). Near to the valley side incomplete mixing of incorporated gravel derived by erosion from higher gravel units causes lithological variations. Down valley there is a clear decrease in the content of exotic vein quartz and quartzite from over 20% and 13% respectively near Reading to 9% and 4% respectively near Rickmansworth. These changes presumably reflect the input from tributaries together with the incorporation of flint from local bedrock and regolith.

As discussed in Chapter 3, the Winter Hill Gravel assemblages are very characteristic, being almost unique in the Thames Valley.

Previous counts from this unit by Walder (1967) and McGregor and Green (1978) have given comparable results. Walder (1967, table X) described counts from gravels beneath the Sealys' (1956) Rassler and 'upper' Winter Hill Terraces and showed that there was no discernible difference between them. There was, however, a marked contrast between this unit and the gravels beneath the 'lower' Winter Hill Terrace (*sensu* Sealy and Sealy, 1956). The latter, flooring the Caversham Channel, has been reassigned to the Black Park Gravel (see Section 2.9) which has a contrasting pebble composition throughout the Middle Thames region. However, the Winter Hill Lower Gravel as used here in the Burnham – Colne Valley area has a composition that is indistinguishable from that of the upper subunit. This strongly suggests that the two subunits should be considered as part of a single member.

As can be seen from Figure 10, the two subunits of the Winter Hill Terrace, mapped by Hare (1947), and the Winter Hill Gravel discussed above, are only found in the Burnham – Colne Valley area. Upstream of Burnham, contrary to the Sealys' (1956) conclusions, only one Winter Hill aggradation and surface can be found. The gradient of this aggradation is 45–55 cm km^{-1} which is comparable to that of both higher and lower gravel members. At Burnham, the two gravel subunits diverge so that the lower continues to fall at the same gradient, while the upper remains almost horizontal at between 79–82 m OD, depending on the amount of post-depositional degradation and modification of the surface. It is not found east of Oak End Wood, Denham. Where sedimentary structures have been seen, for example at Stoke Common, large-scale deltaic-type cross-stratification is present. Sections in both units in the eastern part of the area often record clays overlying the fluvial sands and gravels. While in several cases these deposits may be soliflucted debris or slope wash deposits (see Section 2.7), at Rush Green, overlying the lower subunit, and at Oak End Wood, overlying the upper subunit, silty clay beds occur. At the former the clay was stratified with alternating laminae, a structure very similar to that originating in glaciolacustrine environments.

2.9 Black Park Gravel (Black Park Terrace)

The Black Park Terrace was first recognised as a discrete, well developed stage throughout the eastern Middle Thames area by Hare (1947, pp. 314–16). The ground had previously been included by Wooldridge (1938) in his Winter Hill Terrace, which he acknowledged was complex. Following Hare, Sealy and Sealy (1956) extended the Black Park surface upstream to Henley, but then through the modern valley of the Thames rather than to Caversham where they proposed that it merged with their 'lower' Winter Hill Terrace. Walder (1967, p. 118) also noted this merging of the two terrace stages. However, a reconstruction of terrace gradients using borehole data led Clarke and Dixon (1981) to suggest that the deposits flooring the Caversham Channel, north of Reading, should not be correlated with the 'lower' Winter Hill, but with the Black Park.

Downstream, Gibbard (1974, 1977) showed that the Smug Oak Gravel outwash stream from the Anglian ice in the Vale of St Albans flowed south along a newly formed valley, later to be occupied by the Colne, to join the Thames at Uxbridge during the Black Park Gravel stage. Investigations in the London area led Gibbard (1979) to propose an extension of the Black Park to Richmond, Wimbledon and to Dartford Heath in East London, thus confirming Hare's (1947, p. 315) conclusion that this terrace was the first to be aligned along the present valley of the Thames. Bridgland (1980) extended this line even further downstream to south-eastern Essex, where the river was confluent with the Medway. At Richmond, Gibbard (1979) also demonstrated a confluence with a Mole–Wey river and recognised an upstream equivalent terrace aggradation: the Oxshott Heath Gravel. Recently, Gibbard (1982) has shown that the Kennet and partially also the Blackwater were confluent with the Thames near Henley during the Black Park stage and that the Silchester Gravel stage of White (1902) is its Kennet equivalent.

The correlation of the late Anglian Smug Oak Gravel with the Thames's Black Park Gravel means that the latter provides an important horizon of known age throughout the Thames system.

The type section for this unit is in Black Park at TQ006832 (Gibbard, 1979) where up to 4 m of horizontally stratified, matrix-supported gravel overlies London Clay, below the surface at 62 m OD (Figure 1). Thin tabular cross-bedded sand channel fills within the gravel indicated a palaeocurrent flow towards the east. The well developed terrace surface at this locality prompted Hare (1947) to erect the Black Park stage from this place. Two boreholes penetrated the deposits at Diamond Wood (TQ014832) at 61.5 m OD and at Pinewood Studios (TQ017842) at 62 m OD, and recorded up to 6.6 m of gravel and sand, resting on London Clay. In the Pinewood area the gravels are overlain by 1.5 m of brown pebbly clay probably originating by solifluction (see Section 2.7) or slope wash from the higher ground immediately north-west of the site. The gravels and sand underlying Hare's surface were also found in boreholes for the M25 Motorway near The Clump (TQ024845). However, the surface is heavily degraded here to judge from its form and the presence of only 2.5 m of gravel beneath (Figure 11).

To the east the unit is cut-out by the Colne Valley, but a 'Colne valley equivalent' is present at Uxbridge Common at 61–63 m OD (Hare, 1947). No further outliers of this unit have been found in the Colne Valley as far upstream as Rickmansworth. East of Uxbridge, Hare (1947, p. 315) showed that the terrace is present beneath Hillingdon at 60 m OD. Sections observed by the author at TQ071829 exposed 4.5 m of gravel and sand resting on London Clay. The deposits comprised coarse to medium, often angular, matrix-supported gravel in bar-like horizontally stratified units 30–50 cm thick. These were interstratified with pebbly sand bands, 5–30 cm thick, occasionally showing tabular cross-bedding and sometimes filling shallow channels cut into the gravels. The upper part of the sequence was cryoturbated. A similar thickness of deposits was recorded in a nearby borehole (TQ069826) (Figure 10).

As mentioned above, Gibbard (1979) has traced the unit eastwards to Richmond Park (TQ188731), Kingston Hill (TQ199714) and Wimbledon Common (TQ230725) to link with the so-called 'Kingston Leaf' of Zeuner (1959) and Wooldridge and Linton (1955). Gravels and sands showing similar sedimentary structures at the accordant height were well exposed at Coombe Wood Golf Course (TQ202706). The pebble lithology supported this correlation, but was enriched in Greensand chert and rounded flint incorporated by confluence with the Mole–Wey at this place (Gibbard, 1979, table 2).

The gravel capping Hanger Hill at 62 m OD is too high and probably too far north to be directly correlated with this unit (Allen, 1978). The deposits are absent from beneath the North Circular Road and from a borehole at TQ181819 (41 m OD). Brown (1883, 1887) originally described and illustrated the deposits as being disposed into channels cut into a small outlier of Claygate Beds 1.5 m deep in excavations for a reservoir at TQ182819. They were much disturbed, probably by periglacial cryoturbation processes and contained significant numbers of erratics including various types of Jurassic rocks, Carboniferous Limestone, Greenstone and granites. These rock types can only have been introduced into the area by the Anglian glaciation that reached the Hendon–Finchley district, 8 km to the north-east. The lithology of the gravel suggests that it could represent a remnant of an outwash deposit, although the isolated position prevents correlation with known units. While it might have been deposited during an intermediate stage in the formation of the Thames course through its present valley, it seems more likely that the gravels might represent the first evidence of a southward flowing Brent. This stream must have formed after diversion of the pre-existing Mole–Wey drainage line by the Finchley Till ice (Gibbard, 1979). Such a stream would thus be analogous to the Smug Oak Gravel river of the Vale of St Albans (Gibbard, 1974, 1977). If the deposits were laid down by a stream fed by the Finchley ice, this would explain the erratic content. The waters would presumably also have been confluent with the Thames at the Black Park Gravel stage (cf. Smug Oak Gravel). To achieve this a gradient of 60 cm km^{-1} would be required to link the Hanger Hill and Richmond Park localities, a distance of 10 km. This gradient is very similar to that for the Smug Oak Gravel and it is therefore tentatively suggested that the Hanger Hill Gravel represents a Brent-equivalent of the Black Park Gravel.

West of the type locality the Black Park Gravel is cut away by younger erosion, but it recurs south of Burnham Beeches at Wingroves (SU947845) where old workings exposed gravels and silty sand in places beneath a degraded surface at 61 m OD. The deposits can be correlated with this unit using height and pebble lithology. Their relation to the Winter Hill Gravel to the north and the Boyn Hill Gravel to the south was summarised in a section by Hare (1947, figure 35). This relationship can still be seen and confirms that the base of the Black Park Gravel stands approximately 10 m below that of the Winter Hill and 10 m above that of the Boyn Hill here, allowing for irregularities caused by erosion at the base and subsidence arising from bedrock dissolution.

Continuing to the west, the area underlain by the Black Park broadens north of Burnham, although the terrace surface is much dissected by dry valleys. A ditch section in Lambourne's Wood (SU930847) exposed 2.5 m of horizontally stratified gravel and sand of comparable composition to that at Black Park. Current-bedding in a narrow sand band indicated flow towards the east-south-east. In places, the gravels are highly disturbed by vertical tongues of calcareous light brown silty clay containing small chalk and flint pebbles that appeared to be injected from beneath. These injections closely resembled those seen at the base of other units (e.g. Hill End: see Section 2.5) and may therefore have arisen from loading forming small-scale diapirs or bedrock collapse. The silty material resembled a mixture of local Reading Beds and chalk debris. A borehole north of the wood (SU928853: 64.5 m OD) proved 6 m of gravel and sand overlying Chalk, while a second near Dorney Wood (SU936851) proved over 11 m. This anomalous depth of gravel, compared to those elsewhere, may again result from bedrock collapse. Irregularities in the terrace surface were attributed to solution by Hare (1947, p. 314).

North of Taplow the terrace and underlying deposits are well represented east of the river. The latter were exposed at Wickenden near Taplow (SU907825: 64 m OD) where a gas pipeline excavation showed 3 m of horizontally stratified undisturbed gravel containing a tabular cross-bedded sand band 20 cm thick. The gravel can be correlated with this unit using height and pebble lithology. A borehole near Taplow Lodge (SU912838: 62.5 m OD) also recorded 3 m of gravel resting on Chalk.

The Black Park Gravel is not found again until Medmenham, upstream of Marlow. A degraded outlier at Bolebec's Castle Hill (SU810848: 72 m OD) comprised up to 3.5 m of stratified sandy gravel according to White (1908) and a bevel both here and at neighbouring Danesfield (SU817857) were assigned by Sealy and Sealy (1956, p. 376) to the Black Park Terrace. Although no exposure could be found during the present study, these outliers are correlated with this unit on the basis of their height (Figure 10).

Upstream the unit is absent until Friars Park, Henley (SU754827). Here a road cutting exposed 4–4.5 m of coarse to medium gravel containing tabular cross-bedded sand in minor channel fills 10–15 cm thick below a clear terrace surface fragment at 75–76 m OD. The deposits rest on an irregular Chalk surface arranged as narrow tongues that vertically penetrate the gravel up to 50 cm. Near the base the gravel was rich in large angular, frost-shattered flints, but above it had a composition comparable with that of this unit (see Appendix).

This gravel fragment is only separated from the main mass of gravels flooring the Caversham Channel at the same level to the south-west by a dry valley. The surface developed on these deposits was correlated by Sealy and Sealy (1956) with their 'lower' Winter Hill stage and the gravels beneath were shown by Walder's (1967) pebble counts from eleven sites to be a single unit with a characteristic pebble assemblage. The deposits were exposed in old workings at Highlands Farm (SU743813: 76 m OD) and comprised stratified fine to medium gravel and current-bedded sand, cryoturbated in the upper 40 cm and overlain by 30 cm of faintly banded, red-stained light brown silt. Pebble counts indicate that the gravel should be correlated with this unit. Observations of steeply dipping strata in sections in this pit by R. R. Inskeep (personal communication) and Wymer (1961, 1968) showed evidence for large-scale collapse of the deposits into bedrock solution cavities in places. That at least some of the collapse was post-depositional was shown by a large thickness of the silt, mentioned above, filling a funnel-shaped hollow in the gravels, at least 3 m deep and 6 m wide at one place. The gravels comprised a lower coarse subunit about 3 m thick containing sarsens and coarse flints and a finer upper subunit about 1.5 m thick of medium to fine gravel, both subunits interstratified with current-bedded sand. Sections in this pit have also been described by White (1896), Smith (1917) and Treacher, Arkell and Oakley (1948) in connection with an abundant artefact assemblage that occurs in these deposits (Chapter 8).

Deposits in a nearby pit in the same surface at Herne's (SU743817: 76 m OD) comprised coarse to medium gravel resting again on a surface comprising pillars of chalk between which the lower 1 m of sediments had collapsed into bedrock 'pipes' (Treacher *et al.*, 1948; R. R. Inskeep, personal communication). Two pebble counts by Walder (1967, table IX) confirm the altitudinal evidence for correlation of this gravel with that at Highlands Farm and this unit.

The terrace surface is well developed at 76–77 m around Highland's Farm and Pond Copse (SU734808). Gravel of this unit was again demonstrated by Walder (1967) at the latter place. Boreholes in this surface at Wootten Manor (SU753816: 77 m OD) near Highland's Farm recorded as much as 6.5 m of gravel resting on bedrock. A gravel sample from near the Wootten site was shown by Walder (1967) to belong to this unit. South-west of the Kingshill Wood dry valley the surface is again well developed beneath Crowsley Park at 78–80 m OD. No exposures were available here but boreholes at SU731797 (78 m OD) and at Frieze Farm (SU720817: 80.5 m OD) again recorded up to 6.2 m of gravel and sand overlying bedrock. An old pit at Morgan's Wood (SU721793) near the latter showed 3.5 m of stratified gravel alternating with fine to medium sand bands, the upper 50 cm being cryoturbated. Pebble counts by the author and Walder (1967) from this site support the correlation with this unit.

Continuing towards the south-west, the surface recurs again around Kennylands. No exposure was available in the pit (SU716790) during this study, but R. R. Inskeep (personal communication) observed up to 5.4 m of stratified fine to medium sandy gravel here in 1955. According to Treacher *et al.* (1948), the lower part of the sequence comprised current-bedded sand and included large rafts of Reading Beds. At the northern end of the pit a dome of Chalk rose to within 2.5 m of the surface while in the north-west corner Treacher *et al.* (1948) noted a depth of up to 10.5 m of gravel and sand. As at Highlands Farm the great variation in the deposits' thickness appeared to result from the presence of large-scale solution of the Chalk, although the bedding of the gravels was undisturbed at this site. This suggests that solution occurred during or even prior to gravel and sand deposition. Nearby boreholes record comparably thick accumulations overlying the Chalk: 7.7 m south of the pit (SU717788) and 5.9 m north of the pit (SU711794).

Old sections at Tanner's Farm (SU712777: 80 m OD) were also visited by R. R. Inskeep (personal communication). Here he recorded 4.5 m of stratified gravel overlain by 60 cm of pebbly clay, although Treacher *et al.* (1948) found a maximum thickness of 7.5 m at this site. Walder (1967) obtained a pebble count that again confirmed altitudinal evidence for correlation with the Black Park stage. The terrace surface extends north of this site to Clockhouse Green and, south of the Highdown Bottom dry valley, into the north of Caversham around Emmer Green Golf Course, where there were previously extensive implement-rich gravel workings (Treacher *et al.*, 1948; Wymer, 1968, and references therein). North of Caversham Heights the terrace surface again occurs and extends westwards as far as Chazey Wood where it is cut-out by the modern Thames Valley. A small pit near this place at SU689762 (81 m OD) exposed to 1.5 m of stratified gravel and sand of comparable composition to those of this unit. Pebble counts from sections in this surface at Blagrave Farm (SU696767) and at Farthingworth Green (SU701767) by Walder (1967) also confirmed this to be part of the main gravel spread flooring the Caversham Channel. Sections at the latter site were observed by Blake (1903), Treacher *et al.* (1948) and R. R. Inskeep (personal communication). The sequence was over 3 m deep and comprised stratified coarse to medium sand 5–50 cm in thickness with gravel bands particularly in the upper 2.25 m. Treacher *et al.* (1948) noted the abundance of large boulders of quartzite and sarsen in these deposits. In addition the gravel and sand has been proved in boreholes at Tokers Green (SU701775: 80 m OD) where

Fig. 12. Location plan of the boreholes at Slade Oak Lane (Higher Denham) for the M25 Motorway (source: ERCU). The position of borehole Q sunk for this investigation is also shown.

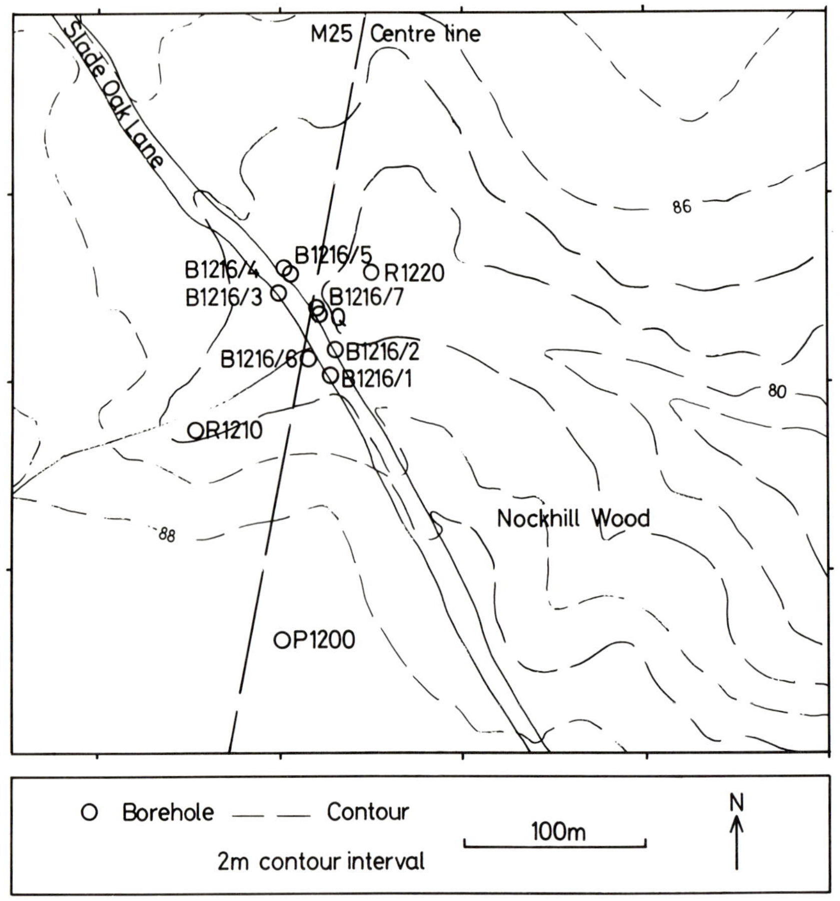

they are 3 m thick, at Chazey Heath (SU695768: 79 m OD) where they are 11 m thick and are overlain by 3.4 m of pebbly clay, and at Chazey Wood (SU687761: 81.5 m OD) where they reach a thickness of 4.7 m.

Pebble assemblages from the Black Park Gravel are characteristic since they are the oldest Thames fluvial gravels in the region that are rich in angular flint (75–89%), poor in rounded flint (3–9%), poor in vein quartz (4–10%) and poor in quartzite (1–6%) (see Appendix). Minor constituents include brown sandstone, Greensand chert (downstream of Henley), Palaeozoic chert, igneous and metamorphic rocks and chalk pebbles. The contrast of this lithology with those of preceding stages is strongly marked and is supported by the multivariate methods mentioned below (Chapter 3). Such a contrast was also noted by Walder (1967, p. 114) in her discussion of the gravels flooring the Caversham Channel compared to those of higher stages. This adds strong support to the correlation proposed here and suggested by Clarke and Dixon (1981).

There is little evidence of down-valley changes in pebble assemblage particularly the frequency of vein quartz and quartzite. This probably arises from incorporation of older quartz-rich gravels and from confluence with the quartz-bearing Smug Oak Gravel river from the Vale of St Albans at Uxbridge. The confluence of the Kennet, partially working Blackwater gravel, with the Thames at Henley (Gibbard, 1982) is supported by the absence of Greensand chert above this place, as also pointed out by Walder (1967).

The resulting gradient for the Black Park Gravel obtained by plotting the localities mentioned above is 50–65 cm km^{-1} and is shown in Figure 10. This figure corresponds exactly with that stated by Hare (1947) and Gibbard (1979).

It would therefore appear that at this stage the Thames was aligned along the floor of the Caversham Channel between Reading and Henley, flowed through the Middle Thames region and, as shown by Hare (1947) and Gibbard (1979), became the first unit to follow the present route through London to Essex (Figure 62b).

The brown silt overlying the gravels and infilling solution hollows at some localities in the Caversham Channel appears to represent a slope wash deposit, derived from local Reading Beds or Clay-with-flints, to judge from its indistinct stratification and variable thickness. The high silt content of this sediment may have been in part derived from loess. Although pebbly clay overlying the gravels near the valley side may be of solifluction origin, as discussed above (Section 2.7), there would seem to be no structural or fabric evidence to suggest such an origin for this material. It clearly post-dates the gravels, but its involvement in collapse structures and in periglacial cryoturbation suggests that it might be of considerable antiquity. The red-staining noted at Highlands Farm may be of pedogenic origin which in future could provide a minimum age for the aggradation.

2.10 Slade Oak Lane deposits

During site investigations for the M25 Motorway, boreholes for a bridge at Slade Oak Lane, Higher Denham penetrated anomalous depths of sediments resting on Chalk. The sediments were sampled in June 1980 by a borehole at point Q (see Figure 12). The complete investigations of the deposits are being prepared for separate publication by P. Gibbard, I. Bryant and A. Hall.

The deposits, proven in a series of boreholes and trial pits, underlie a small, seasonally dry valley trending east to east-south-eastwards (Figures 11 and 12). This valley dissects the plateau, the Misbourne–Colne interfluve, which is underlain by gravels and sands of the Gerrards Cross Gravel member (section 2.7). The surface developed on the gravels is the degraded Harefield Terrace (Hare, 1947). The gravels overlie 14–16 m of Reading Beds clays, which in turn rest on Chalk. The valley is underlain by a very steep-sided, funnel-shaped basin with a minimum diameter of 35–40 m and a maximum observed depth of 37.5 m (Figure 11).

The sequence filling the hollow can be divided broadly into three units as described below.

(a) Basal sands, clays and gravels

This unit comprises a highly complex succession of sorted and non-sorted sediments. The lowest sediment is highly compact yellow sand with brown silty clay partings and is up to 8.20 m thick. The sands are succeeded by as much as 10.80 m of very compact angular to subrounded sandy gravel in a light grey clay matrix. The gravel is rich in frost-shattered, but otherwise unweathered, flint, it also contains rounded and rolled flint, vein quartz, quartzite and sandstone pebbles. The gravels are overlain by predominantly massive light grey, green or buff sandy silty clay with angular pebble concentrations at some levels. Occasional beds of sorted buff sand also occur (up to 20 cm thick). The sediment matrix is very rich in silt (up to 45%). This unit is commonly disturbed by high-angle, near-vertical normal faults with throws of several centimetres.

(b) Organic silty clay
Fossiliferous dark grey to black silty clay overlies the previous sediment. The sediments are finely laminated throughout much of their depth and contain irregularly distributed wood and clasts.

The microstratigraphy of this unit in borehole Q is as follows. Depths are given in metres below ground surface (82.8 m OD).

<12.70 m	Light grey brown silty clay, grades into
12.70–12.81 m	interstratified grey brown sandy silt and dark grey silty clay;
12.81–13.29 m	massive dark grey silty clay with scattered clasts and narrow sand bands;
13.29–13.47 m	black sandy clayey silt with small stones;
13.47–13.67 m	grey clayey silt with small stones, clay clasts and plant remains;
13.67–13.85 m	indistinctly stratified grey clayey silt with clay clasts and plant remains;
13.85–13.89 m	black organic clay mud with plant remains;
13.89–16.51 m	indistinctly stratified dark grey brown clay mud with a few plant remains, wood, small pebbles and grey silt clasts;
16.51–16.58 m	interstratified dark brown silty clay and sand;
16.58–17.03 m	finely interstratified light grey silty clay becoming siltier with depth. Some plant remains. Base very sharp and irregular. Small-scale loading structures at base;
17.03–17.21 m	light grey silty clay with coarse sand interbeds;
>17.21 m	grey silty clay with pebbles and angular flints.

(c) Pebbly silty clay
There is an upward transition from the organic clay muds into light brown silty clay between 12.70 and 12.81 m resulting from interbedding of the sediments. Above 12.70 m the brown silty clay continues up to the surface, becoming progressively more orange brown with secondary iron oxide and black manganese dioxide precipitation above 12.00 m. Pebbly horizons are common. Some fine lamination (laminae 1–2 mm thick) is present, consisting of an alternation of orange sand and grey silt and clay. The deposits are extremely rich in silt (up to 40%) which may represent primary or remobilised loess on the basis of grain-size analyses. In the upper part of the sequence the sediment is highly fissured. Its maximum thickness is 15.15 m.

In order to establish the three-dimensional shape of the basin a geophysical resistivity survey was undertaken. A series of eight traverses using a terrameter with the 'Wenner Configuration' producing a low-frequency, high-voltage current was adopted. These investigations indicated that the deposits are absent to the east and south-east of the borehole sites and therefore that the hollow is an enclosed basin.

Origin of the basin
To judge from its form, position and size it seems most likely that the basin originated by bedrock solution. Its funnel-like shape and proportions closely resemble those of dolines (Sweeting, 1972, p. 44), although it is rather larger than solution features previously described from the Chalk, with the exception of the Breckland Meres. Smaller examples are common in the neighbouring district, and include those around Burnham Beeches (Hare, 1947), from High Wycombe and Beaconsfield (Higginbottom and Fookes, 1971), from the Chiltern Hills (Avery *et al.*, 1982) and from beneath gravel aggradations (see Chapter 5). Sperling *et al.* (1977), in a discussion of Dorsetshire dolines, showed that these structures are concentrated where a thin cover of Palaeogene rocks or plateau deposits overlies the Chalk. This is precisely the situation at Slade Oak Lane. Dolines are thought to form by intense solution activity along joint planes or fissures in the rock.

Materials enter either by an 'hour-glass' effect, by blowing in directly, or by streams flowing into them, and the filling may be highly variable.

Infilling of the basin
The hollow was initially infilled by sorted sand possibly washed in by a stream. The overlying clayey gravel seems to represent detritus that slipped or flowed by solifluction directly into the hollow from the sides during subsidence. The material was probably derived from a mixture of Reading Beds and the overlying gravels together with loess either washed or blown into the doline. The interbedded clay and sand bands may represent periodic inwash or accumulation in standing water. The faulting present suggests that the basin was actively subsiding during this time.

The transition from these deposits into the organic clay muds appears to mark the stabilisation of the basin. Sediments beneath the transition are unweathered and therefore the basin probably never dried-out in the intervening phase. The organic sediments accumulated in a shallow pool into which occasional minerogenic flushes occurred, possibly following storms.

The transition into the overlying silty clay occurred during a period of opening of the contemporary vegetation cover (Chapter 7) and probably therefore resulted from inwash of soil and regolith into the basin. The return to periglacial climates is heralded by the appearance of massive pebbly silty clays of solifluction origin, possibly accompanied by loess deposition.

There is no evidence of subsidence of the hollow after the initiation of organic deposition. The Slade Oak Valley and other valleys in the area (Figure 11) were excavated into the sedimentary fill and therefore post-date it.

2.11 Boyn Hill Gravel (Boyn Hill Terrace)

The Boyn Hill Terrace is arguably one of the most well known features in the British Pleistocene. It has been recognised throughout the Middle and Lower Thames valleys and is regarded as being one of the best-dated terraces in the country.

The term Boyn Hill was first used by Bromehead (1912, p. 74) to distinguish the Thames terrace deposits during the geological survey of the Berkshire – south Buckinghamshire areas. The term Boyn Hill Gravel was used by Dewey and Bromehead (1915, p. 67) who, together with Sherlock and Noble (1922), provided the first systematic descriptions of the terrace deposits and defined a type site now in the suburbs of Maidenhead at Boyn Hill, although the terrace deposits in the area were first mapped by Whitaker (1889). Bromehead (1912) recognised that the Boyn Hill, as he defined it, was a complex unit and even before this Treacher and White (1909) noted that two terraces were present north of Maidenhead. King and Oakley (1936, p. 71) recognised this problem and separated the lower unit as their Iver stage. The latter was incorporated by Hare (1947) into his Lynch Hill Terrace. Hare also examined the Boyn Hill and mapped its occurrence within his area. The work was extended upstream by Sealy and Sealy (1956) who showed that the terrace was aligned along the modern Thames course through Reading rather than via the Caversham Channel (See Section 2.9).

The use of the terms Boyn Hill and Lynch Hill in this study follows Hare's (1947) usage, rather than that of the Geological Survey.

Downstream the terrace was recognised and mapped throughout the length of the Thames Valley. Bromehead (1912) suggested that it was equivalent to the 'uppermost or 100 foot' terrace in the Lower Thames and Dewey and Bromehead (1915) indicated that the implement-rich and fossiliferous deposits at Swanscombe, Kent (Smith and Dewey, 1913) were probably of Boyn Hill age. This correlation has never been questioned since. Wooldridge and Linton (1955), discussing the recognition by Holmes (1892) that gravels of the Lower Thames 'uppermost' terrace overlie Chalky Till at Hornchurch, Essex, suggested that the aggradation must post-date the eastern (Anglian) glaciation. Since the Swanscombe sequence, or at least a part of it, has long been considered to be of Hoxnian interglacial age, it has been taken as the type section for the 'uppermost' or '100 foot' terrace in the Lower Thames (King and Oakley, 1936; Oakley, 1952; Ovey, 1964; Wymer, 1968; Kerney, 1971; Shephard-Thorn and Wymer, 1977). The terrace has therefore been regarded by many as being largely of Hoxnian age throughout the Thames Valley.

The original type localities of the Boyn Hill Gravel at Maidenhead (Figure 1) are difficult to determine from the literature, but according to Lacaille (1961b) and Wymer (1968) following Sherlock and Noble's (1922) mention of the area, the main pits may have been situated around the cemetery (*c.*SU875810). The terrace surface here is well developed at 51–52 m OD. No exposure exists in the area today. Similarly, the sections at the Boyn Hill Church pit (SU878809), Cooper's pit (SU876819), Badnell's pit (SU878814), the Upper Furze Platt pits (SU878826, 876827) and the Shrubbery pit (SU877821) are also unavailable (Lacaille, 1961b; Wymer, 1968). These sections exposed 3.5–9 m of poorly sorted, stratified gravel according to Lacaille (1961b). At Winch's pit (SU872812) Lacaille (1961b) and Wymer (1968) reported 'a layer of large, angular slabs of tabular sandstone', possibly derived from the Reading Beds, within the gravels and sands. A temporary trench in an old pit at SU873813 exposed 4 m of poorly stratified gravel and sand the upper 1 m of which was cryoturbated. This gravel was sampled and will serve as the type section during this study.

No good sections therefore remain in the Boyn Hill Gravel at the type locality. The name has, however, been retained for the sake of clarity and historical precedent.

Only two outliers of this unit occur on the western side of the river near Maidenhead. However, downstream it is better represented. An isolated patch of the gravels is present near Hitcham Church, where the terrace surface was mapped by Hare

Fig. 13. Long profile section showing the deposits correlated with the Boyn Hill Gravel.

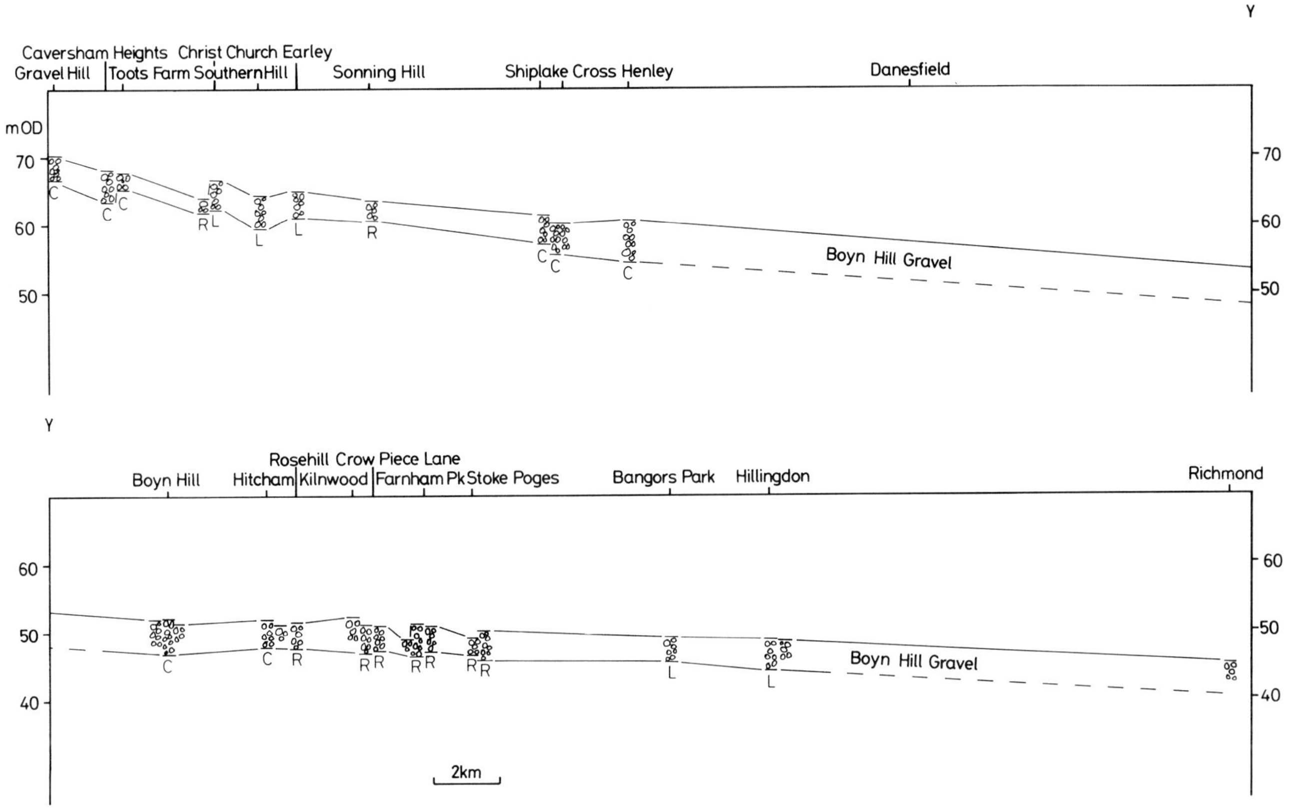

(1947, p. 316) at 52 m OD. Here a borehole in a degraded sloping surface near Taplow (SU914825) recorded 3.7 m of gravel and sand overlying Chalk. East of the Hitcham Park dry valley, an old gravel pit (SU924827: 51 m OD) showed 2 m of poorly exposed gravel and sand. A pebble count reinforces the altitudinal correlation of these deposits with those at Boyn Hill. This pit occurs in the large spread of gravel underlying a well developed although somewhat irregular terrace surface reaching up to 53–54 m OD, north of Burnham (Lacaille, 1939; Hare, 1947). It is probably the same site as that called Britwell by Wymer (1968). Several sections existed in this surface in the past and showed the deposits' variability. According to Hare (1947, p. 318), at Rose Hill (*c*.SU923841) the deposits were 'loamy, with considerable brickearth layers and only occasionally coarse gravels.' However, he continued that 'round Burnham the drift is a coarse, well bedded gravel with some large boulders.' A nearby borehole (SU924838: 53 m OD) recorded 3.7 m of gravel and sand resting on 1 m of Reading Clay: the gravel was overlain by 1.8 m of brown pebbly clay. The slope of the bluff north of this area is formed in part of Reading Beds and these probably provided the source for the interstratified 'brickearth' or pebbly clay material. The stratified deposits occur at the expected height and are therefore correlated with this unit (Figure 13).

Further east, the surface continues south of Burnham Beeches where it is again well developed. Exposures at Kiln Wood (SU943844: 54 m OD) confirmed the presence of 3 m of gravel and sand of this unit, as shown by the pebble lithology (see Appendix). At nearby Wingroves (SU947845), the relationship with the next higher unit, the Black Park Gravel, can be seen (see Section 2.9). The most satisfactory modern section in these deposits occurred at Crow Piece Lane (SU946837: 54 m OD). Here large sections exposed a sequence of 3.5 m of horizontally stratified, frequently massive medium gravels, coarse at the base, containing narrow trough and tabular cross-bedded sand bands up to 15 cm thick. Some of the sand bands contained small chalk clasts. An irregular surface of Reading Beds floored the pit. The gravels were overlain by 1.5 m of brown pebbly clay, the contact between the deposits being heavily disturbed by cryoturbation. Pebble composition of the gravel (see Appendix) corresponds closely with that from Boyn Hill. Palaeocurrent measurements indicated a flow towards east-north-east. This quarry is very close to Deverill's (SU947840) and Cooper's pits (SU945843) (Lacaille, 1961b; Wymer, 1968). The sequences they described closely resemble that from Crow Piece

Lane. Wymer (1968) mentioned several other pits in this altitudinally continuous gravel spread including Burnham Gore (SU924828). Three boreholes in the surface confirm the presence of 3.2–5.2 m of gravel overlying bedrock.

The relationship of the Boyn Hill Gravel member to the lower, Lynch Hill Gravel member was excellently illustrated by a section along Deep Lane, Burnham, in Wymer (1968, figure 83) (*c.*SU954830). Here the former was shown to be separated from the latter by a bluff of bedrock. The base of the Boyn Hill was 7 m above that of the Lynch Hill at this place.

North of Farnham Royal, the terrace surface is again well developed at 51–52 m OD, although it is degraded on its southern side to produce a gently sloping bluff surface termed by Hare (1947, p. 318) the Stoke Park Cut (see below). The gravels and sands of the Boyn Hill member have been recorded in several boreholes in the area. Those at SU961835 recorded 5.2 m of gravel overlain by 1 m of brown pebbly clay, while a borehole in Farnham Park (SU966835: 51 m OD) proved 3.3 m of gravel on Reading Beds. No good exposures existed in this area during this study, but an old pit at SU959828 (Farnham Royal Church) exposed 1.5 m of *in-situ* gravel and sand from which a pebble count sample was taken. The latter (see Appendix) indicates some incorporation of material derived from an older gravel unit, although the gravel is altitudinally equivalent to this member. This pit was also mentioned by Wymer (1968).

On the basis of altitude, the unit is present as narrow spreads at Stoke Poges and Rowley Wood. Boreholes at SU977842 (49 m OD) and SU983840 (61 m OD) proved 2.3 m and 9.1 m of sand and gravel, respectively, resting on Reading Beds and London Clay. In the first borehole the gravel was overlain by 2.3 m of brown pebbly clay, while the latter must certainly include this type of material. A borehole in the north-western corner of Rowley Wood (SU997836: 60 m OD), immediately beneath the London Clay backslope, again recorded interbedded sorted gravel and pebbly clay, the latter of probable mass-flow (solifluction) origin.

East of this area the Boyn Hill Gravel is cut-out by erosion associated with the Lynch Hill member. The Boyn Hill Terrace surface was mapped by Hare (1947) at Bangors Park, Iver. However, no exposure could be found here by the present author, but two boreholes at TQ029822 and TQ035820 (52 m OD) confirmed the occurrence of up to 3.5 m of gravel and sand at the expected height resting on London Clay.

No further outliers of Boyn Hill Gravel seem to occur on the western side of the Colne, except for a marked patch of gravel slightly above the accordant height at 53 m OD north of Broom Warren (TQ024845). The Boyn Hill surface was mapped here by Hare (1947). Boreholes for the M25 Motorway indicate the presence of a well developed terrace gravel unit here (Figure 11) dissected by a small valley. The unit comprises up to 6.5 m of gravel and sand underlain by London Clay and overlain by a brown occasionally pebbly clay that wedges out away from the bounding bluff, towards the south-east. The position of these gravels prevents them from being of Thames origin, since they are separated from patches to the south by the higher, wide Black Park spread. They therefore must be a Colne remnant, equivalent to the Boyn Hill Gravel of the Thames, possibly formed at the Colne–Alderbourne confluence. No exposure of these deposits was available, but since it is not of Thames origin, it should be assigned a unit name of its own. It will therefore be referred to here as the Broom Warren Gravel. In the section (Figure 11) the relationship of this unit to the higher Black Park Gravel can be seen. The separation of the unit bases is 12.5 m.

Very few patches of the Boyn Hill are preserved downstream of the Colne. Allen (1978) suggested that flat surfaces present at 46–49 m OD at Moorcroft (TQ078817) and Hillingdon Heath (TQ079820) may be Boyn Hill Terrace equivalents. Boreholes near the latter locality recorded 7 m of gravel and sand below a surface at 50 m OD. A short section (Figure 14) also recorded the presence of gravel at the accordant height beneath Hillingdon Hospital (TQ069821). In this section, the base of the Black Park Gravel is 11 m higher than that of the Boyn Hill, which is in turn 11.5 m above the base of the Lynch Hill Gravel.

On the north side of the river, deposits on Castlebar Hill, Ealing (TQ171816: 52 m OD) have been mapped as Boyn Hill by the Geological Survey. They comprise 'sub-angular river drift gravel, surmounted by a deposit of rounded pebbles of apparently greater age' (Brown, 1887). Their position on the western side of Hanger Hill suggests that they may be a Brent equivalent rather than a Thames deposit. The 'rounded pebbles' are probably again a mass-flow deposit derived from higher ground comprising Tertiary sands and pebble beds.

South of the river, two outliers in Richmond (TQ185737: 45 m OD) have been shown by Gibbard (1979, figure 5) to fall on the surface gradient of 30 cm km^{-1} for this unit. Similarly, the deposits at Herne Hill (TQ324749: 43 m OD) and Islington (TQ313837: 42 m OD) may be equated with this member. Lack of exposure prevented any correlation of these occurrences using lithological techniques.

Fig. 14. Section from the Hillingdon–Yiewsley area, based on exposure and borehole data (source: IGS).

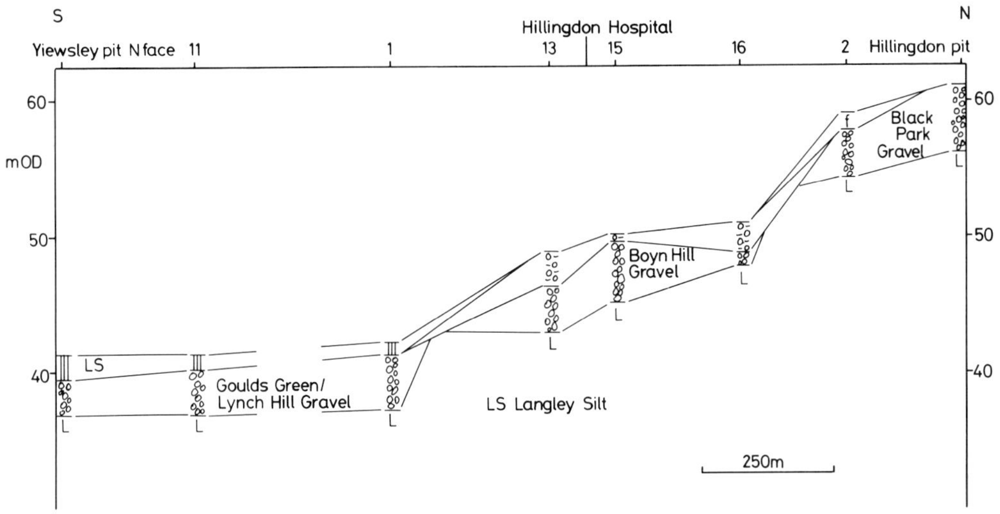

Other gravels mapped by the Geological Survey as River Thames Boyn Hill stage deposits in Central London are too low for inclusion in this unit.

Upstream from Maidenhead, the Boyn Hill is poorly represented. Sealy and Sealy (1956, p. 376) recorded several small terrace relics on the spurs on the north side of the river from Danesfield (SU816845) eastwards at 65 m OD. Westwards they are cut-out by the present valley. Extension of the gradient of 30 cm km^{-1} upstream suggests a height of 58 m OD at Danesfield. It would appear as noted by the Sealys (1956) that these outliers are rather high, especially considering their degraded form. It may be that they represent fans derived from tributary valleys or that they are not remnants of the Boyn Hill, but soliflucted gravel from higher upslope. Lack of sections prevented clarification of this point. A degraded remnant of an altitudinal equivalent of this unit was worked in Henley at SU760817. Bedded gravels up to 6 m deep were described from here by Treacher (in Wymer, 1968) beneath the surface at 61 m OD.

If the Slough area gradient is projected further upstream it intersects exactly with the gravel spread at Shiplake Cross (Figure 13). A section at Shiplake House Farm (SU770788: 60 m OD) displayed 3.5 m of poorly stratified to massive sandy gravel beneath a partially degraded terrace surface. Wymer (1968) reported seeing sections here with dipping strata apparently resulting from bedrock solutional collapse. The Chalk surface beneath was irregular and potholed. According to Wymer, Treacher apparently obtained numerous hand-axes from these gravels. A pebble count sample from the gravels indicates that they are of comparable composition to those at Boyn Hill with which they are correlated. A pit beside Henley Road at SU764783 (64 m OD) was described by Shrubsole (1890) as having sections up to 4 m high of 'sub-angular gravel resting on Chalk.' 'Lumps of conglomerated chalk' were present at the base according to Stevens (in Wymer, 1968). Several boreholes (Figure 13) penetrated the gravels and sands here and recorded up to 4.7 m of sediment resting on Chalk.

South of Sonning Town and into the southern central part of Reading around Christ Church, a partially degraded Boyn Hill surface that projects between the Thames and Loddon valleys was mapped by the Sealys (1956) and Wymer (1968, figures 47 and 60). It is 63 m OD at Sonning Hill, but rises to 65 m at Earley and 67 m at Christ Church, Southern Hill and therefore falls into line with the extrapolated gradient for this unit. A number of sections in this surface have been described by Shrubsole (1890), Monckton (1893), White (1896), Blake (1903) and Wymer (1968), but few were available during the present work. The Sealys (1956) did not stress the importance of these sections which by the presence or absence of southerly-derived Greensand chert record the position of the confluence of the Loddon–Blackwater with the Thames in the area at this stage. This was particularly noted by Blake (1903). The easternmost sections were described by Shrubsole (1890) and Blake (1903) at Sonning Hill (SU754741: 63 m OD) where 3 m of stratified gravel with some clay bands resting on Reading Beds were exposed. Greensand chert was rare at this site. Conversely, beside the Southern Region railway line at Woodley Hill (SU744726: 65 m OD) old pits showed a maximum of 4.5 m of gravel rich in Greensand chert, but lacking quartzite (northerly-derived) pebbles. Temporary building-site trenches at 67 m OD in Beech Lane, Earley (SU747707) exposed up to 50 cm of stratified gravel. A pebble count (see Appendix) showed it to be rich in Greensand chert capped by 70 cm of brown pebbly clay, and therefore probably a Loddon equivalent of this unit. Several small sections of quartz-rich gravel up to 3 m thick were recorded by Shrubsole (1890) and Blake (1903) on Southern Hill (SU725720) that reaches 67 m OD. The gravels vary considerably in thickness reaching up to 4.5 m and resting on London Clay near Christ Church (SU721721). At Edgehill Road, Reading (SU716723), over 2 m of poorly stratified gravel filling channels scoured into the Reading Beds were exposed. A pebble count from these gravels supports correlation with the Boyn Hill member and contains almost no Greensand chert, indicating that the site was located upstream of the Loddon confluence. A borehole at SU734717 (64 m OD) proved 6 m of gravel overlying London Clay and capped by 3.3 m of brown clay. The position of the site, close to the southern bounding London Clay bluff, suggests that the overlying clay was probably derived from this slope.

Deposits altitudinally continuous with this unit occur upstream near Tilehurst (SU669748: 72 m OD) according to Wymer (1968). This could not be confirmed by the author during the present study.

Boyn Hill Gravel deposits are not seen on the south side of the river further west. However, on the north side beneath Caversham Heights a spread of gravel 1.5 km wide by 2 km long occurs. These deposits underlie a well developed although gently sloping terrace surface from 71.5 m OD at Conisboro Avenue (SU701765) to 68 m OD at its southern margin. Numerous old pits were worked in this area and an exhaustive list was presented by

Wymer (1968). The most important implement-rich site was at Toots Farm (SU707753) where Shrubsole (1890) saw 2.5 m of stratified subangular gravel. He also noted that Treacher found a decayed tooth of *Equus* in these deposits. The probability that the surface is degraded at its southern extremity was suggested by Wymer (1968) on the basis of Treacher's (1904) observation that only 2 m of gravel occurred here. Much greater thicknesses of gravel are known from Richmond Road (SU703758) where 4.2 m of gravel rested on Chalk. However, excavations at St Peter's Hill (SU708750) by Treacher (1904) showed that the gravel bedding was much disturbed by collapse into a solution hollow in the underlying Chalk bedrock. The gravel was overlain by light brown pebbly clay at this place. No sections were available here during this study. However, at Gravel Hill near Blagrave Lane (SU690756) a small section in the surface continuous with that beneath Caversham, exposed 2.5 m of bedded medium to fine gravel with a coarse sand matrix. Cross-bedding was visible in one of the sandy horizons that indicated flow towards east-north-east. Chalk was visible in the base of the pit 4 m below the surface at 71 m OD. A pebble count confirmed the altitudinal evidence for correlation of the gravels with this unit.

Pebble count analyses from the Boyn Hill Gravel (see Appendix) show that it comprises 85–91% flint (77–81% angular flint, 5–10% rounded flint), 4–7% vein quartz, 1.5–5% quartzite, 2.5–4% Greensand chert (downstream of Sonning) and less than 1% of other types (excluding the anomalous Farnham Royal count and the Loddon gravel at Beech Lane, Earley, see above). The sample number was, however, rather small because of lack of exposure.

As mentioned above, plotting of the sites gives a gradient of 30 cm km^{-1} (Figure 13) and this was also found by Hare (1947) and Sealy and Sealy (1956). It is somewhat more gentle than that of most higher units. The reason for this is not known.

An erosion feature called the Stoke Park Cut was mapped by Hare (1947, pp. 318–19 and figure 13) east of Farnham Royal. This feature cuts-out the Boyn Hill Terrace and gently slopes down to the Lynch Hill Terrace surface to the south. In one place, a bluff separating this cut from the Boyn Hill surface was mapped by Hare (1947) as a flat terrace-like surface, yet he noticed that it is developed on London Clay with only a few thin patches of gravel in places. Allen (1978) extended this surface into the Yiewsley area where he mapped it as a terrace landform. Sections examined by the present author, for example at Stoke Poges (SU973833) confirm Hare's (1947) view that the surface is not underlain by sorted, fluvial deposits, but by London Clay overlain by 1–1.5 m of brown pebbly clay. The latter was probably remobilised London Clay mixed with pebbles derived from higher gravel spreads and disposed in bands occasionally preserving the original flow-folding in the silty clay matrix. These deposits probably formed by solifluction. For this reason the Stoke Park Cut should not be regarded as a depositional phase of the River Thames or tributaries.

2.12 Lynch Hill Gravel (and in part, Lynch Hill Terrace)

The Lynch Hill Terrace was first defined as a morphologically distinct unit intermediate between the Boyn Hill and Taplow Terraces by Hare (1947, p. 319). This terrace was equivalent to the so-called Iver Terrace of Oakley (in Lacaille, 1936) or the Iver stage of King and Oakley (1936, p. 71). These authors confirmed the earlier observations by Treacher and White (1909) from Maidenhead, and Burchell (1934) in the West Drayton and Yiewsley area, of a terrace unrecognised by the Geological Survey. This stage was frequently referred to as 'Lower Boyn Hill', for example by Lacaille (1939, 1940) or the 'Upper Taplow' Burchell (1934). Wright (1937, p. 126) went as far as calling it the Furze Platt Terrace on the basis of the Maidenhead deposits. These authors not only separated the terrace altitudinally, but also by comparison of artefact assemblages in the underlying deposits.

The deposits underlying the terrace as defined by Hare (1947) include some of the most productive Palaeolithic localities in the country and have been studied intensively by archaeologists for nearly one hundred years (for summary see Wymer, 1968; Roe, 1981). It is from these investigations that most of the site descriptions are derived. However, in parts of the area the terrace surface is underlain by up to three lithostratigraphically distinct units. For the sake of precision, the term Lynch Hill Gravel as defined here, refers only to the fluvial gravels and sands resting on bedrock at the type site and equivalent localities. Overlying units of differing lithology and structure are excluded and will be described separately below.

The type section of the Lynch Hill Gravel is at Lynch Hill, north of Slough (SU950822: 41 m OD). Here 4 m of horizontally stratified sandy medium gravel and interbedded tabular cross-bedded sand resting on Reading Beds were poorly exposed during this study. A pebble count from the gravel is given in the Appendix.

Similar sections were seen by Wymer (1968) at SU952821. Only 500 m to the east is the prolific artefact locality of Baker's Farm (SU958822). This site, now built over, showed a sequence similar to the type section, the gravels being 3.5 m thick here and overlain by 'festoons of white angular gravel in an argillaceous matrix' and 1.5 m of clay silt ('brickearth'), according to Lacaille (1940). A section in the extensive terrace surface at Cocksherd Wood (SU944829) showed poor exposures in 1978 of up to 4 m of stratified gravels of similar composition to those at the type section (see Appendix) overlain by brown pebbly silt. A comparable sequence was recorded by Wymer (1968, figure 83) in Deep Lane (SU954826).

Hare (1947, p. 320) traced the Lynch Hill surface eastwards to a wide relic around Stoke Green (39–40 m OD); however, the presence of gravel here could not be confirmed because of a lack of sections.

The gravels are next exposed south of Iver where the surface is at 36–37 m OD. Large areas have been worked here in the past and were summarised by Lacaille (1940) and Wymer (1968). The deposits in this area, that extends from Mead's Cottages (TQ024805) to the Colne Valley, east of the Iver–Thorney road, were observed in an old pit at TQ036803, at Purser's pit (TQ032797) and in canal bridge parapet excavations (TQ039084). The sequence appears to be consistent across the whole area and comprises horizontally bedded, sandy medium gravel containing isolated, current-bedded sand and pebbly sand channel fills in places. The basal 20–50 cm of gravel is frequently coarse, poorly sorted and rests on irregularly channelled London Clay. The stratified deposits are 3.5–4 cm thick. Palaeocurrent measurements indicate a flow towards the east-south-east. A pebble count from the old pit (Iver) showed that the gravels have a composition comparable to that at Lynch Hill with which this gravel is correlated (see Appendix).

The upper part of the gravel is cryoturbated to a depth of 1–1.5 m. Frost-shattered pebbles, mostly flints and many with vertical long axes, are abundant in this layer, and the gravels are overlain in all sections by red brown clayey silt. In some sections described by Lacaille and Oakley (in Lacaille, 1940) an angular flint gravel with a clayey matrix varying from 0.5–1.5 m in thickness and also heavily contorted by cryoturbation occurred between the stratified gravels and the overlying silts. This probable solifluction deposit (Oakley, in Lacaille, 1940) appears to be consistently present in the area.

Several boreholes penetrate the deposits near Iver. Those for the M25 Motorway proved up to 4.5 m of gravel and sand on London Clay, although most of the material has been removed in gravel workings (Figure 11). A borehole at TQ016811 recorded 6.7 m of gravel and sand overlain by 0.7 m of brown pebbly clay. The gravels of this unit also occur in several boreholes on the north side of the canal at TQ036804 where they are 2.9–4 m thick and are overlain by 1.1 m of silt.

As noted by Hare (1947) and Hey (1976) extension of the gradient of 45–55 cm km^{-1} downstream intersects the deposits present beneath Yiewsley and West Drayton where extensive workings continued until recently. The surface in this area varies between 39 m at Gould's Green (TQ082808) to 34 m at Hayes Town and Dawley (TQ087797). Sections in the area have been described by several authors and were summarised by Dewey and Bromehead (1915) and Wymer (1968). The main sites that have produced extensive artefacts were at TQ071805 and TQ080800 north of the Grand Union Canal. South of the canal at Dawley several old sections also existed at TQ085797, 085804 and 082802 (Monckton, Mangles and Salis, 1904). The most recent study in the area was that of the large Sabey's workings (TQ082803–083807) by Collins and co-workers (1978) north of the canal. These authors provided detailed cross-sections of the site (Collins, 1978, figure 7) and presented complex subdivisions of the deposits. They suggested that two separate 'terrace' gravel aggradations occurred here. The main altitudinal correlative of the Lynch Hill, their Warren gravel, reaches a maximum thickness of 6 m, the upper surface being at about 34 m OD and its base resting on London Clay (Figure 15). The second gravel unit occurs immediately north of the first and is 2.5 m thick with its base at 34.5 m OD sloping uphill northwards to 37.5 m OD near Gould's Green (the Gould's Green gravel). In the sections the two units were continuous, but are now separated by a narrow ridge of London Clay that closely resembles a diapiric structure. Both gravels are overlain by what Collins (1978) called loam and these gravels have been assigned several names in his sections, but with no lithological criteria to support their separation.

Observation of the Yiewsley sections by the writer confirms many of the observations recorded by Collins (1978). Sections in the southern part of the quarry (TQ083803) exposed horizontally bedded sandy gravel with occasional current-bedded coarse sand bands indicating a palaeocurrent flow towards the east. Pebble count samples summarised in the Appendix indicate that this gravel is of a similar composition to that at Lynch Hill. However, sections at the northern end of the pit (TQ084807) exposed up to

Fig. 15. Long profile section showing the deposits correlated with the Lynch Hill Gravel.

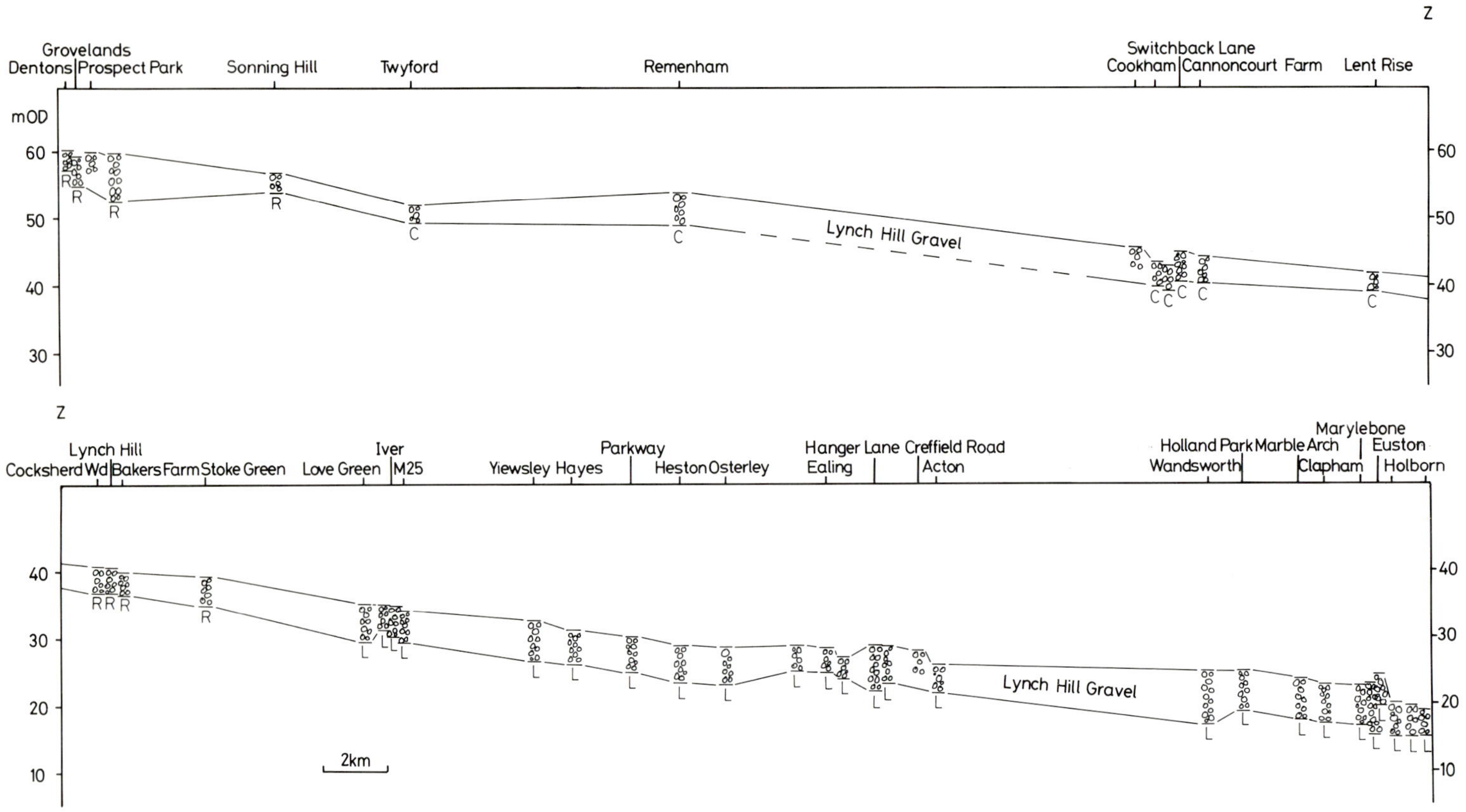

3 m of medium to fine gravel in a coarse sand matrix interbedded with cross-laminated sand bands 10–15 cm thick (Gould's Green gravel). Current-bedding in these sections indicates a flow towards the south-south-east. A pebble count from this section shows that the gravel contains markedly less Greensand chert and slightly more vein quartz than that in the southern sections (see Appendix). This count, although similar to that of the Lynch Hill Gravel (*sensu stricto*) can be compared with that of the Colney Street Gravel flooring the Colne Valley (Gibbard and Hall, 1982 and Table 3). The latter rarely contains Greensand chert and is slightly richer in vein quartz than the gravels of the Thames. In view of the position of the gravel unit, its downslope gradient, pebble composition and palaeocurrent direction, it is strongly suggested that it represents a River Colne-equivalent of the Lynch Hill adjacent to its confluence with the Thames and not a fragment of a higher Thames terrace as suggested by Collins (1978). The term Gould's Green Gravel has been retained for this member. The upper 60 cm of the gravels in these sections was disturbed by cryoturbation and the entire sequence at Yiewsley was overlain by 1.5–3 m of silt ('brickearth').

Gravels 3.5 m thick at the same height as those at Gould's Green occur in boreholes south of Hillingdon Hospital (Figure 14), at Collum Green (TQ069817) and at Yiewsley Nursery (TQ076810) and are equated altitudinally with the Gould's Green member.

The Lynch Hill Gravel can be traced eastwards with few breaks. It is present beneath the EMI factory (TQ094798) where it is 5 m thick. A north-south borehole section along the Parkway, Cranford, and its northward extension to Uxbridge Road, Hayes (Figure 16), recorded up to 6 m of gravel and sand altitudinally equivalent to this unit overlying London Clay. The unit is dissected by the River Crane Valley from the Western Region main line southwards for about 1 km, as shown by the organic floodplain deposits covering thin (2.5 m) gravels. It is, however, clear that in this section the terrace is degraded and the Lynch Hill Gravel unit thins southwards. About 250 m south of the M4 junction (TQ104778) the lower, Taplow Gravel aggradation overlain by silt abuts the higher unit. The base of the Lynch Hill Gravel is 7.5 m higher than that of the Taplow in the section. Numerous boreholes nearby record comparable thicknesses (3–5 m) of gravel and sand of the Lynch Hill member, for example, beneath Nestlé's factory (TQ102793) and Heston Aerodrome (TQ111788 and 118783). Brown (1889, 1896) saw 7.5 m of gravel and sand beneath silt at Southall Gas Works (TQ119797). How-

Fig. 16. Section along the Cranford Parkway and proposed extension to Uxbridge Road, Hayes, constructed using borehole records (source: GLC).

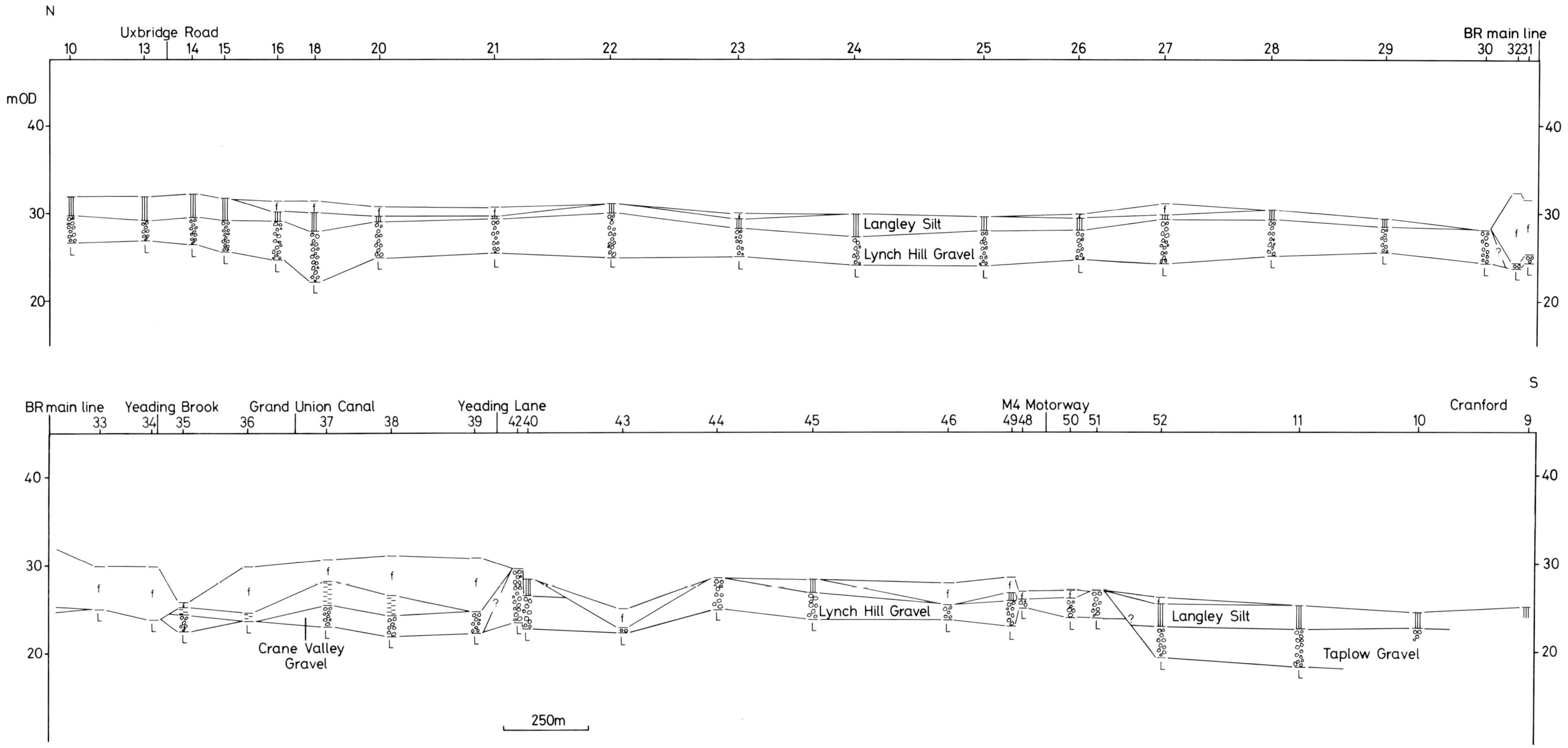

Fig. 17. Section along the M4 Motorway from Langley to Brentford, constructed using borehole records (source: ERCU).

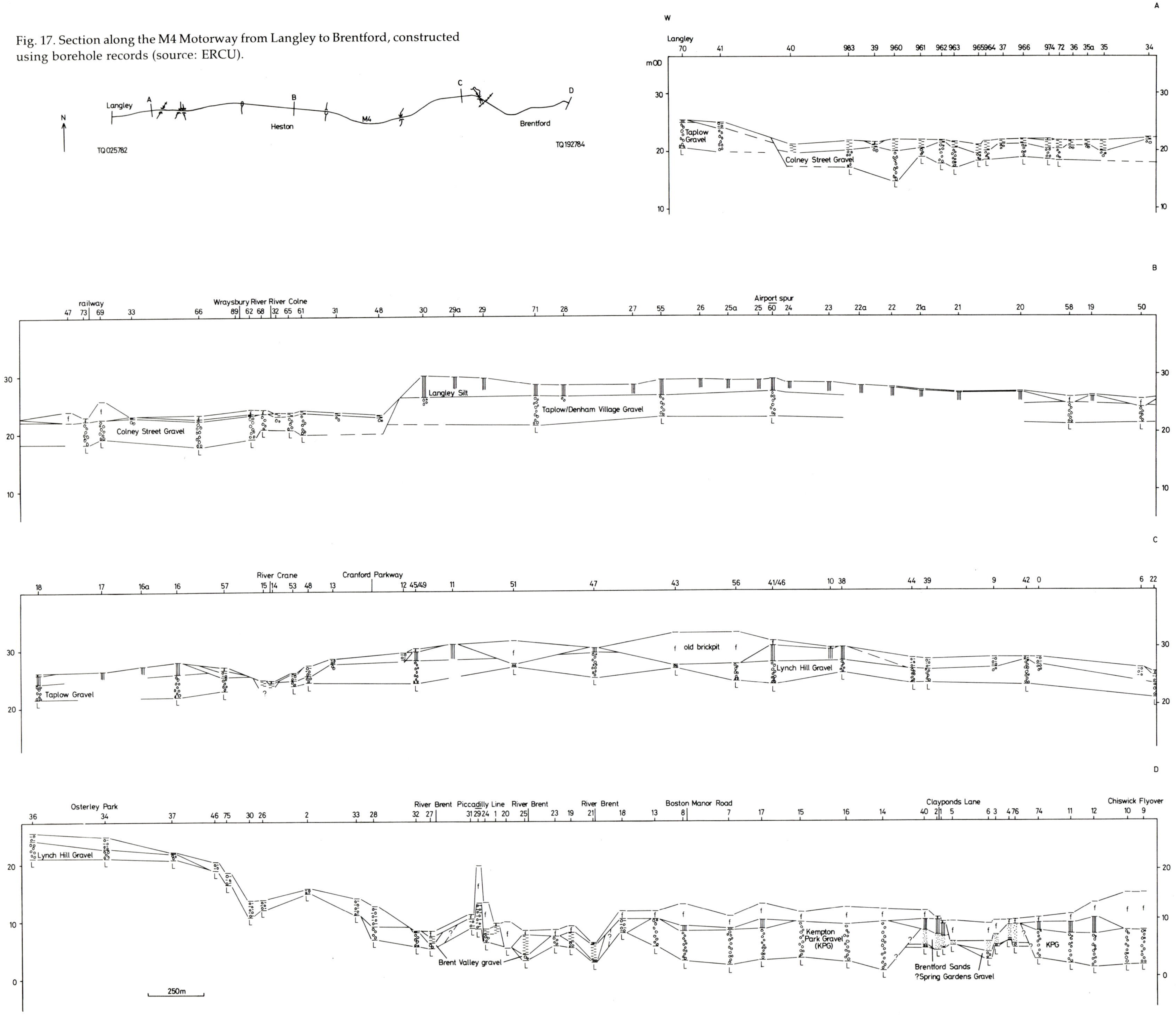

Fig. 18. Section along Hanger Lane, Ealing, constructed using borehole records (source: GLC).

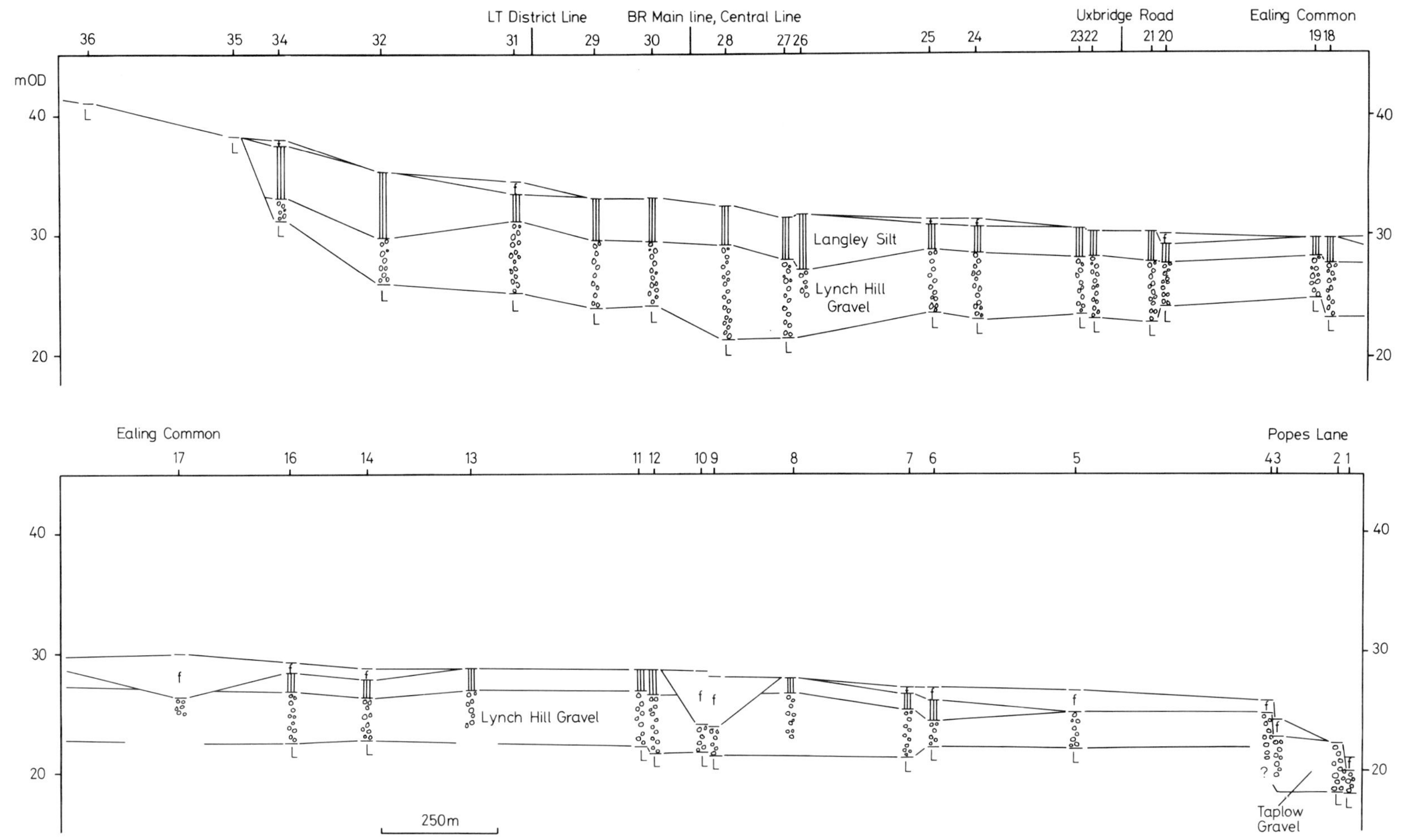

ever, modern boreholes prove only 5 m of gravel and sand here. The section also yielded several bones and teeth of *Mammuthus primigenius* (Bromehead 1925).

For much of its length from the Parkway junction to the Brent Valley, the M4 Motorway is built on the dissected Lynch Hill Terrace (Figure 17). On the basis of altitude the gravels that form the continuous unit beneath Heston and Osterley are correlated with the Lynch Hill member. Wymer (1968) noted the positions of several old pits in the Southall–Osterley area, but no exposures were found during the present study.

The Brent Valley cuts deeply through the terrace and the gravels beneath (Figure 1). The gravels, however, recur on the eastern side beneath Ealing, Hanwell and Acton. Borehole data confirm up to 3.5 m of gravels and sands in an altimetrically equivalent position to this unit at Drayton Green (TQ158807: 29 m OD) at Ealing Fire Station (TQ170805: 31 m OD), at Ealing Polytechnic (TQ177800: 27 m OD) and at the Steyne, Acton (TQ198801: 27 m OD). Apparently degraded remnants occur beneath the Boston Manor – Northfields – South Ealing area and are recorded by boreholes at TQ166793, 171792, 176794 and 172795 (20–25 m OD). The northern limit of the deposits can be determined from the extensive borehole data in the Ealing area; it occurs about 100 m north of the Western Region main line at Ealing Broadway (TQ178809) and it is typically displayed at Woodville Gardens junction with Hanger Lane (TQ186813) in which the gravel is up to 8 m thick (Figure 18). The gravel also underlies Ealing Common where well bedded sandy gravel was observed by both the author and Bromehead (1925). The separation of this unit from the base of the lower, Taplow Gravel is again seen in this section at Gunnersbury Park (TQ191794) to be only 5 m. The unit is overlain by clayey silt over most of the section. Numerous old workings were present in this area; their distribution was shown by Lane-Fox (1872) and Wymer (1968). In nearby Creffield Road, Acton where the surface is 31 m OD, Brown (1886) and more recently G. de G. Sieveking (personal communication) observed over 2 m of bedded gravel and sand beneath clay and silt ('brick-earth') 2.5 m thick at TQ192808 and 195808. Lane-Fox (1872) watched several excavations in the area east of Horn Lane, Acton that showed stratified sand and gravel up to 4 m thick (e.g. Myrtle Road, TQ204803) overlain in the highest places (over 24 m OD) by 'brickearth' (e.g. Chaucer Road, TQ204804). The gravels at Myrtle Road also yielded a tooth of *Mammuthus primigenius* (Lane-Fox, 1872, p. 455).

About 250 m to the east, the deposits are cut-out by later

Fig. 19. Section along the Central Line from Holland Park to Holborn, constructed from borehole records (source: GLC).

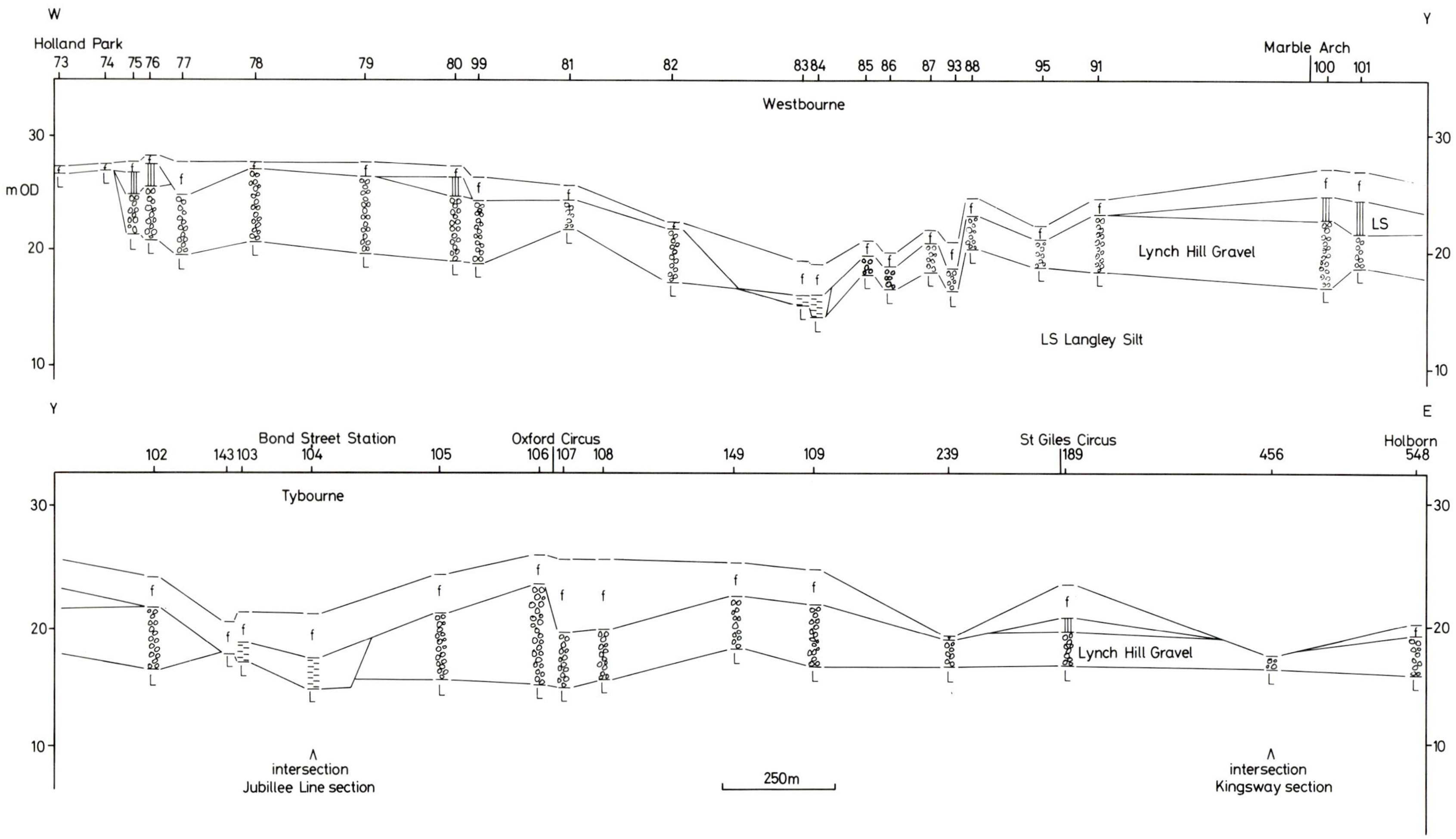

erosion, the gravels being separated from those underlying the 'Upper Floodplain' terrace by a London Clay slope (Bromehead, 1925) approximately along the North London railway line.

The Lynch Hill Gravel is next found beneath Holland Park (TQ252802) where gravels at the accordant height are banked against London Clay beneath Bayswater Road at the Pembridge Road junction (Figure 19). Here the maximum gravel thickness is 5 m, its uppermost surface being at 27–28 m OD, and it is again overlain by silt (Bromehead, 1925). South of Bayswater Road the gravels underlie Kensington Gardens and much of Hyde Park as far south as Kensington Road (Figure 39). Dewey and Bromehead (1921) recorded an exposure near the Albert Memorial (TQ265798: 20 m OD) where the deposits were 8 m thick. The gravels can be followed in the section along the Central Line (Figure 19) almost continuously eastwards below Bayswater Road, Marble Arch, Oxford Street and as far as Holborn (TQ308816). They are dissected by southward-trending stream valleys of the Tybourne between Baker Street and Bond Street (TQ285811), and the Westbourne between Leinster Terrace and Hyde Park Gardens (TQ267807), the latter being the stream dammed to form the Serpentine in Hyde Park. North of this section the gravels (7.5 m thick) have been proved in boreholes for the Westway (M40) at Paddington. The northern limit of the unit can be traced eastwards from here using borehole data. It is present beneath Lisson Grove (TQ273818) and Marylebone Road (TQ274818). In Figure 20 a section along the Jubilee Line from Regent's Park southwards to Piccadilly is shown. Here again the gravel, at the correct height for correlation with this unit (26 m OD), can be seen just north of Euston Road (TQ285822). The stratigraphy remains consistent since the gravel is again overlain by 2.5 m of brown silt at this place. Remains of *Ceolodonta antiquitatis* were found in diggings in the gravels of Regent's Park (Bromehead, 1925). The section (Figure 20) intersects with that in Figure 19 at Bond Street Station (TQ285811). Here again the gravels are dissected in part by the Tybourne Valley and thin markedly southwards being apparently absent beneath Mayfair (TQ285806), although they may have been removed from here. During construction of the tunnel for the Metropolitan Railway a substantial thickness of gravel of this unit was found continuously from Paddington to Euston Square (Whitaker, 1889). The gravels again occur beneath the junction of Euston Road and Tottenham Court Road (Figure 21). According to Bromehead (1925) and Dewey (1926) this gravel spread can be followed as far as the City where it is cut-out by the Fleet Valley. The former author reported several sections within this spread

Fig. 20. Section along the Jubilee Line from Regent's Park to Whitehall, constructed using borehole records (source: IGS).

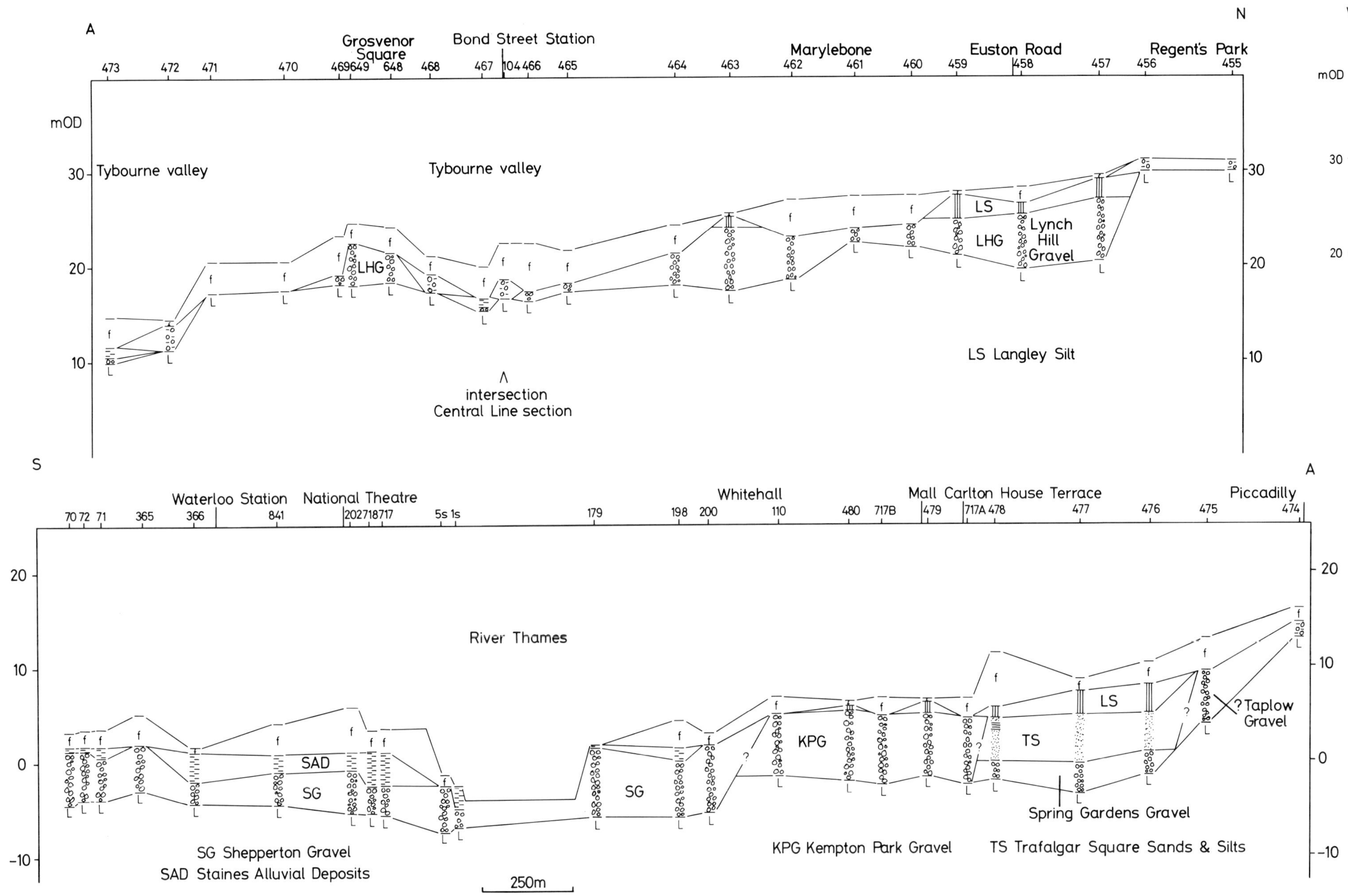

Fig. 21. Section beneath the junction of Euston Road and Tottenham Court Road, constructed using borehole data (source: GLC).

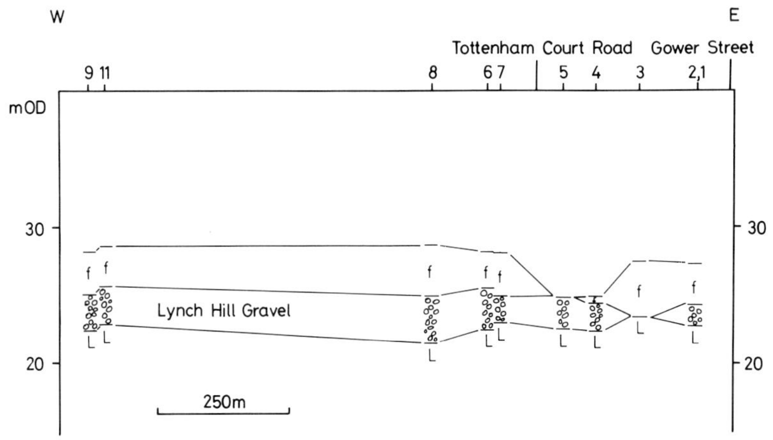

from Bayswater to Holborn where gravel has been observed. The numerous borehole records support his statements.

Modern sections in these deposits are very temporary; however, the writer observed 4.5 m of horizontally stratified sandy gravel containing shallow pebbly sand channel fills and overlying London Clay in foundation excavations beside Tottenham Court Road (TQ297813). Substantial foundation works in Shaftesbury Avenue (TQ299809) exposed 2.5 m of gravel and sand above London Clay and overlain by 0.3 m of brown clay. Excavations in Hyde Park near the Serpentine (TQ281801) were recorded by Dewey (1926). These sections exposed almost 7 m of sandy gravel resting on London Clay at the accordant height for correlation with the unit. The basal 1 m was coarse and yielded several implements together with a well preserved tooth of *Bos primigenius/Bison priscus*. At the southern end of a 13.5 m long trench the gravel thickness greatly increased to over 12.2 m and suggested to Dewey (1926) that a separate, lower aggradation was present. This lower unit may be referable to the Taplow Gravel (see Section 2.13), however, it would seem to be rather high. The position of the site suggests that the gravel may represent the buried valley side of the Westbourne.

The abundance of palaeoliths from this member in Central London, especially beneath Holborn and the West End (Wymer, 1968), adds support to the height-based correlation presented here. Southwards, the gravel almost reaches the river. Immediately north of the Strand (*c.*TQ310812) Hudleston and Price (1873) noted that in the Law Courts excavations it varies from 3 to 4 m in thickness, but is absent beneath the Strand itself. The remains of *Bos primigenius/Bison priscus*, *Capreolus capreolus* and *Equus ferus* were found in the workings. An outcrop of London Clay in places overlain by brown pebbly remobilised clay separates this gravel from lower spreads to the south (Figure 22) and can be traced from the Strand to beneath the Haymarket, Lower Regent Street and St James's Street, south of Jermyn Street (Dewey, 1926).

South of the Thames two large gravel outliers fall altitudinally into this unit. They are those beneath Wandsworth and Clapham Commons. Boreholes for the West Cross Ringway south of Wandsworth Bridge (Figure 23) confirm the presence of gravels at the accordant height beneath Wandsworth Common (TQ267742: 25–31 m OD). The gravels occur immediately beneath the surface, but are degraded at the northern end. They are up to 7 m thick. A borehole in the Prison (TQ267718) supports these records. Dewey and Bromehead (1921) noted that in the Southern Region railway cutting (TQ272744) the gravel includes extensive bands

Fig. 22. Section beneath Kingsway, from Holborn to Waterloo Bridge, constructed using borehole data (sources: GLC, IGS).

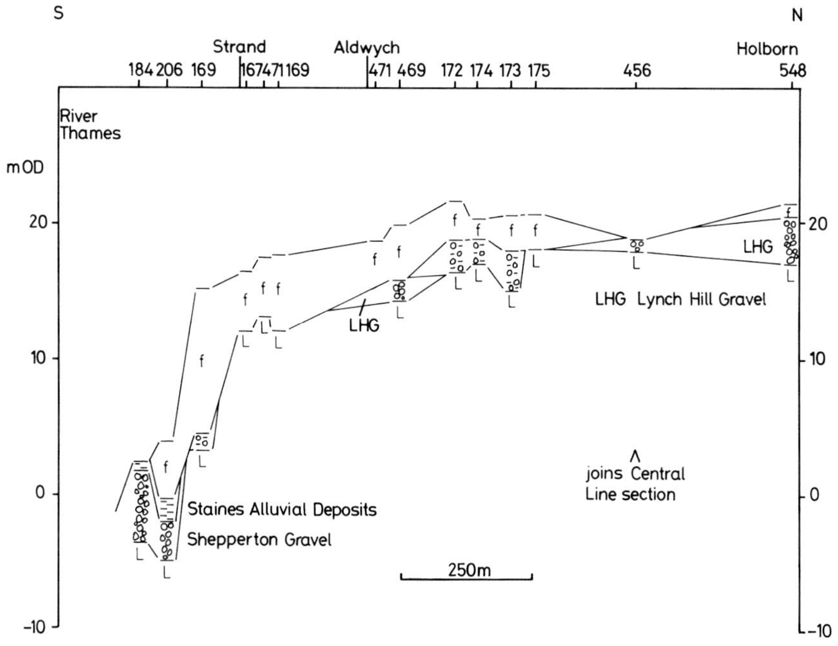

Fig. 23. Section along part of the West Cross Route ('Westway') between Wandsworth Common and Battersea Bridge, constructed using borehole data (source: GLC).

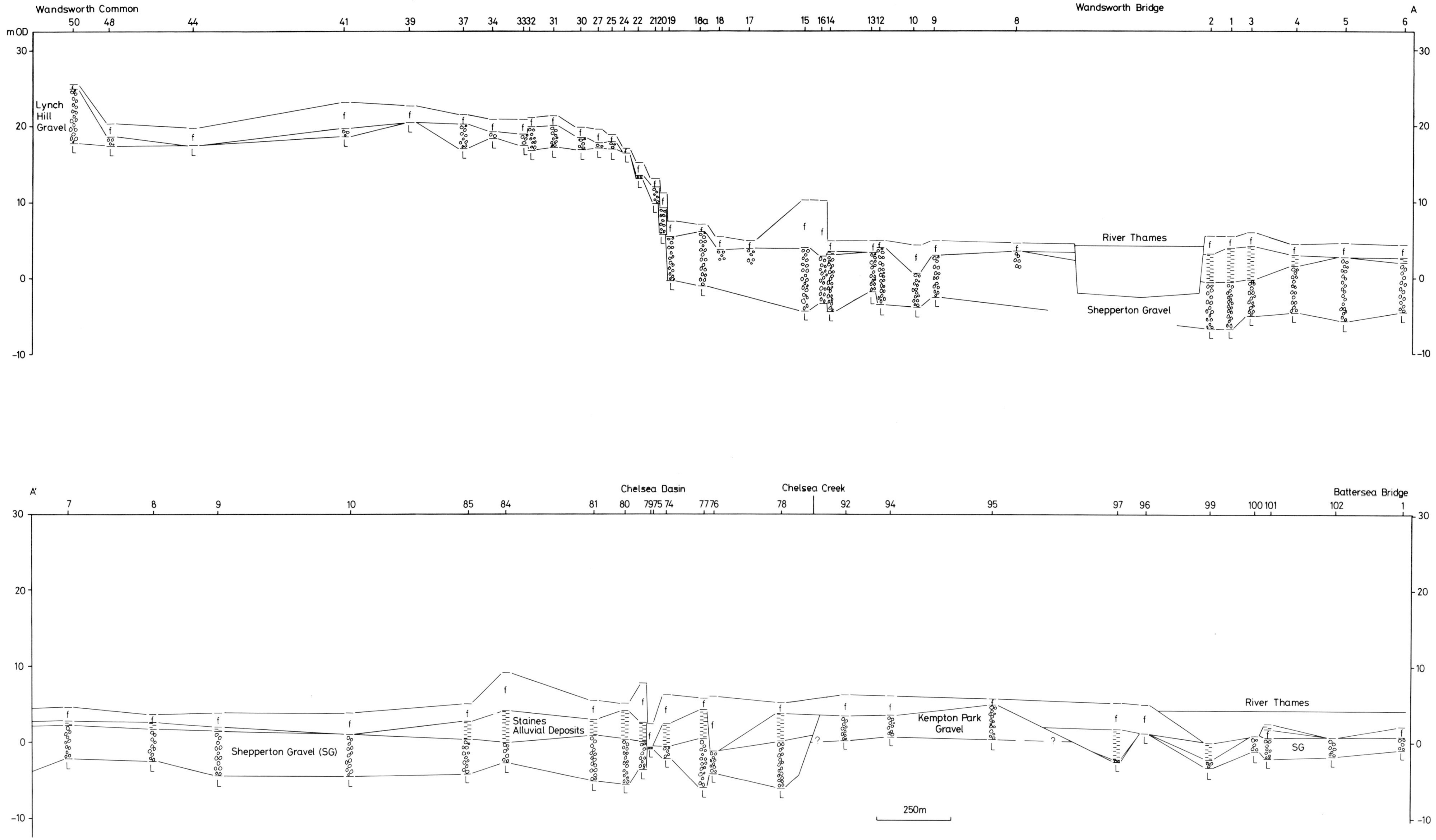

Fig. 24. Section from Wandsworth to Clapham Common, constructed using borehole records (sources: IGS, GLC).

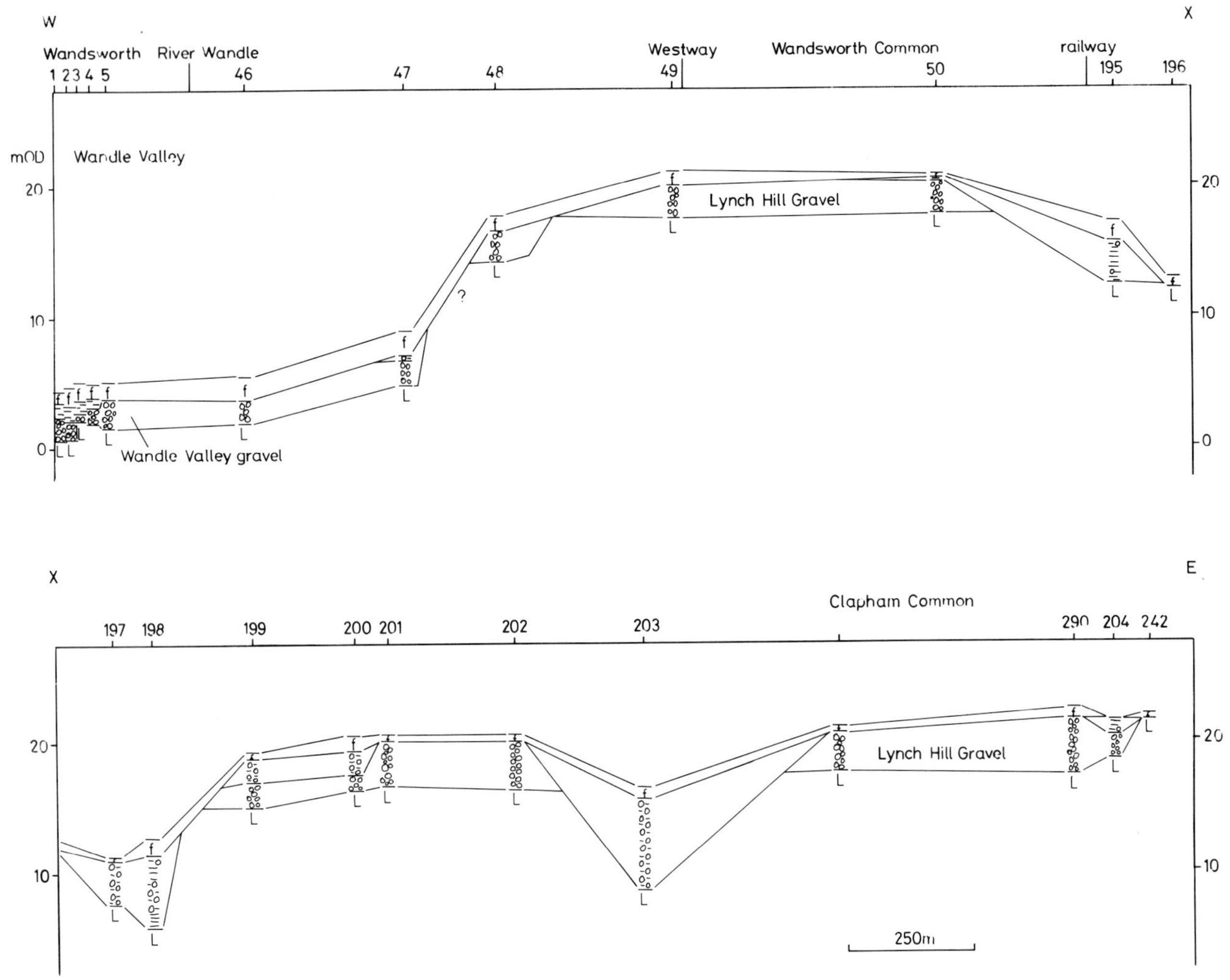

of sand (Whitaker, 1889). In a pit beside Nightingale Lane bridge (*c.* TQ275736: 24 m OD), Jones saw about 3 m of gravel, in which the uppermost 1.8 m was much disturbed. Between the gravel and the London Clay he noted some black clay containing *Equus* bones (in Whitaker, 1889). Dewey and Bromehead (1921) showed that the outlier extends southwards continuously, parallel to the River Wandle as far as Tooting Bec Common (TQ291739). The gravels of the Wandle comprise only flint, Greensand chert and occasional ironstone pebbles (Hinde, 1897; Dewey and Bromehead, 1921) and can therefore be separated from those of the Thames. On this basis they suggest that the gravels beneath the northern part of Wandsworth Common are of Thames origin, while further south the gravels are a Wandle equivalent, i.e. the gravels underlying the Fairfield terrace (Peake, 1982). By the same criterion the similar gravel patch beneath Clapham Common is also a Thames deposit equivalent to the Lynch Hill member. It was described by Prestwich (1857, p. 10) as consisting 'of angular flints with their edges slightly worn, and of flint pebbles perfectly rounded imbedded in siliceous sand, mixed with a small proportion of clay.' He also noted the presence of Greensand chert, a few quartz, reddish sandstone (quartzite) and slate pebbles. This composition is certainly in keeping with that known to occur upstream and is confirmed by Dewey and Bromehead (1921). A borehole section across the northern part of the outcrop mapped as Taplow Terrace by the Geological Survey (Figure 1) suggests that, as at Wandsworth, the gravels of this member occur beneath a degraded surface. In 1806, Jones (in Whitaker, 1889) saw a gravel pit section in Victoria Rise (Road), Clapham (*c.*TQ286755), that showed irregularly bedded gravel and sand, 4.25 m thick, upon a 'wavy' surface of London Clay. The gravel can be traced southwards, however, and on the west side of Clapham Common (TQ283748) where the surface is 28.5 m OD, 5.5 m of gravel and sand are overlain by 0.75 m of clayey gravel and 0.3 m of made ground. On the eastern side (TQ291749) only 3.5 m of gravel is overlain by 1.5 m of similar materials. At the north-western corner of the common (TQ294753: 24 m OD) up to 3.5 m of gravel is banked against a wall of London Clay that forms the southern margin of the gravel unit channel (Figure 24).

Immediately upstream of the Lynch Hill type site the terrace surface mapped by Hare (1947) at 42–43 m OD continues south of Burnham village. No exposures were found during the present study, but extensive workings at Lent Rise (SU925818–927817) were described by Lacaille (1940). He noted that up to 2.5 m of bedded, poorly sorted gravel rested on an undulating Chalk surface and was overlain by greyish-buff silt ('brickearth') at this site. The upper part of the gravel was disturbed by cryoturbation here. He illustrated the gravel being penetrated 'seemingly from below by irregular pinnacles (up to 2 m high) composed of disintegrated Chalk' (Lacaille, 1940, p. 249). These diapir-like structures may have resulted either from downslope movement of the gravel incorporating weathered Chalk or by differential pressure resulting during permafrost development or decay.

Between Maidenhead and Cookham a number of shallow dry valleys dissect the terrace surfaces and it is on the interfluves between them that the full thickness of deposits can be seen. North of Maidenhead, the gravels of this unit and the overlying terrace reappear at Furze Platt. The surface here reaches 45–46 m OD and formed the Lower Boyn Hill Terrace of Treacher and White (1909). The sections at Cannoncourt Farm (SU878831) were famous for their prolific hand-axes. Lacaille (1940) described and illustrated the sections that comprised on average 2–3 m of bedded sandy gravel with occasional large flints and erratics resting on Chalk and capped by beds of buff silt. This author also noted cryoturbation in the upper part of the gravel. The bedded gravel here together with that at nearby Switchback Lane (SU879837: 45 m OD) can be correlated with this unit on the basis of its height and pebble assemblage (see Appendix). The sections at Switchback Lane exposed up to 4 m of very poorly sorted sand and gravel overlying an irregularly channelled and blocky, weathered Chalk surface. The upper 1 m of the gravels was cryoturbated and overlain by 20 cm of very clayey, probably soliflucted gravel.

A section at Lower Mount Farm, Cookham (SU881842: 46 m OD) in the Lynch Hill surface was described by Wymer (1968, figure 76). Here a sequence similar to Switchback Lane was present, but with 3 m of disturbed, possibly soliflucted sandy gravel resting respectively on horizontally bedded chalky gravel, current-bedded pebbly sand, sandy gravel and Chalk bedrock. The total thickness of stratified gravel was 4 m. Older workings at Danesfield pit (SU878852) exposed altitudinally equivalent gravels to this unit that yielded a significant number of rolled artefacts (Treacher, 1896; Sherlock and Noble, 1922).

Between Cookham and Henley, only one outlier of gravel that can be correlated with this unit using height and pebble composition, can be identified, that at Remenham (SU775844). Here the Lynch Hill Terrace surface was mapped by Sealy and Sealy (1956) at 55 m OD. Sections in this surface (SU774841) were described by White (1899, 1908) and Wymer (1968). Exposures in this pit were still visible in 1979. The deposits comprised horizontally bedded fine to medium sandy gravels containing occasional shallow channel fills of fine sand. The upper 1 m of the gravel was cryoturbated, beneath it was undisturbed except by two large hollows filled with brown pebbly clay. These features, up to 2 m wide and 3 m deep, penetrated the entire thickness of the bedded deposits and extended into the Chalk beneath. The form of these hollows suggests that they are solution features. They extend almost to the ground surface, but are overlain by 1 m of pebbly clay of probable solifluction origin. Two pebble counts were made from the bedded gravel. The markedly higher frequency of Greensand chert present in the lower sample in comparison with that in the upper sample (see Appendix) may arise by incomplete mixing of material from the nearby Loddon–Thames confluence.

The deposits are next present in the Sonning–Twyford area. On the eastern side of the Loddon at Twyford, two large outliers occur, the more northerly beneath the Floral Mile Nursery (SU792774) and the second beneath Ruscombe (SU796763). The deposits at Ruscombe in Priors pit (SU795762: 52 m OD) were described by Shrubsole (1890) and Treacher and White (1901) as being only 1 m thick and comprising 'sub-angular to rounded flints with a number of red and brown Bunter quartzite pebbles and rounded pieces of Greensand chert.' In places, on the north side of the pit, a distinctly southern (Loddon) type pebble assemblage was overlain by gravel with a northern (Thames) suite. Modern sections in this pit, now a builder's yard, were examined. They exposed 2.5 m of gravel and sand, a pebble count from which showed an assemblage that supports the altitudinal correlation of the deposits with this member. Several other pits have been dug into these gravel spreads and are listed in Wymer (1968). He, however, equated the truncated Ruscombe outliers with the Lynch Hill Terrace, presumably on a relative height basis. As can be seen in Figure 15, the deposits here fall into line with the Lynch Hill Gravel by extension of the unit gradient (45–55 cm km^{-1}) into this area. The gravels at Twyford are 2.75 m thick and overlie Chalk at SU790772 (50 m OD).

West of the Loddon Valley, in the Reading area, this unit is poorly preserved. East of the town, around Sonning, the surface gradient intersects with the fragments south of Woodley and Sandford (SU757738: 58 m OD) and near Earley Station (SU746724: 59 m OD), both of which are probably degraded Loddon-equivalents of this unit (cf. Wymer, 1968). Sections in this area mentioned by Blake (1903) and Wymer (1968) showed stratified gravels composed only of flint and Greensand chert, i.e. a Loddon-type assemblage. However, a section of 3 m of clayey gravel at Sonning Hill (SU752748: 57 m OD) described by Shrubsole (1890) may be a Thames-equivalent to this unit.

Sealy and Sealy (1956) recognised only one Lynch Hill Terrace surface patch in the Reading area that they felt was of Thames origin, that at SU702728 (61–62 m OD) at and east of Prospect Park. Two old sections were known in this area, Grovelands (SU693732: 59 m OD) and Denton's (SU693723: 60 m OD). Both pits produced artefacts and a vertebrate fauna was found at the former, from the base of the gravels (Stevens, 1882; Shrubsole, 1890). According to Wymer (1968), the Grovelands pit was excavated into the bluff between the Lynch Hill and lower, 'Taplow' (Reading Town) surfaces. The gravel (up to 4 m thick), although stratified in part, was clearly only partially decalcified since Shrubsole (1890) saw lenses of chalky gravel at the base beneath clayey (soliflucted) gravel at the top. At Denton's, Smallcombe (according to Wymer, 1968) also recorded up to 4 m of stratified gravel and sand resting on Reading Beds, but Wymer (1968) found only 3 m of disturbed contorted gravel here. Modern exposures are rare in the area, but an old pit face in Prospect Park (SU692723: 60 m OD) exposed over 2 m of finely bedded sandy gravel overlain unconformably by 20–30 cm of cryoturbated brown clayey gravel. A pebble count sample from the bedded gravel supports the height correlation of these deposits with the Lynch Hill unit, and palaeocurrent measurements indicate a flow towards due east. A nearby borehole beside the Bath Road (SU707728: 60 m OD) recorded 7.5 m of gravel and sand resting on Reading Beds.

The pebble count samples from the Lynch Hill Gravel (see Appendix), show remarkable uniformity. They comprise 85–94% flint (80–87% angular flint, 4.5–9% rounded flint), 2.5–7% vein quartz, 1–4% quartzite and 0.75–7.5% Greensand chert. The counts also contain minor quantities of other lithologies. The significance of these counts will be discussed in Chapter 3.

As stated above, plotting of the localities described gives an average gradient for the Lynch Hill Gravel of 45–55 cm km^{-1} (Figure 15) which is of similar magnitude to that obtained by Hare (1947) and Sealy and Sealy (1956).

2.13 Taplow Gravel (and in part Taplow Terrace)

As early as 1856, following the discovery of vertebrate remains at Taplow Station pit, Prestwich recognised that the deposits here represented a younger accumulation than those on higher ground to the north (Figure 1). However, it was not until 1912 that Bromehead proposed the term Taplow Terrace for the deposits. Dewey and Bromehead (1915) were the first to use the term Taplow Gravel following their mapping of the Windsor–Chertsey area. The station pit type locality was well established when Sherlock and Noble (1922) published their memoir for the Beaconsfield Geological Survey sheet, which includes the site. Subsequent authors, including King and Oakley (1936), Oakley (1937), Hare (1947) and Wymer (1968, 1977b), have accepted its status.

The 'brickearth' that overlies the Taplow Gravel over large parts of the area has generally been regarded as being 'intimately associated' with the gravel 'so that it may be regarded as immediately following in time the underlying gravel to which it belongs' (Sherlock and Noble, 1922, p. 42). The present author would dispute this statement and cannot therefore accept that the terrace surface and the gravel are synonymous. As a consequence in this description the term Taplow Gravel (*sensu* Dewey and Bromehead, 1915) will be used. The overlying 'brickearth' accumulation will be described separately below (Section 2.16). The Taplow Terrace, like the other terraces in the Middle Thames, was mapped morphologically between the Colne Valley and Bourne End by Hare (1947). His work was extended upstream to Reading by Sealy and Sealy (1956), and downstream to Gunnersby by Allen (1978).

Excavations in the 1840s by the Great Western Railway for a station at Taplow (SU919816) initiated gravel extraction that seems to have continued sporadically into the 1930s. Descriptions of the sequence exposed in the station pits vary slightly, but most authors agree that the gravels were 4.5–6 m thick, poorly sorted, but stratified. They were underlain by weathered Chalk bedrock comprising 'lumps of Chalk and broken flint nodules in a chalky paste' (Oakley, 1937), termed 'Coombe Rock' by Breuil (1934). About 1 m of silt ('brickearth') locally capped the sections (Wymer, 1968). The ground surface here is at 32–36 m OD. The stratigraphy described was confirmed by the author's observations at the station pit (SU916816). The 'brickearth' was 1 m thick and rested on gravels and sands the upper part of which was

Fig. 25. Sections in the Slough–Windsor area. The upper two are orientated north–south, along the A332 road into Slough centre; the lowest section runs west–east. All were constructed using borehole data (sources: IGS, SBC).

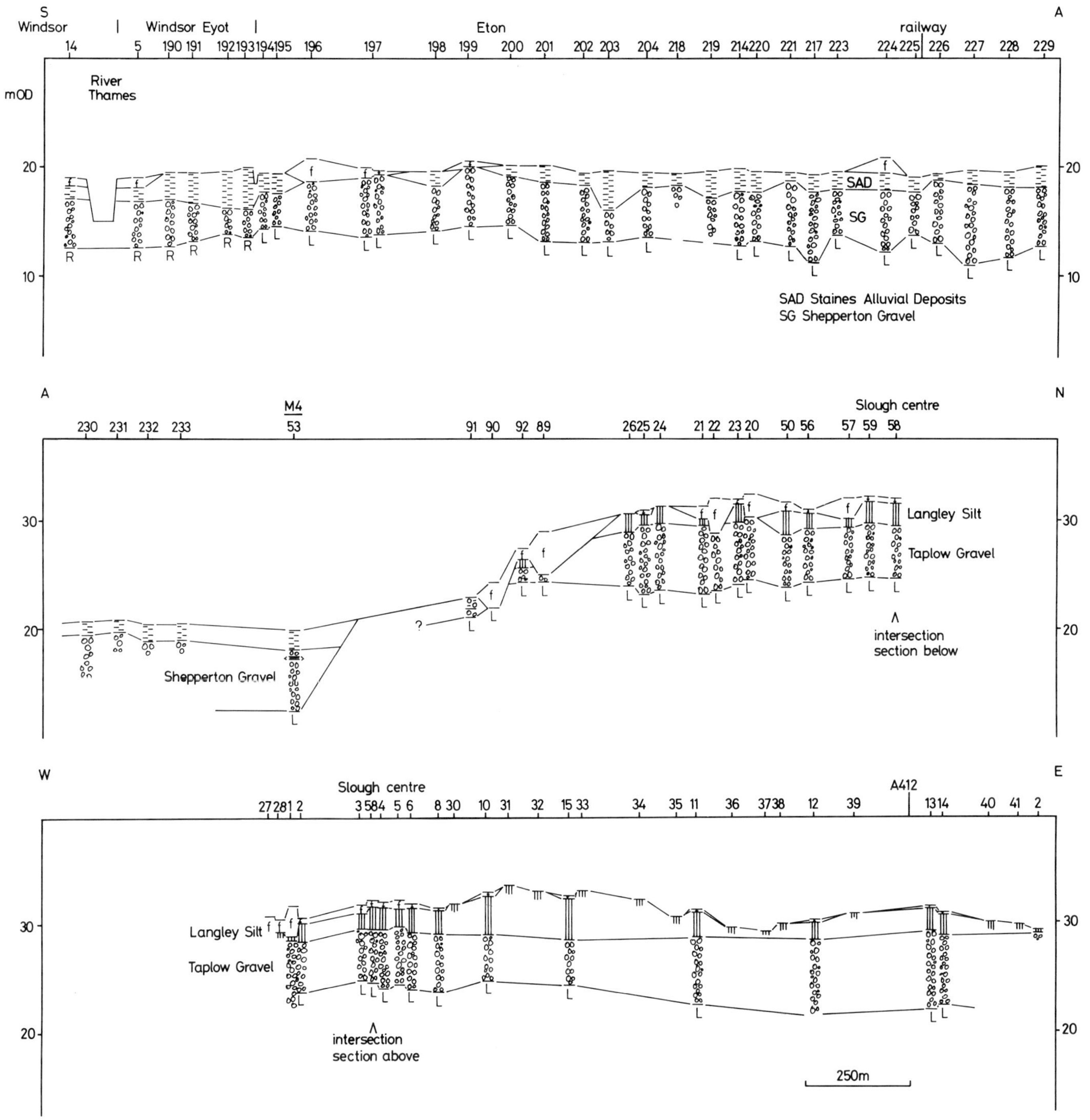

cryoturbated to a depth of about 1 m; beneath, the gravels were indistinctly, horizontally stratified. Narrow tabular cross-bedded sand units within the gravel indicated a palaeoflow towards the east-south-east. A pebble count from the gravels is given in the Appendix. Bones found by workmen at or near the base of the gravels included mammoth, woolly rhinoceros and the first musk ox (*Ovibos moschatus*) recorded in Britain (Owen, 1856; Oakley, 1937). A number of rolled artefacts have also been found in the gravel. A comparable sequence to that present in the pit was found in a borehole at SU913816.

The main mass of Taplow Gravel extends downstream from the type area almost without a break as far as the Colne Valley. The terrace surface is only dissected by minor transverse tributaries, but is degraded on its south side associated presumably with downcutting and slope degradation during younger phases.

Close to Taplow, at Lent Rise an exposure beside the railway bridge (SU926815: 31 m OD) showed over 5 m of current-bedded sand and gravel overlain by about 2 m of silt ('brickearth'). Cross-bedding in the sand indicated a flow towards the east-north-east. Altitudinal correlation of these gravels with those at Taplow is supported by a pebble count.

Much of Slough is built on the Taplow Terrace (Hare, 1947). Sections through the deposits beneath the town (Figure 25) confirm the presence of up to 7 m of gravel and sand resting on an irregular surface of Reading Beds. The gravels occur at the expected height of between 24 and 30 m OD for correlation with that at Taplow (Figure 26). The upper surface of the gravel is distinctly more regular than the base and is overlain by the Langley Silt Complex ('brickearth'). The down-valley gradient can also be discerned. The transverse section shows the relationship of the Taplow Gravel to the lower-lying floodplain deposits. The base of the Floodplain gravel (Shepperton Gravel) here is 14.5 m OD, while that of the Taplow is 22 m OD, i.e. a difference of 7.5 m. The degraded nature of the surface towards the southern margin of the terrace is also well illustrated (see below).

Dewey and Bromehead (1915) recorded sections at two old pits in the area: one between Chalvey and Salt Hill (*c.*SU968802) and a second at Cippenham (*c.*SU951804: 30 m OD). Both exposed up to 3.5 m of 'gravel with current bedded seams of sand and sandy clay.' At both sites the gravel produced molars of mammoth and rested on Reading Beds clay.

A large flooded abandoned working at Langley (TQ002800: 28 m OD) exposed 2 m of sand and gravel beneath 2 m of brown 'brickearth' (Langley Silt), although Wymer (1968) recorded over

Fig. 26. Long profile section showing deposits correlated with the Taplow Gravel.

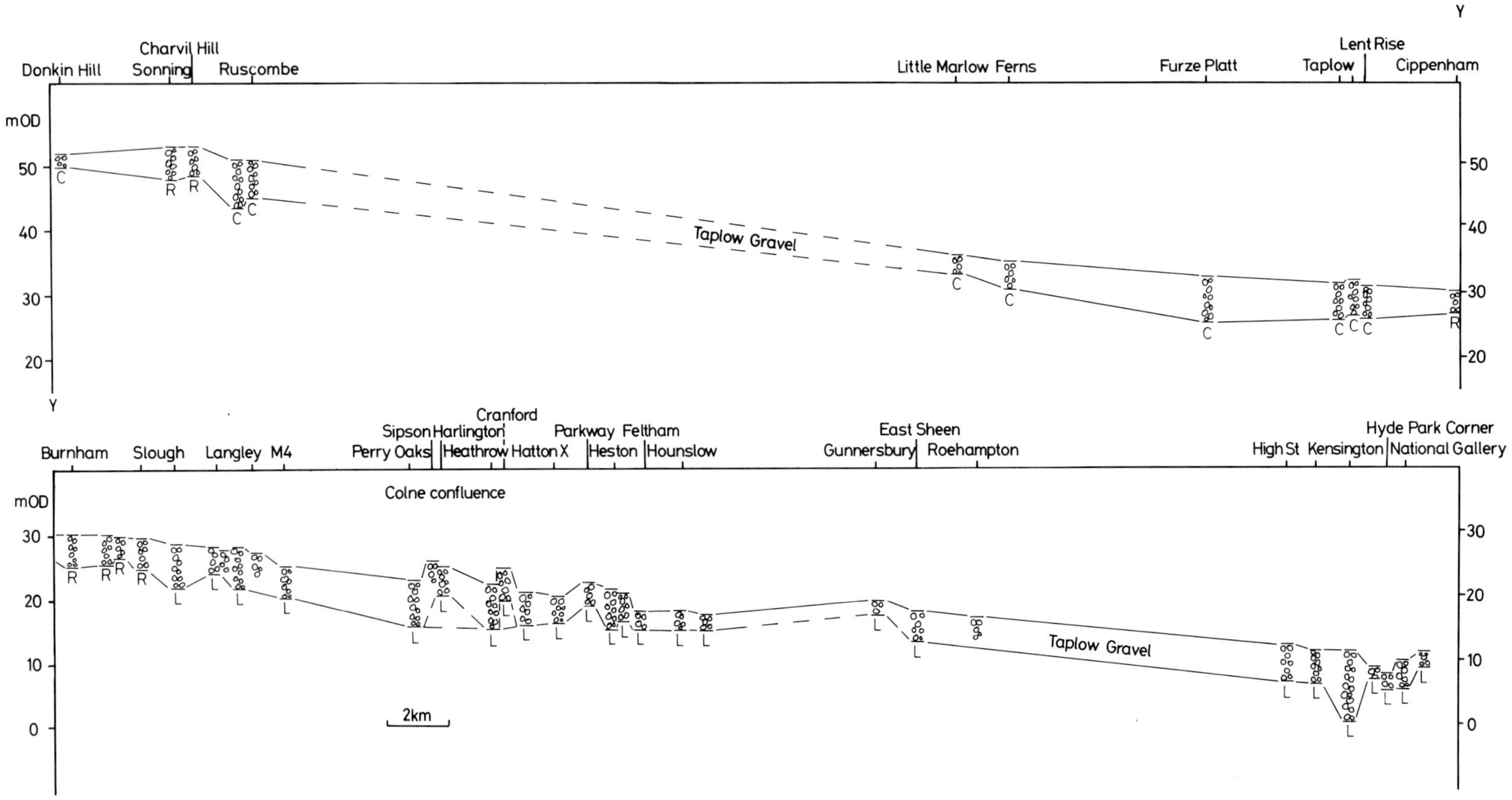

3 m of gravel at this site. Fine gravel bands interstratified with tabular cross-bedded sand beds up to 20 cm thick form the deposit. Current-bedding in the gravel and sand indicated an eastward flow and a pebble count demonstrated that this gravel is equivalent to that at Taplow. The upper 60 cm of the gravel and sand unit exhibited many stones with vertical long axes and in one place a small ice wedge cast, 40 cm deep and 20 cm wide, penetrated the gravel from the junction with the silt above. The contact between the two units gently undulated across the pit. North of the site four boreholes confirm the occurrence of gravel in the same stratigraphical position. Immediately north and east of this area the land rises to the Lynch Hill Terrace surface and the vertical separation of the bases of the two aggradations is here about 8.5–10 m.

South of Langley, the deposits have again been penetrated by several boreholes. The section along the M4 (Figure 17 and Gibbard and Hall, 1982, figure 5) shows a maximum of 5 m of gravel and sand in an altitudinally equivalent position to that at Langley. Nearby boreholes in Tower Way (TQ014783) and Langley School (TQ005787) demonstrated 5.6 m of this member above Reading Beds. The boreholes in all these three localities occur in the area mapped by Hare (1947) as a minor sub-facet of the Taplow Terrace, about 3 m below the general level. The sub-facet was called the 'lower Taplow Terrace' by the Sealys (1956). As shown by the borehole sections this surface does not seem to be developed on a separate aggradation, but merely in those areas where the 'brickearth' overlying the gravel is thin or absent. This presumably results from slope degradation during a post-'brickearth' period. The recognition of this surface as representing a significant separate stage is therefore unjustified.

The Taplow Gravel is cut-out by the valley of the Colne (Figure 17), but the terrace and its underlying deposits are present again on the eastern side beneath Harmondsworth and Heathrow Airport as far as the valley of the River Crane (Hare 1947, Allen 1978). In this area the terrace is extremely wide reaching a maximum of 8 km between Cranford and Feltham. Four large gravel workings were present in the area north of Heathrow during this study, at Sipson (TQ070782 and 077781: 29 m OD), Harlington (TQ085785: 28 m OD) and Cranford Park (TQ094776–091783: 27 m OD). The sections all exposed stratified sand and gravel overlying London Clay and capped by silt ('brickearth') up to 2.5 m thick. At Sipson, only 2.5 m of horizontally bedded medium gravel interstratified with tabular cross-bedded sand laminae up to 70 cm thick were exposed above water level. Palaeocurrent

Fig. 27. Long profile section showing the deposits correlated with the Denham Village Gravel and its confluence with the Taplow Gravel beneath the Heathrow Airport Spur, constructed using exposure and borehole data (sources: IGS, ERCU).

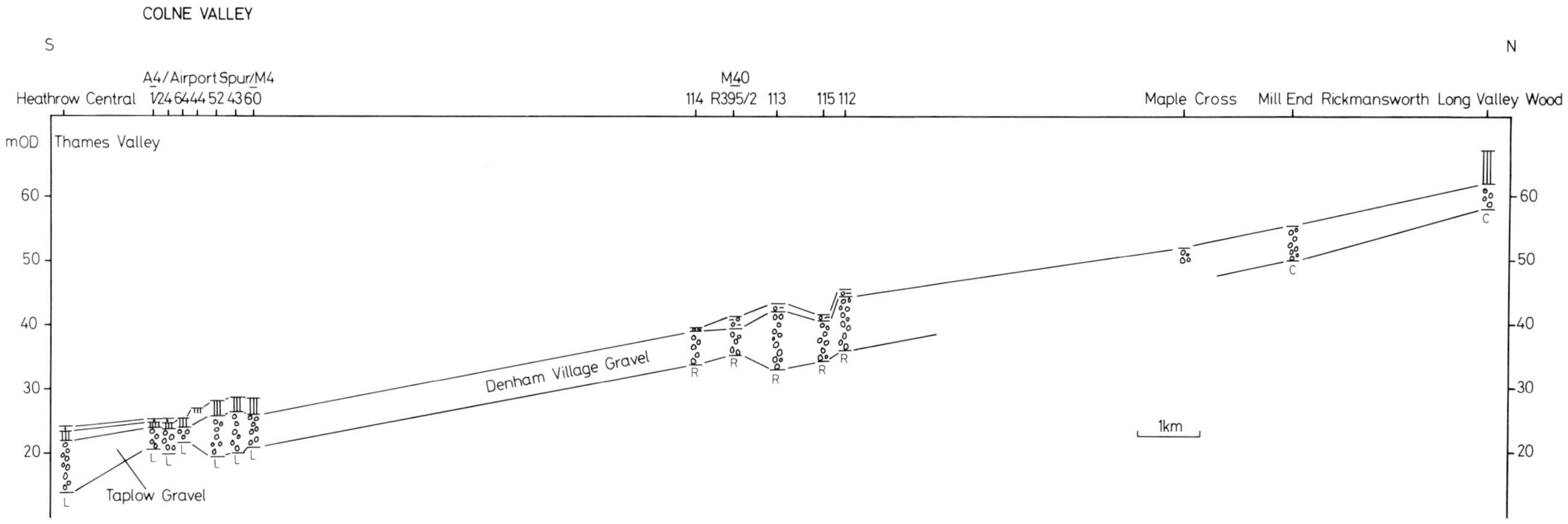

Fig. 28. Section along the Piccadilly Line extension from Heathrow Central to Hounslow West Station, constructed from borehole records (source: IGS).

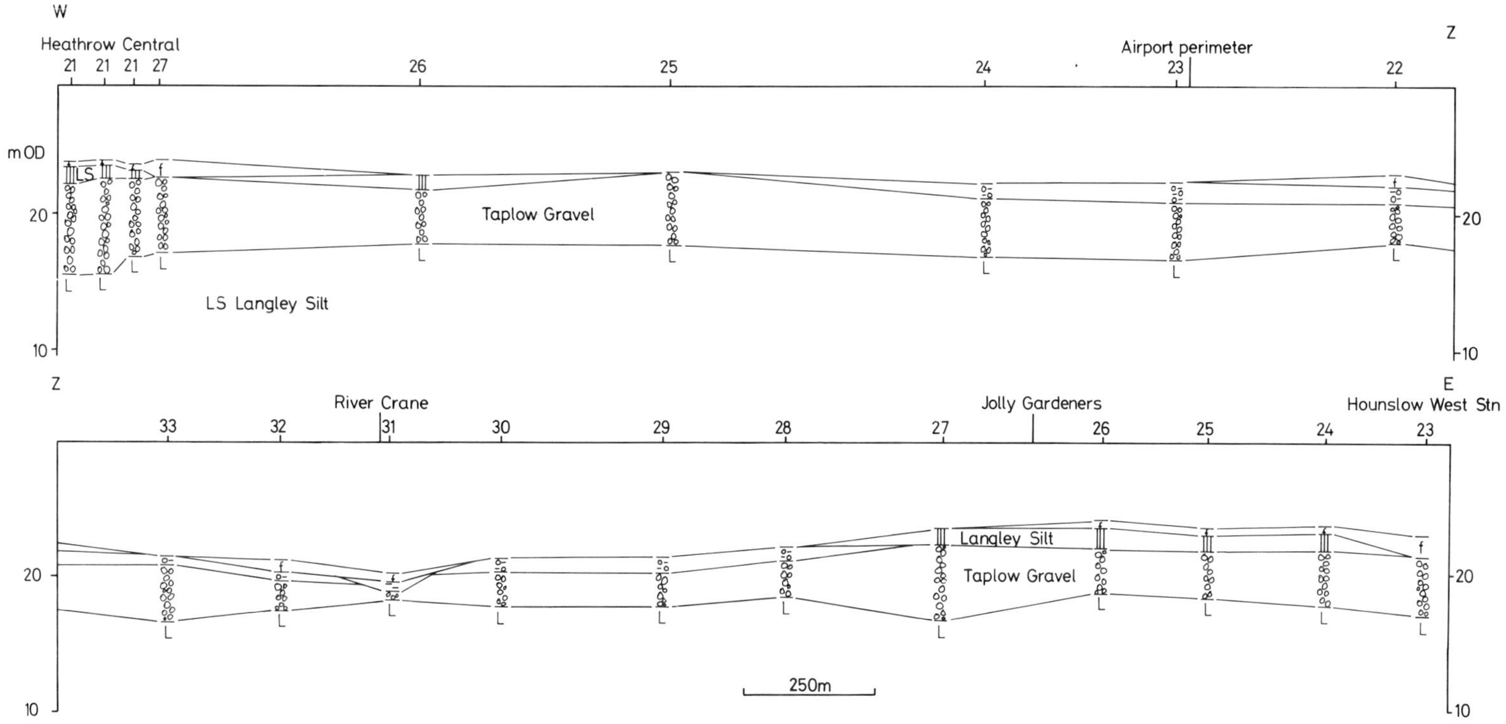

directions here indicated a flow towards the south-east to east-south-east. At Harlington a very similar sequence was exposed, and palaeocurrent measurements from here indicated a flow towards the south-east. The Cranford Park sections showed up to 4.5 m of bedded gravels and sands. Here a lens containing alternating 1 cm thick fine to medium sand and silt laminae filling a cross-cutting channel-like feature was present in one place. Elsewhere narrow silt bands were found interbedded with the sand. The gravel was calcareous in many places at this site and included a number of small rounded chalk pebbles. Here the palaeocurrent measurements indicated a flow towards the east-south-east.

At several horizons within the gravels at Cranford thin, fine gravel bands (10–20 cm) showed wave-like pendants of vertical stones with a wavelength of 7–10 cm. The tops of these micro-involution-like structures were removed by erosion at the base of the overlying gravel bed. A number of flint pebbles in the involuted bands were frost-shattered. It would seem reasonable to interpret such a structure as a product of short subaerial exposure of these beds to frost processes resulting in a type of 'arctic structure' soil. A substantial double ice wedge cast 1 m wide penetrated the gravels to a depth of 2 m from the gravel/'brickearth' interface at Cranford. Elsewhere the gravels were strongly involuted in their upper 60 cm.

Pebble count samples (see Appendix) from Sipson, Harlington and Cranford support altitudinal evidence (Figure 26) for correlation of the deposits with those at Taplow.

A section along the M4 Airport Spur (Figure 27) from the motorway (TQ074785) to the Heathrow entrance beneath the A4 (TQ075771) shows the transverse slope on the deposits in the Sipson area. The gravels, which are 6 m thick beneath the M4 (Figure 17), thin markedly southwards and are only 4 m thick beneath the airport entrance. However, at Heathrow Central (TQ076759: 24 m OD) the gravel thickens to a maximum of 7 m (Figures 27 and 28) and comparable thicknesses (up to 5.4 m) occur at Heathrow North (TQ086768).

The difficulty of separating the Lynch Hill and Taplow aggradations in the West Drayton – Harmondsworth areas has been discussed previously (e.g. Allen, 1978). However, the borehole data clearly show that the gravel deposits are separated altitudinally. The base of the Lynch Hill Gravel (Section 2.12) at Yiewsley is 29 m OD, while the base of the Taplow Gravel at the M4 is 18.25 m OD falling southwards to a minimum of 16 m OD at Heathrow Central. The upper surfaces of the aggradations can be

Fig. 29. Section from Chelsea to Kensington, constructed from exposure and borehole records (sources: GLC, IGS).

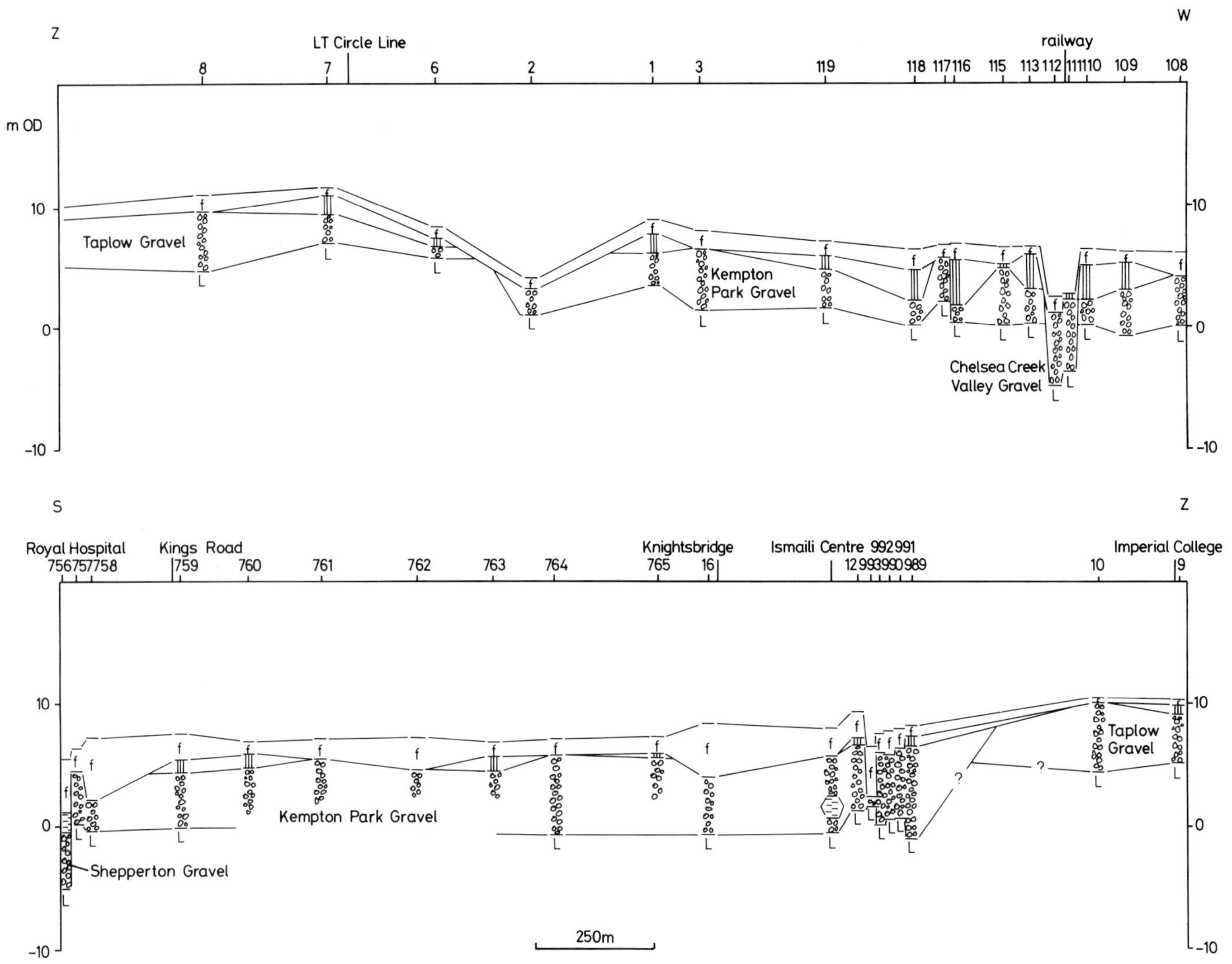

similarly separated. The problem of distinguishing between the aggradations on the ground arises in part from the masking of the gravel surface by overlying 'brickearth'. The upper surface of the Taplow Gravel at the M4/Airport Spur junction is 26.5 m OD, a marked break of slope is present immediately north of this junction at TQ074787 where the ground rises steeply to the Lynch Hill Terrace. The Taplow Gravel surface falls southwards to Heathrow Central where it is 23.75 m OD, i.e. a drop of 2.75 m in 1.6 km or 180 cm km^{-1} (assuming that the surface was not degraded before the 'brickearth' was deposited). Such a gradient is almost exactly that of the Colney Street Gravel, flooring the present Colne Valley. In view of the position of the Sipson–Harmondsworth area, the gradient on the deposits, and the south-eastward palaeocurrents from the exposures near Heathrow, it is thought highly probable that the gravels were deposited at the Colne confluence with the Thames in this area.

South of Heathrow Central the present surface appears to be gently sloping so that the 'brickearth' 'wedges-out' as a result of post-depositional slope degradation (Figure 29). This section reaches its most southerly point at Hatton Cross (borehole 23: TQ094753) and thence continues eastwards. As can be seen, apart from the valley of the River Crane, the Taplow Gravel forms a continuous spread in the area. The section along Cranford Parkway (Figure 16) close to Cranford Park quarry illustrates the junction between the Lynch Hill and Taplow Gravels (see also Section 2.12). The gravel, at the accordant height, abuts the degraded Lynch Hill member 250 m south of the M4 junction and is recorded in boreholes as far south as Cranford High Street (TQ106773). South-east of Feltham a marked break of slope separates these deposits from the Kempton Park Gravel beneath the Isleworth ('Upper Floodplain') Terrace. This has been illustrated by Gibbard *et al.* (1982, figure 5). An almost identical feature is shown in Figure 37 from the A316 at Twickenham (TQ148737) and the break of slope can be followed northwards just west of Mogden Sewage Works (TQ151753).

East of the Crane, the Taplow Terrace narrows markedly, but the gravel is present beneath Hounslow Heath (TQ121749: 21 m OD) where it was worked in the 1950s. Numerous boreholes here prove up to 4 m of gravel and sand beneath 0.75 m of clayey gravel. At the nearby Barracks (TQ119756: 22 m OD), 6.25 m of gravel is present. A series of trial holes between here and Hounslow West Station showed a maximum of 6 m of gravel beneath 'brickearth', according to Dewey and Bromehead (1921). In Hounslow High Street (TQ144759: 18 m OD) 3.5 m of gravel is present and at Whitton (TQ148749: 16 m OD) 3.8 m of gravel rests on London Clay. The gravel recorded in all these boreholes is the altitudinal equivalent of that beneath Heathrow and therefore that at Taplow (Figure 26).

Downstream from Hounslow the terrace feature continues to narrow as it is cut-out by younger deposits. East of the Brent no morphological expression of the Taplow Gravels can be discerned. However, Allen (1978) traced a morphological feature which he equated with the Taplow Terrace, upstream along both sides of the Brent Valley. The borehole section along the M4 does not intersect one of the patches, but a borehole at Boston Manor (TQ161790: 21 m OD) recorded 2 m of gravel that may be a Brent-equivalent of this unit. It is, however, possible that this material may represent degraded Lynch Hill Gravel, a wide spread of which occurs immediately north and east of this locality.

At Gunnersbury, the Taplow Gravel appears to be almost entirely removed, but the section along Hanger Lane and Gunnersbury Avenue (Figure 18) shows 4 m of gravel at the accordant height at Popes Lane (TQ191794: 20 m OD).

South of the Thames, patches of deposits equated with the Taplow Terrace on the basis of their height were mapped by Bromehead (1912) and Dewey and Bromehead (1915, 1921). Those associated with the Chertsey Bourne, the Wey and the Mole are not Thames deposits as shown by their gravel composition and therefore will not be considered here. However, at East Sheen, Dewey and Bromehead (1921) described gravels beneath Sheen Common (*c.*TQ199146: 18 m OD) and beneath Roehampton Club grounds (*c.*TQ218749: 17 m OD) possibly representing in part a confluence of the Beverley stream with the Thames. No exposure of this material was available during this study and therefore it can only be equated altitudinally with this unit.

Probably equivalent deposits outcrop downstream beneath Kensington High Street (TQ255796: 10 m OD), south of Kensington Gardens, beneath Queen's Gate (TQ264793) and eastwards to Knightsbridge (Figure 29) where the gravel is cut-out by the Westbourne Valley. A borehole at the corner of Knightsbridge and Kensington Road (TQ277796: 11 m OD) proved an anomalous depth of 11.5 m of gravel and sand. Boreholes at Hyde Park Corner (TQ283797: 14 m OD) showed 1–2 m of gravel beneath as much as 6 m of made ground, while at the western end of Piccadilly (TQ286799) 2.5 m of gravel occurs at the expected height. As noted above, excavations near the Serpentine were examined by Dewey (1926). These may have exposed the junction of the Taplow and Lynch Hill Gravels (see Section 2.12), the equivalent of the former being over 12 m thick here, although some of the upper part of the deposit may have been soliflucted fluvial deposits. Elsewhere the boundary of these two members is impossible to determine on the ground in Central London because of slope degradation and surface modification by fill. It must, however, run south of Kensington Gardens, north of Kensington High Street and towards Holland Park, to judge from the deposit's altitudinal relationships.

The deposits are again dissected by the Tybourne Valley, but east of this feature 5.8 m of gravel and sand at the appropriate height occur in the section along the Jubilee Line near St James's Palace (Figure 20; borehole 475, TQ292802). The base of the deposit is 7.5 m above that of the Spring Gardens Gravel (Section 2.14) here. The Taplow Gravel is probably again encountered beneath the National Gallery (TQ300805) (Figure 33). Here it is much thinner, reaching a maximum of 4 m. The gravels in this area seem to have been heavily degraded during subsequent phases and are absent completely 500 m further east in the Kingsway – Waterloo Bridge section (Figure 22).

On the south bank of the Thames opposite the Taplow type section, a spread of gravel has been mapped as Taplow by the Geological Survey and by Hare (1947). Comparison of the elevation of the type deposits between 27.5 m (base) and 31–32 m (top of gravels) with those south of Bray (e.g. SU922776), 18.6 m (base) and 24.4 m (top of gravel), strongly suggests that the correlation is incorrect. The gravel spread south of Bray is equated below with the Kempton Park Gravel (Section 2.17).

Between Maidenhead and Cookham Rise, Hare (1947) mapped a series of patches of Taplow Terrace on the interfluves between the eastward-trending dry valleys. A large exposure east of the railway line at Furze Platt (SU884833: 33 m OD), showed 7–8 m of deposits. These comprised poorly sorted, medium to fine chalk-rich gravel at the base overlain by tabular cross-bedded sandy gravel bands alternating with silty sand up to 20 cm thick. The upper 3 m of the sequence consisted of horizontally bedded gravel interbedded with coarse sand. The upper 75 cm of the gravels were strongly involuted and were overlain by 50 cm of brown silty 'brickearth'. Palaeocurrent measurements indicated a flow towards the south-south-east. Two pebble count samples from the gravel demonstrate that this unit should be correlated with that at Taplow. The gravels rested on a very uneven surface that consisted of irregularly sized blocks of Chalk and fresh flint in a finely comminuted chalk matrix ('rubble chalk'). The amplitude of the surface was 2–3 m at this section.

The Taplow Terrace is next encountered upstream between Marlow and Bourne End where it was mapped by Sealy and Sealy (1956) at 36–40 m OD. Although the surface is dissected by dry valleys and degraded by solutional collapse, it is still very marked in this area. A series of fans emanate from the dry valleys, e.g. Marlow Bottom (SU858878), according to the Sealys (1956, p. 377) and grade into the terrace gravel. These fan gravels comprise 'angular to subangular Chalk flint'. Sections in the main terrace at Ferns pit (SU883883: 38 m OD) show a sequence exactly like those described from here by Treacher (1916), and noted by Sherlock and Noble (1922) and Wymer (1968). They show horizontally stratified interbedded medium gravel and coarse sand bands, the latter being tabular cross-bedded in places. This unit was 4 m thick. This accords with Sherlock and Noble's (1922) observations. They noted that the lower 1 m of the gravels were very chalky. Above the stratified deposits, a bed up to 3 m thick, of coarse to medium angular to subangular flint gravel with a large proportion of quartz pebbles in a clayey matrix with occasional thin seams of clay occurs. The base of this bed is almost horizontal and in places is erosional since it truncates involutions in the upper part of the gravel beneath. The upper bed is thickest in the northern face and thins rapidly southwards. Wymer (1968) followed Treacher (1916) who suggested that it was derived by solifluction of older quartz-rich gravel from higher up the slope to the north. The present author sees no reason to question this interpretation. In places, interstratified between the stratified gravel and the soliflucted bed, was a band of brown loam 50 cm thick containing small pebbles with their long axes orientated vertically. The Sealys interpreted this band as being equivalent to the 'brickearth' of the type area. A pebble count from the lower gravel at Ferns pit (see Appendix) again supports the height correlation of the stratified gravel with that at Taplow. Extension of the gradient of 45–55 cm km^{-1} gives a surface at 34.5 m at Ferns, i.e. within 1 m of the upper surface of the stratified deposits. Part of a mammoth molar was found in the lower gravel here by Treacher (1916). Three boreholes in the terrace surface near Marlow confirm the gravel pit sequence.

Upstream the deposits are not present until the Sonning–Twyford area, where they appear to form remnants on either side of the Loddon Valley. The main outliers at Ruscombe and Twyford mapped as Taplow Terrace by the Sealys (1956) appear to represent eroded remnants of the Lynch Hill Gravel (Section 2.12). However, gravel and sand at SU796762 (51 m OD) 4.9 m thick, and at SU793757 (51 m OD) 7.6 m thick, are too low for inclusion in the Lynch Hill and may represent Taplow-equivalent deposits. A pebble count from a ditch close to the latter site, supports this correlation.

On the western side of the Loddon Valley, Sealy and Sealy (1956) mapped Taplow Terrace at Charvil Hill, but no exposures could be found here by the present author. However, gravel and sand resting on bedrock are found in two boreholes. At Sonning (SU762754: 53 m OD) the unit is 4.9 m thick and beside the Bath Road (SU768752: 53 m OD) it is 3.9 m thick. A section at Charvil Hill pit (SU766753) showed 'well stratified gravel and a little current bedded sand' (Blake 1903). Shrubsole (1890) noted some 'thin layers of clay here' and a 'lump of mottled clay about 4 feet [1.25 m] in diameter, embedded in the gravel.'

The extension of the Taplow Terrace into the Loddon Valley to include the extensively worked outliers at Norris' Green (SU776745) and Woodley Aerodome (SU775733) would seem problematical. The surfaces of these deposits occur at 45.5 m and 47.5 m OD, respectively, and are underlain by stratified gravels. The Sealys (1956) assigned these patches to the 'lower' Taplow Terrace. However, as shown above, this so-called 'lower' sub-facet is only an erosional feature present in the Slough area on the Taplow Gravel. Although they may be interpreted as degraded remnants of a previously thicker aggradation, these deposits are markedly lower than the Taplow Gravel at Charvil Hill and have therefore not been included in this unit (see also Section 2.15).

For the same reason the deposits mapped as Taplow in central Reading by Wymer (1968, figure 47) and by the Sealys (1956) as 'lower' Taplow cannot be included in this unit. However, Wymer (1968) recognised a so-called Henley Road Terrace in the area south of Caversham. The well developed terrace surface here was assigned to the 'upper' Taplow by Sealy and Sealy (1956). Wymer's (1968) type section in the degraded slope of the deposits at Donkin Hill (SU724752: 52 m OD) has been developed, but sections were still available in 1978. Here the author found 2.5 m of current-bedded gravel and sand filling a channel 4 m wide cut into sand over 1 m thick, the base of which was not seen. A palaeocurrent slightly south of east was measured from the channel fill, and a pebble count suggests that the height correlation with this member is correct. The upper 1 m of the deposits was strongly cryoturbated. Shrubsole (1890) and Wymer (1968) stated that here the sandy gravel rests on weathered Chalk and that the remains of a molar and tusk of mammoth have been found from this pit.

The pebble counts from this unit (see Appendix) show that the gravel consists of 89–94% flint (85–91% angular flint, 2.5–7.5% rounded flint), 1.5–6.5% vein quartz, 1–3.5% quartzite, and 1–5% Greensand chert (downstream of the Loddon confluence). Very small numbers of pebbles of other rock types occur, but are exceedingly rare. The significance of the counts will be discussed in Chapter 3.

The downstream gradient of this member, as mentioned above, varies between 35 and 60 cm km^{-1} (Figure 26), but is of the same order as that for other units and that determined by Hare (1947) and Sealy and Sealy (1956) on purely morphological grounds.

Colne Valley (Denham Village Gravel)

The gravels beneath Harmondsworth and the M4 were previously assigned solely to the Thames Taplow Gravel stage by Dewey and Bromehead (1915), Hare (1947) and Gibbard and Hall (1982). However, if they are accepted as representing a Colne–Thames confluence (as mentioned above), it might be expected that remnants of a Colne-equivalent aggradation could be found upstream. Terrace gravels were recognised in the Colne Valley by Howe and Skeats (1903) and Sherlock and Noble (1922), but were not correlated with the Thames sequence. Hare (1947) mapped a terrace remnant that he assigned to the 'Colne–Taplow' between the Alder Bourne and Misbourne confluences on the northern end of which Denham village is built (TQ041863). The dissected surface of the outlier extends for 2 km along the valley at 39.5–45 m OD, i.e. about 10 m above the present floodplain. There are several records of boreholes in this surface, but a transverse section along the M40 (Figure 30) demonstrates that the gravel is up to 6 m in thickness. According to Squirrell (1974, pp. 120–2), the gravels consist of subangular to rounded flint with a little quartz and quartzite. They are overlain by as much as 3 m of brown pebbly clay, probably of solifluction origin. The gravels beneath Harmondsworth occur between 21 m (base) and 26 m OD (top) in Figures 17 and 27. If the gradient of 175–180 cm km^{-1} is extended upstream, the Denham–M40 gravel outlier falls precisely at the expected height (Figure 27). The deposits which may tentatively be termed the Denham Village Gravel (type section: M40 borehole section, TQ033857–042838) are therefore thought to represent the Colne Valley equivalent of the Thames's Taplow Gravel.

Further upstream two important terrace remnants occur. The first on the outside of the northern bend in the Colne Valley, extends from Maple Cross (TQ033927) to beneath Rickmansworth and was mapped by Sherlock and Noble (1922). This gravel outlier was worked at Mill End (*c.*TQ045943: 56.5 m OD) where highly disturbed, implement-rich red gravels and sands up to

Fig. 30. Section along the M40 Motorway in the Denham – Rush Green area, constructed from borehole records (source: ERCU).

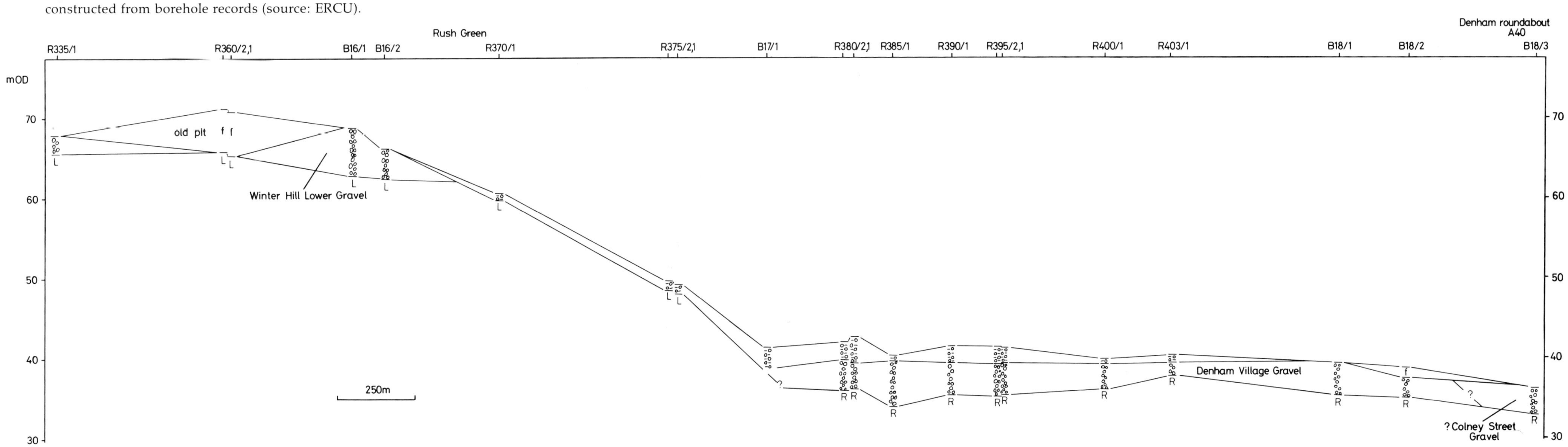

Fig. 31. Location map showing the position of exposures and borehole sections in the Trafalgar Square area. Sections: I: Jubilee Line (Figure 20); II: Shaftesbury Avenue to Whitehall (Figure 33); and III: Charing Cross to St James's (Figure 33).

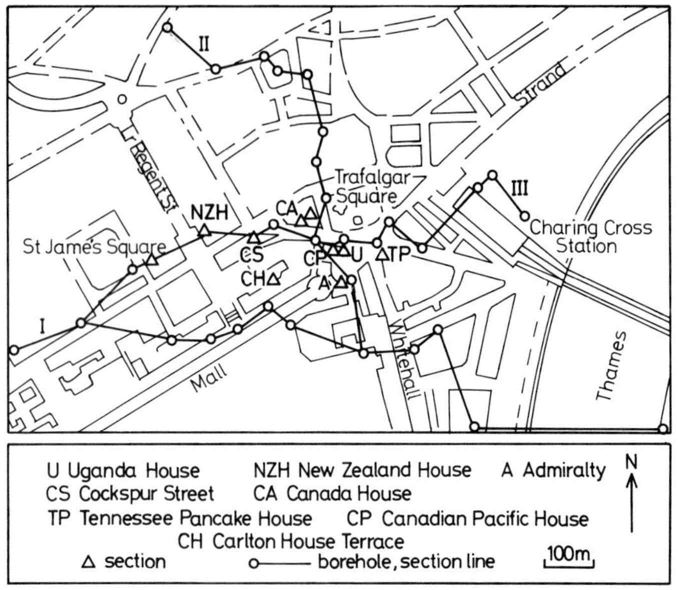

Fig. 32. Summary profiles of the main exposures recorded in the Trafalgar Square deposits, compiled in part using descriptions supplied by A. J. Sutcliffe and A. Currant.

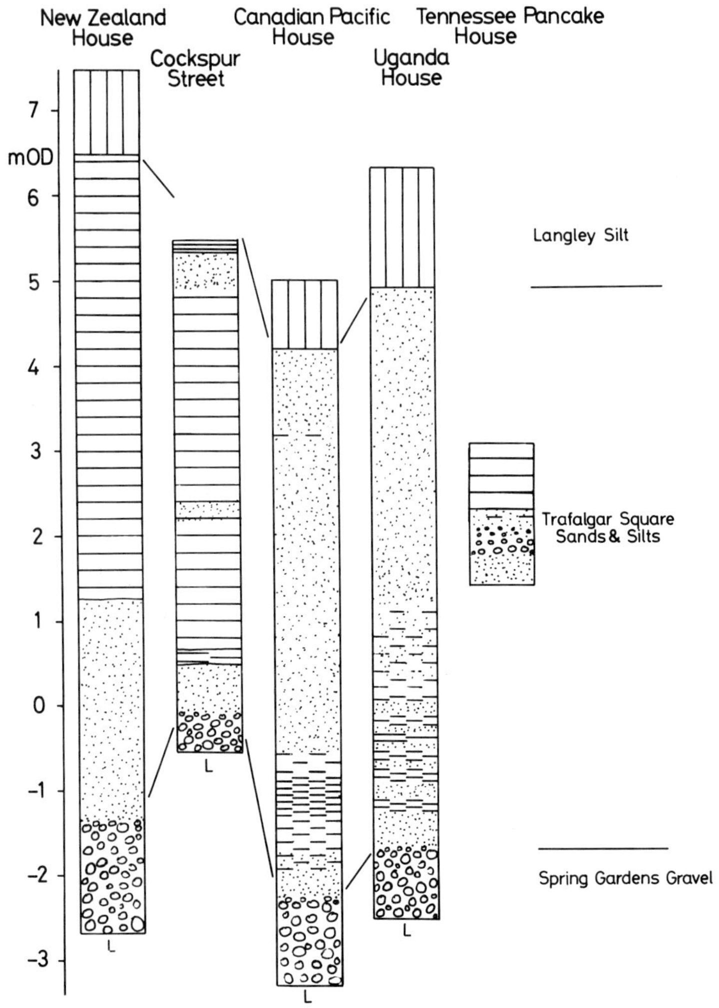

5.5 m thick were exposed in the early part of this century (Woodward, 1909; Smith and Dewey, 1915; Sherlock and Noble, 1922). The disturbance of the gravels, thought by Smith and Dewey (1915, p. 221) to be due to removal of frozen sand, may more probably be interpreted as resulting from decalcification by solution. This would also account for the very irregular Chalk bedrock floor recorded by these authors at the same site.

East of Rickmansworth, on the eastern bank of the River Gade, was the site of workings in Long Valley Wood (Croxleyhall Wood) (TQ077949). As already mentioned above (Section 2.8), sections here originally described by Smith and Dewey (1915) and Sherlock and Noble (1922), yielded very large numbers of Palaeolithic artefacts at the turn of the century. Although the pit still exists the implement-rich gravel and sand deposits have been completely removed; however higher level, probably Winter Hill Gravel is still exposed. The section illustrated by Smith and Dewey (1915, figure 3) strongly suggests that the lower gravels represented a younger aggradation that rested on a highly irregular Chalk surface at 59–60 m OD. The overlying gravels and sands were about 1–3 m thick beneath 5 m of clayey silt ('brickearth'). The upper surface of this lower gravel unit was at about 62 m OD.

Extension of the gradient on the Denham Village Gravel 7.5 km and 11.5 km, respectively, to the Rickmansworth and Croxleyhall Wood spreads shows that they fall within the expected height range of the unit (Figure 27) and are therefore tentatively correlated with it. This correlation should, in future, be evaluated using gravel lithology.

2.14 Trafalgar Square Complex

The earliest description of fossiliferous deposits in the environs of Trafalgar Square was published in the *Gentleman's Magazine* in 1758 when elephant remains were reported from sand beneath Pall Mall and St James's Square (for locations see Figure 31). Excavations for Drummond's Bank in Trafalgar Square produced a large fauna that was published in the *English Mechanic* (1882). The first systematic description of sections was made by Abbott (1892) who watched excavations for the Admiralty Building in Spring Gardens. These sections not only produced a number of vertebrate fossils including hippopotamus, but also plant and molluscan fossils. During the 1920s and 1930s further discoveries were made beneath Lloyds Bank in Cockspur Street (unpublished), Rex House in Lower Regent Street (unpublished) and the Canada Sun Life Building, Cockspur Street (Bate, 1937). Sections in these sediments were almost certainly seen by Bromehead (1925, p. 48) on the south side of St James's Square. The foundations for Uganda House, on the corner of Trafalgar Square and Spring Gardens in 1957–58 produced sections which were studied and later published by Franks *et al.* (1958) and Franks (1960). The richly fossiliferous deposits yielded large vertebrates including hippopotamus and straight-tusked elephant. Terrestrial and freshwater Mollusca, beetles, pollen and plant macrofossils were also recovered. Both the faunal and botanical evidence indicated an interglacial environment of Ipswichian age.

Subsequently, excavations have taken place at several sites including New Zealand House, Lower Regent Street; Pall Mall; St James's Theatre; the British Council Building, Spring Gardens; Cockspur Street (Laing's site) and most recently at the Tennessee Pancake House, Whitehall and Canadian Pacific House, Trafalgar Square. Almost all the recent sites were recorded by members of the British Museum (Natural History) staff, A. Currant and A. J. Sutcliffe in particular, and some of the descriptions mentioned here are based on their unpublished observations.

In addition to the open sections, a large number of site investigation boreholes have been sunk in the vicinity and are included here.

The results of new investigations at the Canadian Pacific House, Tennessee Pancake House and other localities are in the course of preparation. The work presented here will therefore take the form of a review for regional stratigraphical purposes rather than a detailed site report. Two new lithostratigraphical units can consistently be recognised in the vicinity:

(i) Spring Gardens Gravel
(ii) Trafalgar Square Sands and Silts

Both units have been studied at the Uganda House (TQ30008037) and adjacent Canadian Pacific House (TQ29958037) sections and it is proposed that these sites are jointly considered as the type sections. The sections are summarised in Figure 32 together with additional new sections at New Zealand House, Cockspur Street and the Tennessee Pancake House. For the sake of simplicity the units will be described in stratigraphical order.

Spring Gardens Gravel

Beneath Uganda House and Canadian Pacific House a bed of red brown stratified, coarse to fine subangular gravel and sand up to 1.4 m thick overlies London Clay at −2.3 to −3.3 m OD. The longitudinal and transverse sections (Figures 20 and 33) through the deposits show that the gravel underlies almost the whole Trafalgar Square – St James's area as a continuous spread. The

Fig. 33. Sections showing the disposition of sediments in the Trafalgar Square area. Upper section: Shaftesbury Avenue to Whitehall (intersects with Jubilee Line section at borehole 110); lower section, Charing Cross to St James's. Both sections were constructed using exposure and borehole data (sources: GLC, IGS).

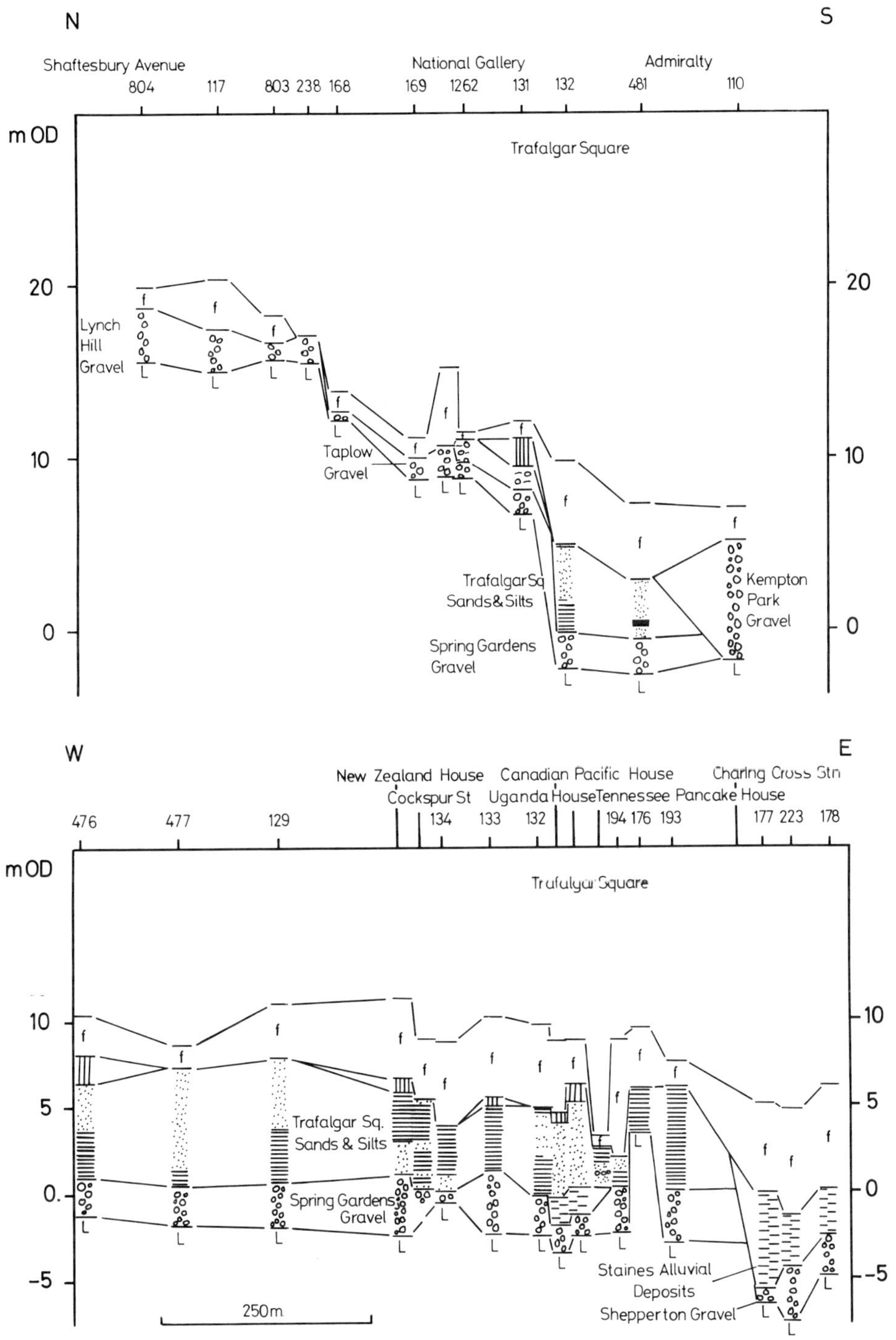

maximum recorded thickness is 4.5 m at New Zealand House where stratification and sand bands were present in the deposit. Here the gravels were penetrated by small, pinnacle-like diapirs of London Clay over 2 m long. Elsewhere the base of the deposit appears to lie directly on a gently undulating bedrock surface. The unit is absent in borehole 176 (Figure 33). Similarly, beneath the Canada Sun Life Building, Bate (1937) remarked on the marked change of gravel depth from 12.2 m below the surface at the southern end of the site to 5.2 m at the northern end. At New Zealand House, Sutcliffe saw a marked rise in the London Clay surface from −2.4 m OD on the south side beneath Pall Mall to 1.8 m OD on the north side adjacent to Her Majesty's Theatre, Haymarket. At this section the gravel was banked against the London Clay rise and was virtually absent on the higher surface. This rise presumably marks the northern limit of the unit which is therefore present beneath the southern side of Trafalgar Square and south of a line running from Pall Mall across the centre of the Square and east to Charing Cross.

The southern limit of the gravel is problematic. The transverse borehole sections (Figures 20 and 33) indicate that the Kempton Park Gravel abuts both this and the overlying Trafalgar Square Sands and Silts unit and rests directly on bedrock. It is, however, possible that the gravel is preserved as eroded remnants beneath the Kempton Park member in places. Some evidence suggesting that this may not be true comes from Abbott's (1892) observations. In the Admiralty foundations he recorded mammalian remains found in the basal gravel. They included *Bos primigenius, Cervas elephas, Mammuthus primigenius, Equus caballus, Hippopotamus amphibius, Rhinoceros* sp. and a *Microtus* sp. This fauna appears to be mixed including, for example, mammoth and horse with hippopotamus which, although they may have co-existed in later parts of the interglacial, probably did not do so at earlier stages (cf. Stuart 1974, 1977, 1982). The faunal remains at the Admiralty should perhaps be interpreted as derived by erosion of the Trafalgar Square Complex deposits during the Kempton Park phase. It remains unclear exactly what Abbott saw since the sections would probably have exposed the boundary between the Trafalgar Square and Kempton Park members. Abbot certainly saw and sampled the Trafalgar Square Sands and Silts since Mollusca undoubtedly from these deposits collected by him are preserved in the Geological Museum. In view of this Abbott's observations should perhaps be regarded with some caution.

Trafalgar Square Sands and Silts

At the type sites the Spring Gardens Gravel was overlain abruptly by a complex fossiliferous sequence (Figures 32 and 33). At Uganda House, the lower part of the deposits consisted of 1.5 m of alternating grey clayey silt and brown sand bands 15–30 cm thick, some horizons being rich in plant remains and containing Mollusca. This bed was overlain by 3.3 m of brown sand that was silty at the base, cross-bedded at some horizons, and contained occasional shelly bands. The lower part of the sand was again rich in plant remains. A narrow sandy gravel horizon 20 cm thick succeeded the sand and was in turn overlain by 30 cm of greenish-grey, well sorted shelly sand from which several vertebrate bones were found. Further shelly sand about 1 m thick was separated from that below by a narrow silt band 15 cm thick. It therefore appears possible to subdivide the sequence broadly into two groups of strata: a lower subunit of alternating clays, silts and sands and an upper subunit of sand, frequently shelly and occasionally cross-bedded. An almost identical sequence occurred beneath Canadian Pacific House. The lower subunit comprised grey clayey silt and included a dark grey organic silty clay band at −0.8 m OD. The upper part of this subunit seemed to grade into the overlying sands. At both sites the section was capped by up to 1 m of mottled grey brown clayey silt 'brickearth' with a marked erosional base.

At the Tennessee Pancake House section, a slightly different sequence was present. Here about 20 cm of coarse, shelly gravel rested on an erosional surface of cream–yellow sand at 1.75 m OD (Figure 32). The gravel was overlain by 60 cm of grey to pale buff clayey silts rich in Mollusca at some levels. To judge from the generally fining upward sequence the deposits appear to represent a channel fill, the channel having been excavated into the sand of the upper subunit in the nearby Uganda House section.

As can be seen from the plotted sections (Figures 20, 32 and 33), the type section subunits can be recognised at some sites and in some boreholes. However, the sequence is clearly very variable. At Cockspur Street, the lower 4.5 m comprised a sequence of grey brown clayey silt laminated in places and containing abundant shells and occasional plant remains. Two organic mud horizons were present, a lower bed 17 cm thick near the base and a second 30 cm thick higher in the section. At New Zealand House the deposits comprised a basal sand 2.5 m thick overlain by an alternating sequence of silt and sand bands 30–40 cm thick, followed by about 3.5 m of green grey silt. Brown silty clay ('brickearth') capped the section here, again resting on an eroded surface of the underlying silt. The deposits therefore show a considerable facies variation.

The plotted sections (Figures 20, 32 and 33) suggest that the lower subunit at the type section occurs in a depression in the underlying gravel surface. If this is correct it may represent the filling of a channel either cut into gravel or inherited from the gravel aggradation phase. The gradation from initial fine sediment deposition in the deepest hollows into coarser sediments showing current structures must be contrasted with the apparently opposite occurrence at sites such as New Zealand House or the almost total absence of current-bedded sand at Cockspur Street, for example. This suggests that main flow channels were migrating during the aggradation of the sequence. The palaeobotanical sequence represented at the Canadian Pacific and Tennessee Pancake House sections is considered in Chapter 7.

Brentford Sands

A strikingly similar succession to that at Trafalgar Square occurs upstream at Brentford (Figure 17). Here a series of fossiliferous deposits containing many vertebrate fossils in common with those at Trafalgar Square including hippopotamus and straight-tusked elephant occur in a geographically analogous position. Abundant Mollusca and plant remains are again present and the weight of evidence suggests that they are of Ipswichian age (Zeuner, 1959; Kerney, 1959). The deposits were originally described by Trimmer (1813) from Gunnersbury Park (*c.*TQ186785: 11 m OD) and near Boston Manor Road (*c.*TQ172781: 12 m OD) and were subsequently seen by Morris (1838) at Brentford Waterworks (TQ184781), by Zeuner (1959) and Kerney (1959) at Beecham's House (TQ181782) and by A. J. Sutcliffe (personal communication) beneath an extension to Beecham's House (TQ180783). The sequence is currently being reinvestigated and will be the subject of a future publication.

The sequence at Brentford according to Trimmer (1813) also included a basal gravel unit that overlies London Clay. At Gunnersbury Park, the deposit comprised 'an heterogeneous mass of clay, sand, and gravel' at the base, was 0.6–3 m thick and rested on London Clay at about 3 m OD. According to Trimmer's (1813) description the gravel here contained the remains of large vertebrates. It is likely, however, in view of the site stratigraphy, that these remains were in fact not from the gravel but from the interface with the overlying loam. At Boston Manor Road, the gravel again rested on London Clay. On the basis of these descriptions and the gravel's stratigraphical position and height (Figure 17) the Brentford gravel is tentatively equated with Spring Gardens Gravel. Such a correlation gives a gradient of 50 cm km^{-1} on the unit. This is comparable with those for both higher and lower gravel members and suggests that the correlation is correct.

According to Trimmer (1813) the Gunnersbury section exposed a laterally impersistent 'peat . . . a few inches thick' that rested on the gravel beneath. This bed was overlain by 30–150 cm of shelly sandy loam which yielded vertebrate material. 'A few inches' of sandy gravel occurred above the loam and was overlain by a further sandy loam over 2 m thick. Work currently in progress indicates that the upper 2–2.50 m of sediment is an unfossiliferous brown silty clay, probably a solifluction deposit derived from higher ground north of the site. This overlies a sandy, fine, angular gravel 20 cm thick beneath which a buff silty sand occurs.

A comparable sequence was present at Boston Manor Road (Trimmer, 1813). Here the basal gravel was overlain by 0–3 m of sandy loam that contained vertebrate remains and Mollusca. Above was a bed 1–2 m thick of fossiliferous sand becoming a sandy gravel at the base. This was in turn overlain by 2–3 m of brown silty clay.

The similarity of sediment types, the suggestion of a parallel sedimentary development, the presence of animal and plant remains and their altitudinal and stratigraphical relationships suggest that the deposits at Brentford may be equated with those at Trafalgar Square and they are therefore correlated. This correlation is supported by the biostratigraphical vertebrate and molluscan (Kerney, 1959) evidence.

2.15 Reading Town Gravel

In the Reading area there is a well developed spread of gravel deposits that occur in a position intermediate in height between the modern floodplain and the Taplow Gravel. These deposits in central Reading were assigned by Wymer (1968) and by the Sealys (1956) to a 'lower' Taplow stage. However, as shown above, this discrete unit is too low for inclusion in the Taplow member. The intermediate position of the deposits is precisely that of the Kempton Park Gravel in the West London area. It therefore seemed most reasonable initially to attempt to correlate the Reading Town Gravel with the Kempton Park Gravel. However, as will be discussed in Section 2.17, the latter cannot gen-

Fig. 34. Sections showing the disposition of deposits in the Reading area. Upper section: aligned approximately north–south; lower section showing the long profile of the Reading Town Gravel (source: IGS).

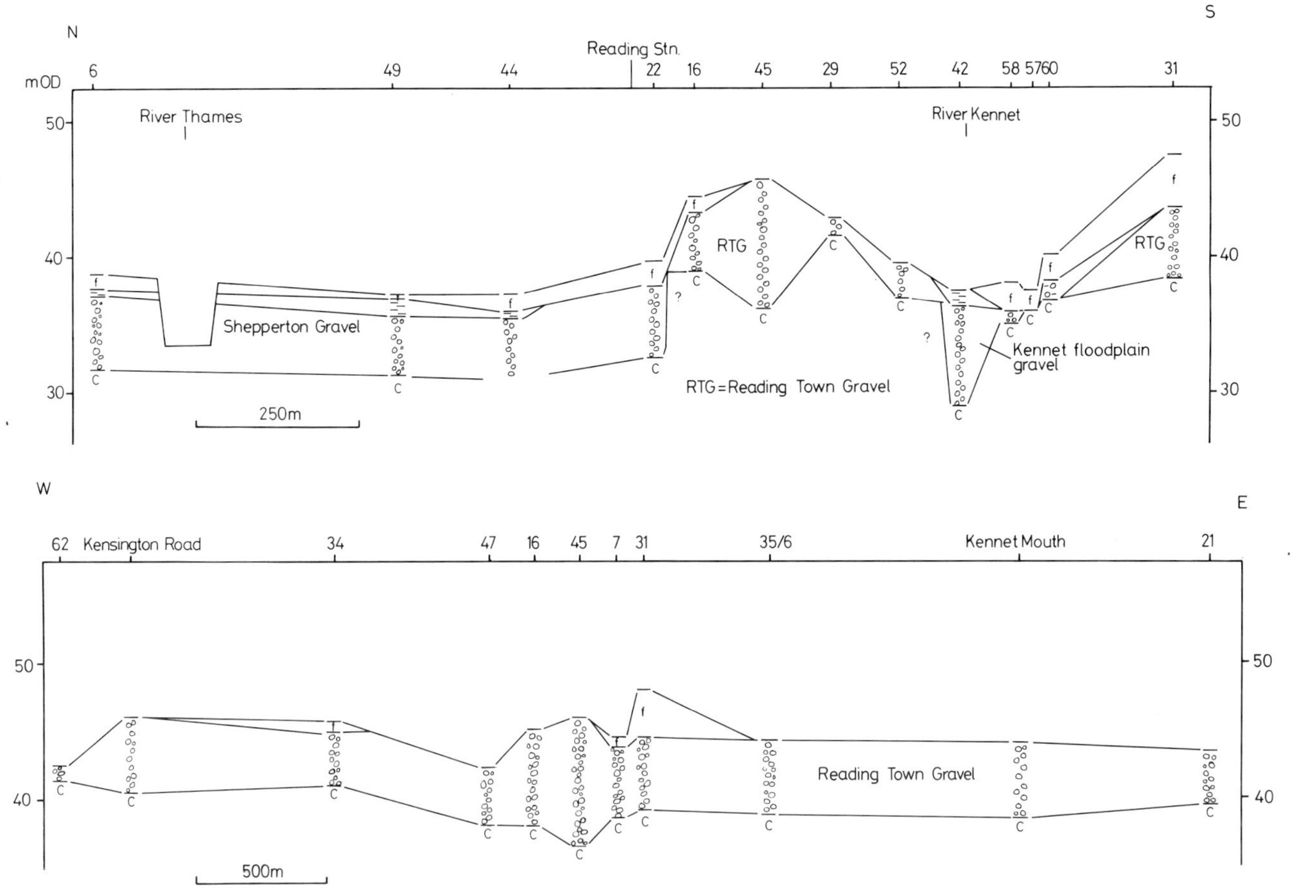

erally be separated from the floodplain upstream of Chertsey except where isolated outliers are found but little above the floodplain surface. By contrast, the deposits in the Reading area occur 6–10 m above the local floodplain surface. A marked steepening of the gradient of the sediment body would be required to link these units if they are lateral equivalents. This seems unacceptable in view of the consistent gradient of 30–40 cm km^{-1} on the Kempton Park Gravel downstream. For this reason the aggradation will be referred here to a separate unit, the Reading Town Gravel. Stratigraphically it appears to be intermediate in age between the higher, and therefore probably older Taplow Gravel and the lower and therefore presumably younger Kempton Park Gravel. Its relation to the Langley Silt Complex cannot be assessed at present since their distributions do not overlap.

Unfortunately, the gravel was not exposed during this study so that descriptions presented are taken from previous publications. The lithological content of the gravel cannot therefore be assessed at present. Inter-site correlation must be based only on altitude. The terrace surface developed on the gravel cannot be considered as Taplow (Section 2.13) and therefore a new term, the Reading Town Terrace should preferably be used.

The Reading Town Gravel occurs in two main spreads on which much of the centre of the town is built. The first, downstream of the Coley Gap, extends beneath the London Road as far as New Town (Figure 1). Here the surface is at 44–46 m OD. The deposits were exposed in the Kennet Mouth pit (SU733737: 44 m OD) which for present purposes may be regarded as the type section for the Reading Town Gravel, where Shrubsole (1890) recorded 4.9 m of gravel and sand. According to Blake (1903), the well stratified gravel contained some pebbles of chalk and fragments of Eocene shells, as well as numerous pebbles of quartz and quartzite and 'bits of fossils probably from the Oolites. The bulk of the deposit was as usual formed by sub-angular flint'. Shrubsole (1890) noted the occurrence of fragments of mammalian bone, but could only identify mammoth. Stevens (1882), however, also recorded horse as well as further mammoth remains from this site. Several excavations in the gravel spread are mentioned by Blake (1903, pp. 76–7) all of which showed the gravel and sand to be 3–5 m thick.

Boreholes available for the area confirm the observations of these authors. At the extreme eastern end of the outlier 3.8 m of gravel are present above Chalk (SU742736: 44 m OD). A similar thickness is present west of Wokingham Road (SU731731: 46 m

OD), while near Coley Gap, borehole 31 (Figure 34), records 5 m of gravel and sand.

One further locality should be mentioned in connection with this gravel spread; the Redlands quarry, which was located immediately west of the Erleigh and Addington Road junction (SU732727: 48 m OD). The sequence, exposed in the late nineteenth century, was described in detail by Poulton (1880) and was summarised by Wymer (1968, p. 161). The Pleistocene deposits here comprised broadly two aggradations, a lower bed of fluviatile sands and clays, 2.75 m thick and locally disturbed, capped by an upper bed of clayey gravel, 3.6 m thick, of probable solifluction origin. The latter was thought by Wymer (1968) to have been derived from the higher Boyn Hill Gravel immediately to the south. The lower bed was well stratified and was probably formed of sediment and fossils reworked from the Reading Beds. It contained trunks of trees (*Pinus* sp.) possibly also derived from the Tertiary rocks, but equally possibly contemporary with the deposits. Vertebrate remains including mammoth, ox, horse and rhinoceros were found together with fragmentary shells. The upper gravel bed also contained vertebrate material of the same species. However, they were distinctive in having been broken rather than water worn. Palaeoliths were also found in the upper gravel.

The description of these deposits shows that in spite of their altitudinal equivalence to the Reading Town Gravel, they are lithologically distinct and cannot therefore be correlated with this unit. The presence of fossils in the sand and clay bed led Wymer (1968) to speculate that they were of interglacial fluvial origin, blanketed by later solifluction deposits. While they need not strictly be of interglacial age, the deposits are clearly distinct and appear to represent a remnant of a separate aggradation termed here the Redlands Sand. Their relationship to the Reading Town Gravel can be little clarified at present. However, the slightly higher elevation of the Redlands Sand deposits suggests that the Reading Town Gravel abuts them; the nature of the contact possibly being analogous to that between the Kempton Park and Trafalgar Square members beneath the Admiralty (Figure 20). Further investigation of the Redlands Sand deposits would be valuable.

West of the Coley Gap, the deposits of this unit continue. According to Wymer (1968, figure 47) this westerly spread can be followed from near Reading Prison to near Norcot Farm (SU685740). The section (Figure 34) shows that the Kennet Valley merely dissects the sediment body at Coley. The base of the deposits is rather irregular resulting from localised scouring or bedrock collapse that gives rise to the apparently anomalous thickness of gravel, i.e. 9.3 m (borehole 45). Numerous boreholes in central Reading record gravel and sand at the expected height; for example at SU716736 (44.5 m OD) 4.2 m of gravel were recorded beneath made ground, and at SU706737 (45 m OD) 3.5 m of gravel was present. Shrubsole and Whitaker (1902) described an exposure in Kensington Road (SU698734: 46 m OD). Here 5.5 m of stratified gravels and sands resting on Chalk were visible. Blake (1903) mentioned that the gravel consisted of many small chalk pebbles, subangular flints often of large size and a few large quartzites. Of particular interest is the occurrence of pieces of broken bone and several teeth of hippopotamus found by Shrubsole and Treacher. Blake (1903) again recorded several other localities that exposed gravel of comparable thickness in this spread.

No further deposits equivalent to this aggradation are known to occur in the Middle Thames Valley. However, extensively worked outliers of gravel and sand, again equated with the 'lower' Taplow by Sealy and Sealy (1956) and Thomas (1961) occur in the Loddon Valley at Norris' Green (SU776745) and Woodley Aerodrome (SU775733). The upper surfaces of these deposits are well developed at 45.5 m and 47.5 m OD, respectively. The Norris' Green site is now infilled, but some gravel and sand that may have been *in situ* was found. However, at Woodley extensive gravel sections occur beside the perimeter road. Here 2.5–3 m of gravel resting on Reading Beds were exposed, the upper 1.5 m of which was involuted. The deposits comprise horizontally bedded fine to medium gravel present as discrete lenticular bodies, but also in parts as continuous bands, individual units 25 cm in thickness. A pebble count (see Appendix) indicates that the gravel comprises rounded flint and Greensand chert, i.e. it is of Loddon type. Tabular cross-bedded sand bands and lenses up to 5 cm thick usually occurred irregularly interstratified with the gravels. Palaeocurrent measurements from the sand bands indicated a flow towards the north-north-east. The gravel and sand was locally overlain by a 25 cm thick bed of brown pebbly silt of possible solifluction origin derived from high ground to the east. The palaeocurrent measurements and the pebble counts confirm that the deposits were deposited by the Loddon, upstream of its confluence with the Thames. Since no type section has been defined for the gravel of this stage in the Loddon Valley, the Woodley site might be considered. If so the deposits would best be termed the Woodley Gravel.

If the gradient of the Reading Town Gravel (55 cm km^{-1}) is extended downstream from New Town to the Loddon confluence, it converges precisely with that on the surface of these outliers (125 cm km^{-1}) at 41.5 m OD. The Woodley Gravel is therefore considered here to be the Loddon equivalent of the Reading Town member in the Thames sequence.

2.16 Langley Silt Complex ('brickearth')

The term 'brickearth' has been applied in the Thames Valley to any fine-grained sediment that was suitable for brick-making. As a result it is of extremely variable character (Bromehead, 1925). It varies from a sandy silt to little more than recycled London or Reading clay (Dewey and Bromehead, 1915, 1921; Bromehead, 1925). In places it is stratified, while it is predominantly massive over most of the study area. It often contains artefacts (cf. Wymer, 1968) and is generally unfossiliferous, but occasionally yields faunal or floral remains.

As early as 1864 the sedimentary characteristics of the dominant sandy or clayey silt 'brickearth' led Prestwich to compare it with the loess of Central Europe. These comparisons have continued up to the present day (Wooldridge and Smetham, 1931; Breuil, 1934; King and Oakley, 1936; Oakley, in Lacaille, 1936; Dalrymple, 1960; Wymer, 1968, 1977b; Catt, 1977, 1978; Lill and Smalley, 1978). Although Hollin (1977) favoured a tidal origin for this material, it is now generally accepted that these clayey silts originated as a combined loessic and waterlain deposit. However, the exact mechanism of deposition has given rise to much controversy. For example, Wymer (1968) favoured a colluvial and fluviatile origin. Catt (1977) preferred reworking of loess combined with overbank deposition ('flood-loam' of earlier workers). In contrast, the writer, following Dalrymple (1960), concluded that loess deposition could have been contemporary with fluvial overbank accumulation (Gibbard, 1974, 1977).

Not all of this material has the same origin however. As well as loessic material, the name 'brickearth' has been applied to demonstrably solifluction deposits, stratified estuarine deposits (especially in the Lower Thames cf. Kennard, 1944; Hollin, 1977) and colluvial sediments. Because of its polygenetic origin it is found in apparently conflicting stratigraphical contexts.

'In Middlesex [the deposits] cover indifferently all the three terraces' according to Pocock (1903). The main mass overlies the Taplow and Lynch Hill Gravels, but a thin cover is also present on the lower, Kempton Park Gravel in some places in West London. In the past, the 'brickearth' was often regarded as being 'intimately associated' with the gravel on which it rested (e.g. Sherlock and Noble, 1922) by analogy with the alluvium of the modern floodplain that overlies the Shepperton ('Lower Floodplain') Gravel. However, since the 'brickearth' can generally be shown to be a cold stage deposit, it is certainly not directly comparable to modern interglacial alluvium.

The word 'brickearth' cannot therefore be used as a precise lithological term. For this reason the term 'silt' will be used here to describe the silt-rich sediments that form the predominant lithology. The single name 'Langley Silt Complex' is proposed to describe the main thickness of deposits that are thought to be in their original position since there is reason to believe that the internally complex spreads might represent a single aggradation (see below). This formal name should ideally be restricted to *in situ* material only, but in practice difficulties may arise in deciding whether sediment is *in situ* or not. Such a decision must be based on an investigation of the sedimentary structures at each site. Where material has been soliflucted or washed onto lower units the problem of its precise stratigraphical position may be difficult to determine. Naturally, the material must postdate the youngest aggradation on which it lies unless post-depositional redistribution can be demonstrated. The relation of the Langley Silt Complex to deposits of known age should clarify its stratigraphical position. However, the possibility of multiple cycles of deposition must also be borne in mind.

The approach adopted here for the correlation of this unit is largely descriptive, supported by selected analyses of grain-size and heavy-mineral assemblages. Some of the analytical work was undertaken by J. Catt (Rothamstead). Ultimately, between-site correlation will only adequately be achieved by using a combination of approaches including sedimentology, heavy-mineral assemblages, stratigraphical position, palaeontology (if any) and interpreted origin. Such a correlation, supported by thermoluminescence dating (Wintle, 1981; Wintle and Huntley, 1982) now beginning to be applied, will help elucidate the problem in the future. TL dates obtained from four sites in west Middlesex and south Buckinghamshire by A. Wintle (personal communication) indicate that the main mass of Langley Silt was deposited around 17 000 BP, i.e. during the Late Devensian.)

The large gravel working at Langley (TQ002800: 28 m OD) is proposed as the type section for this unit. Here the silt is 2 m

Table 1. *Particle-size analytical results from a selection of nine samples of sediments of the Langley Silt Complex, kindly supplied by J. A. Catt.*

Size: Phi (ϕ) divisions	μm equivalent		Cranford Park (1.0 m)	Cranford Park (2.0 m)	Langley (0.2 m)	Langley (0.9 m)	Yiewsley Lynch Hill (1.0 m)	Yiewsley Lynch Hill (2.0 m)	Iver	Thorney Lane, Iver	Burnham, Taplow
0 to −1	1000–2000	Sand	0.7	1.0	3.6	0.0	0.2	0.8	1.5	1.6	2.0
+1 to 0	500–1000	Sand	1.9	1.8	5.0	0.3	0.4	1.3	1.4	3.0	3.5
+2 to +1	250–500	Sand	8.8	8.9	13.9	1.5	2.1	4.7	4.0	7.8	6.3
+3 to +2	125–250	Sand	11.7	10.4	14.9	2.3	3.8	5.7	4.2	8.2	4.3
+4 to +3	63–125	Sand	9.7	12.2	15.2	6.5	6.1	9.9	5.7	9.4	3.9
+5 to +4	31–63	Silt	11.2	13.8	17.5	25.6	26.8	10.8	24.5	19.2	21.4
+6 to +5	16–31	Silt	9.1	6.8	7.2	23.2	17.4	8.7	18.8	14.9	20.7
+7 to +6	8–16	Silt	6.3	4.0	3.4	11.4	9.5	7.4	10.1	8.1	9.5
+8 to +7	4–8	Silt	3.8	3.5	2.3	5.1	4.7	5.7	4.7	4.5	5.1
+9 to +8	2–4	Silt	3.0	3.6	2.0	2.8	3.1	4.8	3.2	3.9	4.1
>9	<2	Clay	33.8	34.0	15.0	21.3	25.9	40.2	21.9	19.4	19.2

Table 2. *Heavy-mineral analytical results from the 16–63 μm (coarse silt) fractions from the same samples of Langley Silt Complex sediment as used for particle-size analyses in Table 1. An additional sample from Langley Silt at Trafalgar Square (Canadian Pacific House) is included. The light fractions are given as percentages (%), the heavy fractions as per thousandths (‰). Analyses by J. A. Catt.*

	Cranford Park (1.0 m)	Cranford Park (2.0 m)	Langley (0.2 m)	Langley (0.9 m)	Yiewsley Lynch Hill (1.0 m)	Yiewsley Lynch Hill (2.0 m)	Iver	Thorney Lane, Iver	Burnham, Taplow	Trafalgar Square
A. Light fraction										
Quartz	70.9	77.3	80.3	76.3	78.4	80.6	80.1	81.3	78.8	80.0
Alkali feldspar	21.5	16.4	12.5	16.1	15.5	12.1	14.6	14.6	18.9	16.8
Flint	1.0	0.5	1.5	2.1	1.5	1.8	1.1	0.5	0.9	1.3
Muscovite	6.0	5.2	3.7	3.8	3.4	2.9	3.1	1.9	0.7	1.7
Glauconite	0.6	0.6	2.0	1.7	1.2	2.6	1.1	1.7	0.5	0.2
B. Heavy fraction										
Epidote	92	106	326	338	363	311	330	352	368	309
Zoisite	21	28	16	15	18	24	24	28	22	19
Zircon	555	430	208	177	196	262	140	212	262	178
Tourmaline	73	62	44	31	25	40	37	26	56	106
Chlorite	—	17	146	113	78	113	104	134	39	160
Biotite	1	—	1	11	—	1	3	2	—	—
Garnet	72	120	60	77	58	83	90	40	51	50
Green hornblende	40	67	85	132	117	52	146	110	55	48
Brown hornblende	8	13	2	2	—	1	3	5	—	1
Tremolite/actinolite	8	13	26	15	28	13	24	11	9	17
Yellow rutile	44	44	36	34	49	46	39	36	83	37
Brown rutile	26	35	13	10	19	20	17	9	23	12
Red rutile	2	1	—	—	1	—	—	2	—	—
Anatase	6	8	15	23	30	18	14	26	19	33
Brookite	9	6	3	2	2	4	4	2	6	3
Straurolite	24	30	4	—	2	2	3	—	4	6
Kyanite	15	12	2	—	6	6	1	5	3	4
Andalusite	1	2	—	—	—	—	1	—	—	—
Sphene	3	1	1	—	3	3	3	—	—	5
Spinel	—	1	1	—	—	1	—	—	—	1
Augite	—	—	—	2	1	—	1	—	—	—
Hypersthene	—	—	—	—	—	—	—	—	—	1
Apatite	—	—	8	13	2	—	15	—	—	9
Collophane	—	1	3	5	2	—	1	—	—	1
Monazite	—	2	—	—	—	—	—	—	—	—

thick and rests on the gravel beneath with a sharp, slightly undulating base. The basal 75 cm of the deposit consists of a buff clayey silt with a 10 cm thick sand band at the base. The silt contains scattered subrounded to angular flint or rare quartz pebbles many of which have vertical long axes. Above, a light brown silt band 50 cm thick is present and passes upwards into a 75 cm thick chocolate brown clayey silt almost totally free of pebbles, but showing a marked vertical hexagonal, columnar-jointing pattern. The three transitional sediment zones present in the section appear to result from increased post-depositional weathering and decreasing moisture content up the section. The entire deposit is decalcified and shows small irregular 'worm-like' tubules throughout. Grain-size and heavy-mineral analyses are shown in Tables 1 and 2 respectively. Oakley (in Lacaille, 1936) attributed Levalloisian implements to the silt in these sections. The 'laminated brickearth' seen by Oakley was not found by the writer.

There is an almost continuous spread of the silts overlying the Taplow Gravel in the Langley area (Figure 25). Near Langley Station (TQ013798) clayey silt showing the same colour variation as that at the type section was seen by Howe (in Dewey and Bromehead, 1915). North of the type section, four boreholes record variable thicknesses of the deposit. At Parsonage Farm (TQ016805: 29.5 m OD) the unit is over 4.4 m thick where it is banked against the bluff formed by the higher, Lynch Hill Gravel. A shallow augered transect in this area showed that the silt above the Taplow is continuous with that overlying the Lynch Hill Gravel in the Iver area.

The silt complex around Iver was described by Oakley (in Lacaille, 1936) as a 'red–brown structureless loam' containing occasional strings and festoons of angular, frost-shattered flints. He also noted 'the characteristic vertical fissuring' which it develops on drying. Old pit sections at TQ036803 exposed 70 cm of this deposit and a sample taken at 10 cm above the base was analysed and gave the results shown in Tables 1 and 2. Although it was generally decalcified, Oakley saw some places where deep festoons in the gravel contained a yellow-orange calcareous silt which was undoubtedly the same deposit as the red brown sediment. The base of the sediment again rests on cryoturbated Lynch Hill Gravel or in places on the soliflucted Iver Clayey Gravel. In Purser's pit (TQ032797) and elsewhere the red brown silt was overlain by up to 60 cm of grey buff structureless calcareous clayey silt. The latter was separated from the former by a marked hiatus. The hiatus suggested to Oakley that emplacement of the grey buff silt post-dated the decalcification and reddening of the subunit beneath. The local setting suggests that the grey buff silt could conceivably represent later soliflucted material derived from a higher London Clay slope north of the sites. This deposit has not been seen by the writer. The silt is recorded in several boreholes near Iver, and a small section nearby (TQ040796) exposed 40 cm of pebbly red–brown silt overlying sandy gravel. The analytical results of a sample from this section (Thorrey Lane, Iver) are shown in Tables 1 and 2. The silt is also present in boreholes for the M25 Motorway (Figure 12).

East of the Colne Valley, thick silt deposits are again present in the Yiewsley–Harlington area (Dewey and Bromehead, 1915). The gradient on the uppermost surface of the aggradation, although much has now been removed, at the Colne confluence is 25–30 cm km^{-1}. This gradient intersects with the deposits at Yiewsley described by Collins (1978) and observed by the writer where the main workings exposed 3 m of deposits overlying the Lynch Hill Gravel (TQ083803: 37 m OD). Here the basal 1.20 m of the unit were stratified comprising a partially fining upward sequence of ripple cross-bedded sand 10 cm thick at the base overlain by alternating silty sand and brown silt bands up to 10 cm thick. Grain-size analyses from these bands suggest that the sediments were predominantly waterlain and therefore probably of colluvial origin. A palaeocurrent measurement from the cross-bedded sand indicated flow towards due south and suggests that these deposits were derived from the higher ground north of the site. Above the stratified sediments, decalcified, massive brown clayey silt containing occasional pebbles and again showing the characteristic prismatic structure, occurred. In one part of the section a gravel band 20 cm thick (1 m above the base of the silt), possibly the Stockley gravel of Collins (1978), included a small ice wedge cast (50 cm long and 15 cm wide) at its base. At Yiewsley, the silt unit could be traced continuously to the north face (TQ084807). Here 1.5–2 m of massive brown silt was present. In these sections the silts lacked the basal stratification and rested on the gravels beneath with a sharp junction. Collins (1978) proposed several names for this 'loam', but provided no criteria to support their separation. Indeed, the sections seen by the author showed that the deposit in the north face is the same as that overlying the stratified sediments in the south face. Such a similarity is supported by the grain-size and heavy-mineral analyses (Tables 1 and 2). At Hayes and Dawley, the deposits appear to be a continuation of those at Yiewsley. Southwards the deposits form a continuous sheet as far as the M4 Motorway and bury the 'bluff' formed by the Taplow Gravel abutting the Lynch Hill (Section 2.12).

The section along the M4 Motorway shows a maximum of 3 m of this unit almost continuously overlying the Taplow and Lynch Hill Gravels between Harmondsworth and Osterley (Figure 17). It may also be followed along the spur from the M4 to Heathrow Airport entrance (TQ075771) where it is 1.5 m thick.

The Geological Survey mapped the silt as a continuous spread underlying the area as far south as Longford and Cranford. Sections at four localities north of Heathrow exposed the silt complex overlying Taplow Gravels at Sipson (TQ070782 and 077781), Harlington (TQ085785) and Cranford Park (TQ094776–091783). At Sipson the silt was 2–2.5 m thick and consisted of partially stratified fine gravel and silt at the base. Above, it was massive, light brown in colour and characteristically columnar jointed. At one place, on the north side of the pit (TQ077783) an eastward-trending channel-like feature 1.5 m deep and 3 m wide filled by the silts was present in the gravel surface. This feature was not found elsewhere. A similar lithological sequence was present in the Harlington and Cranford Park sections, although the sediment was much more clay-rich and darker in colour at the latter site. Grain-size and heavy-mineral analyses from these sites are given in Tables 1 and 2.

The boreholes for the Piccadilly Line extension to Heathrow Central (Figure 28) confirm the presence of up to 1.25 m of the Silt Complex beneath the Heathrow Central area, being cut-out immediately to the south. The sediments recur near the junction of the Great South West Road and Bath Road where they are 1.5 m thick. The Cranford Parkway section (Figure 16) again records a nearly continuous spread of the silts from Uxbridge Road, Southall to the Western Region main line (TQ105794). Here the deposit has been removed by the later River Crane and by gravel exploitation. However, near the M4 Motorway the silts reappear and can be followed onto the Taplow Gravel at the southern end of the section almost without a break. Boreholes on both sides of the Crane Valley record the occurrence of the silt resting on Lynch Hill Gravel, for example beneath the Nestlé factory (TQ101792), at Southall Gas Works (TQ116797) and beneath Heston Aerodrome (TQ112784).

The deposits can be followed therefore from Southall and Hayes, where they were extensively worked in the past, south to Hounslow reaching a line that runs parallel to the Staines Road. Dewey and Bromehead (1915) noted that trial holes between Hounslow West Station and Hounslow Barracks encountered

0.6–2.5 m of silts, resting on gravel. The silts were, however, absent beneath Hounslow Heath, 600 m south of the barracks. A number of old workings in the Osterley area yielded artefacts and were summarised by Wymer (1968).

The Brent Valley dissects the deposits east of Osterley, but they are present again on the opposite side forming a spread that extends from beneath Hanwell to Ealing and beyond to Acton. In the Hanwell–Ealing area the Silt Complex is again found above gravel. Between Drayton Green and Hanwell stations several boreholes show that the deposits vary considerably in thickness. At the former site (TQ156806) they are 3 m thick and were described as 'brown clayey silt with stones', while at the latter (TQ158807) the basal 1 m was laminated. Similar brown clayey silt occurs beneath Lammas Land (TQ172797) and Ealing Polytechnic (TQ177800). Near Ealing Broadway Station, boreholes beside the Western main line record 1.6 m of brown sandy clay (TQ178809), but this material together with the gravel beneath are absent immediately north at TQ177810. The sequence in this area is best illustrated by the section along Hanger Lane (Figure 18). Here the Silt Complex comprises an almost continuous stratum that thickens northwards towards the London Clay slope of Hanger Hill. Between the District Line and Woodville Gardens (TQ186813) the deposits are over 5 m thick and were described as 'brown silty clay with stones.' South of the District Line they become more uniform in thickness, so that beneath Ealing Common they are only 1.5–2 m thick and are silt rich (cf. also Bromehead, 1925). Further south the deposits thin and finally wedge-out beneath the Popes Lane – Hanger Lane junction (TQ191794).

Approximately 600 m east of Hanger Lane, the temporarily exposed Creffield Road section of Brown (1886) occurred (TQ192808: 31 m OD). The sequence here seems to have consisted of a basal 'sandy loam' about 0.3–1 m thick overlying the Lynch Hill Gravel. The bedding plane between the gravel and the loam yielded abundant Levalloisian artefacts. This led Brown (1886) to suggest that it represented a buried land surface or floor. Unbedded brown 'brickearth' 60–70 cm thick was present above the loam and this in turn was overlain by 60 cm of 'trail' (soliflucted clayey gravel) showing involution structures. The 'brickearth' horizon also yielded some artefacts. More recent excavations by G. de G. Sieveking (personal communication) nearby (TQ195808) revealed a comparable sequence. Here the deposits comprised a partially stratified sandy silt.

Several sections in the Acton area were observed by Lane-Fox (1872) during house building. In Chaucer Road (TQ203804) he saw 1.2 m of clayey silt containing two narrow seams of white sand. The silt deposits are not preserved east of Acton, since they have been removed by later erosion.

The Silt Complex recurs in the Central London area near Holland Park (Figure 19). Here, in spite of exploitation over a long period by man for brick-making, some boreholes record its presence above Lynch Hill Gravel. Beneath the junction of Bayswater Road and Pembridge Road (TQ252802) 1.8 m of 'brown clayey silt with occasional stones' are present. A similar thickness was found at Queensway (TQ259806), while at Marble Arch (TQ278810) up to 2.75 m of 'brown stony silty clay' was present in a similar stratigraphical position. At St Giles Circus (TQ301813) only 1.25 m of clayey silt is present. The Jubilee Line section (Figure 20) again shows 'brown silty clay with pebbles' overlying gravel beneath Euston Road (TQ285821) and Marylebone (TQ286816).

Sections beneath Endsleigh Gardens and environs near Euston (TQ297825) described by Hicks (1892) exposed dark grey silty clay (60–80 cm) containing plant and vertebrate remains resting on London Clay and overlain by 1–2 m of sandy gravel, 0.3–0.6 m of silty sand and up to 3 m of yellowish-brown clay with flint pebbles. The lower silty clay, gravel and sand deposits which are separate from, and probably later than, the local Lynch Hill Gravel appear to fill a northward-trending valley cut into the London Clay. The overlying yellowish-brown sediment appears to consist of redeposited clay and contains sand, 'bluish streaks due to percolation of water saturated with lime' and flint pebbles, which Hicks (1892, p. 460) thought was 'mainly derived from local London Clay bedrock.' His description suggests that the material originated by solifluction. It is unclear therefore whether these yellowish-brown sediments should be equated with the Silt Complex or not.

The silts are, however, absent over large parts of Central London being present only as isolated outliers. Beneath Knightsbridge, for example (Figure 39), they rest on both Lynch Hill and Taplow Gravels, and beneath Trafalgar Square (TQ300805) 'brickearth' rests on Taplow Gravel. It is, however, possible that some of these occurrences may represent pebbly clays of solifluction origin and not silts of the type found upstream. In such cases the clays have no particular stratigraphical significance and it should not be assumed that they are equivalents of this unit.

Upstream from the Langley type section the Silt Complex continuously underlies Slough (Figure 25), where it is generally

2–4 m thick, but wedges-out southwards presumably as a result of later erosion (see also Taplow Gravel, Section 2.13). It is upon this material that the Taplow Terrace surface is developed at 24–30 m OD. From Slough town centre the silt can be traced northwards beneath the Trading Estate onto the higher Lynch Hill level. Sections at Baker's Farm (SU958822), illustrated by Lacaille (1940), exposed up to 1.5 m of red brown clayey silt resting on gravel. North-west of this site, similar silt overlies gravel in Cocksherd Wood (SU944829), while excavations in Deep Lane (SU954826) recorded by Wymer (1968) showed a comparable sequence with the silts banked against the London Clay bluff. Continuing upstream the silts were recorded by Lacaille (1940) at Lent Rise (SU925818) beneath the Lynch Hill Terrace. Here they were grey buff in colour and included 'loops of angular flints . . . in the lower part'. About 350 m south, sections in the Taplow Terrace surface beside Burnham railway bridge (SU926815) exposed brown clayey silt showing typical columnar jointing and tubular structure. Analyses of a sample from here are shown in Tables 1 and 2.

The silts again occurred immediately beneath the surface at the Taplow type site where they were about 1 m thick. The basal part here contained occasional isolated angular flint pebbles. A neighbouring borehole at SU913816 showed a comparable thickness of the deposit.

Above Taplow, only the Cannoncourt Farm section (SU878831) appears to include an horizon of material comparable to that included here in this unit. Sections described and illustrated by Lacaille (1940) included about 2 m of 'tenacious sandy brickearth (clayey silt), the lower part occasionally containing shattered flints . . . [and in places] . . . filled pockets descending to the bench [bedrock surface]'. The silt deposit here locally overlies a 25–30 cm thick layer of clay-rich, frost-shattered angular gravel thought by Lacaille to be of solifluction origin.

Deposits that can be assigned to this unit do not apparently occur further upstream. The reason for this is obscure and will be discussed below.

The grain-size analyses of the samples from this unit (Table 1) indicate that the sediments generally consist of well sorted silt with minor sand and clay components. A secondary maximum in the fine sand range (2–3 phi) is present in some samples. The silt maximum is in accord with the findings of earlier workers. It has often been stated that the silt is of loessic origin on the basis of its abundance, the sediments' characteristic tubule-like internal structure, decalcification and presence of calcareous nodules (loess pupchen), the massive appearance of the deposit and the vertical, hexagonal, columnar dessication cracks. The bimodality of samples implies that the sediments are not pure loess, with the possible exception of Langley (0.9 m), but have been mixed to a greater or lesser extent with the waterlain fine sand component. The extent to which water has played a part in deposition is difficult to determine, since the content of the sand component is variable. However, this is not the only indication of water activity.

Although the deposits are generally massive, they often include pebble horizons one to two clasts thick. These occasionally thicken to form shallow, channel-like lenses presumably resulting from low-energy stream washing of the sediment. The clasts may have been transported by rolling across the sediment surface. The rare sand bands probably originated in a similar way. Where gravel bands were exposed to periglacial processes ice wedges grew in the sediments, the casts of which have been found, albeit rarely. However, the probably unstable nature of these fine sediments possibly prevented the large-scale formation of such features. The presence of undoubted slope wash, usually at higher localities near the bluff or back-slope of a gravel aggradation, is shown by the stratified or laminated sediments found at the base of the complex, in places. Grain-size analyses of samples from the stratified horizons clearly indicate that the current-bedded bands consist of medium sand, and the finer bands consist of clayey silt. The stratification indicates periodic downwash of sediment resulting from alternating higher and lower energy flow depositing coarser and finer sediments respectively.

The lack of bedding in the main mass of the silts implies that deposition occurred slowly in fine increments either of loess locally remobilised and redistributed by sheet wash and/or, particularly at lower levels, by overbank stream deposition of fine sediments mixed with contemporarily deposited or possibly reworked loess.

The silt mineralogical analyses (J. A. Catt, personal communication) show that almost all the samples have a similar composition (Table 2) and therefore source. The assemblages include high frequencies of zircon, chlorite and green hornblende and closely resemble those known to be of Devensian age. The Cranford Park samples include staurolite, kyanite and andalusite and high frequencies of garnet and zircon possibly derived in part from the Reading Beds or perhaps Clay-with-flints. This material must have been derived by reworking of the parent rock; however, possible sources do not occur near this locality today and this therefore poses an interpretational problem. The solution to this is unclear at present.

2.17 Kempton Park Gravel ('Upper Floodplain' or Isleworth Terrace)

The term 'Floodplain gravel' appears to have been first used in a formal sense by Bromehead (1912, p. 75), who applied it to the 'lowest deposits, but little above [and underlying] the modern alluvium.' This definition included 'all river gravels younger than those of the Taplow' (Dewey and Bromehead, 1915, p. 76). The term replaced the 'Lowest Terrace or Valley gravels' used in earlier Geological Surveys (e.g. Whitaker, 1889; Blake, 1903). When initially defined the Floodplain gravels were not subdivided. However, following the mapping of the South London area (sheet 270), Dewey and Bromehead (1921) proposed a two-fold division into Upper and Lower Floodplain gravels separated by an erosional episode. They recognised this subdivision downstream from Brentford, but pointed out that it was not always possible to separate the units. King and Oakley (1936) discussed extension of the 'Upper Floodplain' terrace deposits downstream and into the Lea Valley. Upstream from Brentford the single Floodplain gravel remained undivided until Zeuner (1959) suggested that the 'Upper Floodplain' terrace could be found as far west as Chertsey, and this has been reaffirmed recently by Gibbard *et al.* (1982).

Beyond Chertsey the 'Upper Floodplain' terrace was not found during geomorphological mapping by Hare (1947) or Sealy and Sealy (1956). However, recently Clarke and Dixon (1981) attempted to show that the Floodplain deposits upstream of Windsor could be divided into two subunits using widely spaced borehole data without field control. These authors, however, did not allow for irregularities in base level caused by solution of Chalk bedrock when establishing their correlations and therefore their results must be regarded with caution.

Until very recently, the interglacial and pre-interglacial deposits at Brentford and Trafalgar Square were regarded as underlying the 'Upper Floodplain' terrace. Zeuner (1959), for example, thought that this terrace surface marked a sea-level of last interglacial age. However, as shown above (cf. Gibbard, 1981) the interglacial deposits are eroded remnants unrelated to the aggradations upon which the terrace is developed.

Numerous fossil discoveries have been made from deposits beneath the terrace. The earliest published finds are those from Kew Bridge by Morris (1850), which included mammoth, woolly rhinoceros and reindeer, as well as Mollusca. From Acton and Chiswick, Lane-Fox (1872) found a similar vertebrate fauna with hippopotamus, the latter possibly derived from local interglacial strata. In 1891, Leeson discovered saiga antelope (*Saiga tartarica*) at Twickenham and soon after Leeson and Laffan (1894) found a vertebrate fauna including horse, reindeer and bison associated with an organic deposit rich in plant and molluscan fossils. More recent discoveries have included mammoth (from several sites including Earls Court and South Kensington), woolly rhinoceros, reindeer and bison (see Chapter 6).

The first fossiliferous deposits within the gravels to be studied in detail were those at Isleworth. Here organic sediments, rich in beetles, plants, Mollusca, Ostracoda and vertebrates, were discovered in 1956 (Coope and Angus, 1975; Sutcliffe and Kowalski, 1976; Kerney *et al.*, 1982). The fossil assemblage, particularly the beetles, indicated that the deposits accumulated during an interstadial which was shown by a radiocarbon date of $43140 \pm {}^{1520}_{1280}$ BP (Birm – 319, Shotton and Williams, 1973) to be the Middle Devensian Upton Warren Interstadial. A similar deposit found at Kempton Park that also contained plants, beetles, Mollusca and Ostracoda gave a radiocarbon date of 35230 ± 185 BP (Q-2019, Gibbard *et al.*, 1982). A third site, at the Ismaili Centre, South Kensington, again yielded members of all these fossil groups and has also provided comparable dates (Coope, Gibbard, Hall, Preece, Robinson and Sutcliffe, unpublished). These three localities therefore span the Upton Warren Interstadial interval (Figure 35). The results led Gibbard *et al.* (1982) and Kerney *et al.* (1982) to assign the gravel and sand in which the organic deposits occur to the Middle Devensian substage. The aggradation has been estimated to have begun at 45000–44000 BP and ended at about 34000–30000 BP on the basis of the dates so far obtained.

As a result of the stratigraphical studies at Kempton Park, Gibbard *et al.* (1982) proposed a new lithostratigraphical member, the Kempton Park Gravel. This unit includes the gravels and sands together with organic deposits as beds exposed at this or other equivalent sites. The type section at Kempton Park (TQ118703: 12 m OD) was described in detail by these authors. It consisted of 4 m of current-bedded gravel and sand overlying London Clay. The composition of the gravel based on four samples is given in the Appendix. Palaeocurrent measurements from the sand horizons indicated a flow towards east-north-east. In the northern part of the quarry a fossiliferous grey clayey silt bed of vari-

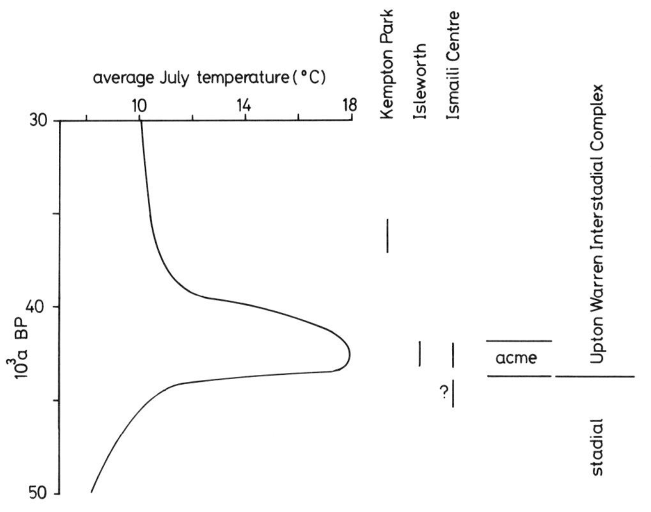

Fig. 35. Diagram showing the climatic changes in lowland England during part of the Middle Devensian substage (after Coope, 1977). The interpreted time spans of the three sites described in the text are plotted.

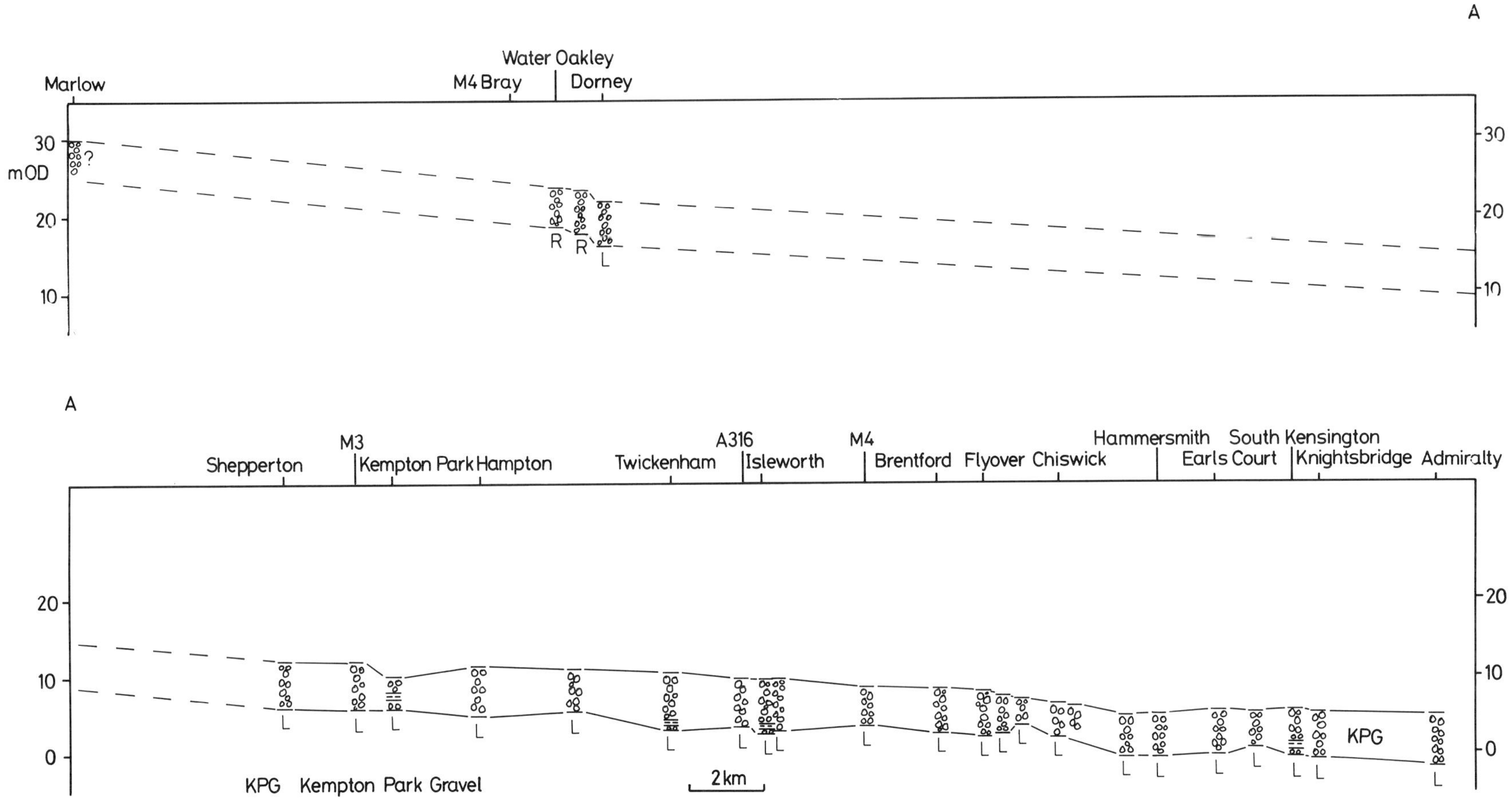

Fig. 36. Long profile section showing the deposits correlated with the Kempton Park Gravel.

able thickness was present in the gravels and appeared to fill a depression that originated as a shallow channel. The section was capped by 2.6 m of brown clayey silts.

Immediately north of the type site the relationship of the Kempton Park Gravel to the Taplow Gravel could be seen in a section along the M3/A305 (erroneously called the A316 in the original publication) road from Sunbury to Hanworth (Gibbard *et al.*, 1982, figure 4). Beneath this road the base of the Taplow Gravel at 16 m OD was clearly separated from that of the Kempton Park member at 6–7 m OD. South of the site there is a marked break of slope on the south side of the A305 where the ground drops to the modern floodplain at 9–10 m OD, underlain by the Shepperton Gravel (Section 2.18).

Downstream this unit is cut-out at Hampton (TQ137696) where the modern floodplain abuts the Taplow Gravel. However, the deposits recur on the eastern side of Hampton and form a continuous spread upstream as far as Twickenham (cf. Whitaker, 1889). Gravel 6.5 m thick at the expected height (*c.* 11 m OD) was found in boreholes beneath Hampton centre (TQ141696). Gravel overlying London Clay was seen in the cutting along the Strawberry Hill – Sunbury railway line when it was dug (Whitaker, 1889).

The sewer trench sections observed by Leeson and Laffan (1894) at Strawberry Hill, Twickenham (TQ155724–160725: 11 m OD) showed a dark blue silty clay bed 0.7 m to over 1 m thick overlain by 'coarse reddish yellow gravel' and underlain by 'gravel in every respect similar to that above.' These deposits are the altitudinal equivalents of those at Kempton Park (Figure 36). At two places in the trench London Clay was seen below the gravel. The fossil vertebrate remains, already mentioned, were found mostly in the silty clay, but some were present in the gravel immediately above. The vertebrate remains were presumably derived from the organic bed below. Other boreholes and excavations in the vicinity have not found the silty clay, but only gravel and sand. For example, at Twickenham Station (TQ160737: 10 m OD) 2.5 m of gravel and sand were present beneath made ground. Leeson and Laffan (1894, p. 459) noted that sand at *c.*TQ162731 produced rhinoceros bones 6 m below the surface and horse remains 2.4 m above them. The saiga antelope skull was discovered in sand beneath gravel at Orleans House (TQ170736) (Leeson, 1891), and stratified gravels at Twickenham Green (TQ153729: 10 m OD) contained a bison bone (A. J. Sutcliffe, personal communication).

The wide gravel spread is dissected by the shallow River Crane

Fig. 37. Sections showing the disposition of deposits beneath the A316 (above) and in the Parson's Green – Fulham area (below).

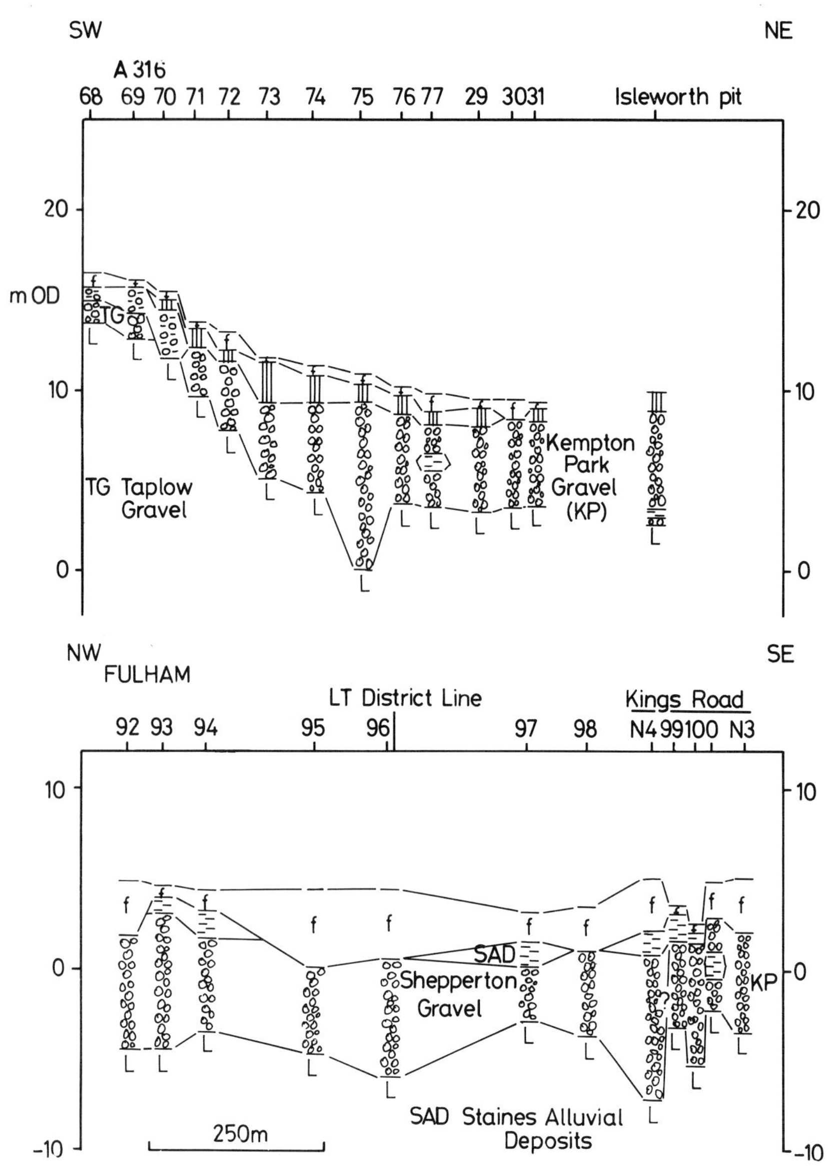

Valley, but continues to the north where it forms a conspicuous feature as far as the Brent Valley. A short section along the A316 between Whitton and Twickenham (Figure 37) again demonstrates the relationship of this unit with the Taplow Gravel. The section closely resembles that along the M3/A305 (A316), and shows that the Kempton Park Gravel is on average 5 m thick and abuts a steeply rising surface of London Clay. The unusual thickness of the gravels in borehole 75 (8 m) suggests that a scour hollow is present in the bedrock surface here. In contrast borehole 77 proved a 1 m thick bed of grey silty clay containing plant fossils within the gravel sequence. Above the gravel and sand is a bed of brown sandy clay. It is thickest adjacent to the terrace bluff and thins markedly eastwards. According to the geological map the deposit forms a continuous cover banked against the London Clay slope. The Isleworth site (Willment's pit) was about 850 m north-east of this section. In the 1950s, excavations at this site (TQ158746), now built over, exposed a basal red gravel resting on an undulating London Clay surface overlain by finely laminated, fossiliferous dark grey silty clay. Above this, gravel and sand occurred, the base of which was irregularly channelled into the silty clay. The lower 1.75 m of the gravel contained several narrow impersistent silt bands one of which contained Mollusca. Above, current-bedded pebbly sands gave way to horizontally stratified gravel towards the top. The uppermost gravel and sand contained a number of post-depositional ice wedge casts, the largest of which were over 1 m wide at the top and penetrated the full thickness of the deposits, into London Clay. A brown sandy silt ('brickearth') deposit capped the section (Coope and Angus, 1975; Kerney *et al.*, 1982). No samples of gravel were available during this study, although altitudinally these deposits are equivalent to those at the type section (Figure 36). The extent of the organic bed is not known. It does occur in an adjacent borehole beside Mogden Lane (TQ157748), but is absent beneath Mogden Sewage Works (TQ155749), the sequence there consisting entirely of gravel and sand.

East of the Brent Valley the deposits recur beneath the southern half of Brentford. A borehole at TQ178780 (10 m OD) proved at least 4 m of gravel and sand overlain by 1.5 m of brown pebbly sandy clay. At Kew Bridge, 1 km further east (TQ190782), the railway embankment sections recorded by Morris (1850) showed 5.1 m of stratified sandy gravel and included a clay band 2.3 m above the base. It was overlain by 1.2 m of brown sandy clay. The basal gravels yielded the bones and molluscs already mentioned. The upper 1.8 m comprised cross-stratified sand with fine gravel, while below gravels predominated. A borehole in the neighbouring railway yard (TQ189782) showed a comparable sequence of 5 m of gravel in a sand matrix resting on London Clay. A pebble count from this borehole is given in the Appendix and suggests that correlation of these gravels with those at Kempton Park is acceptable.

The neighbouring section along the M4 Motorway (Figure 17) also records gravels and sand that are a continuation of this unit. The section also illustrates the nature of the relationship of these deposits to the interglacial sediments described above (Section 2.14). The line of the section appears to cross the junction of the deposits twice; first near the Ealing and Great West Road intersection (*c.*TQ179785) and again near the railway bridge (TQ191785). In both cases the gravel and sand abuts the earlier sediments that occur at approximately the same level. The deposits recorded in the section are also found at several sites locally. In nearby Gunnersbury (TQ196784: 9.5 m OD) 5.3 m of gravel and sand overlie London Clay and are in turn overlain by 1.2 m of brown stony sandy clay.

In the 1860s Lane-Fox (1872) saw sections in the terrace surface at Gunnersbury. The first at Brown's Orchard (*c.*TQ202787: 7.5 m OD) and the second at Chiswick Row (*c.*TQ201785: 9 m OD) both showed a broadly similar sequence of stratified gravel and sand, 3 m thick, resting on London Clay. The lower part of the gravels at both sites contained vertebrate remains. At the former site these included an apparently 'mixed' fauna of 'cold' and 'warm' taxa (see below). This apparent mixing almost certainly results from reworking of the local interglacial sediments in the vicinity. At both sites the fluvial deposits are overlain abruptly by brown sandy silt up to 1 m thick. The structures present in this horizon at Brown's Orchard drawn by Lane-Fox (1872) closely resemble those of cryoturbation.

A comparable sequence continues from the Gunnersbury–Brentford area to beneath Chiswick (Whitaker, 1889). A borehole at Chiswick Town Hall (TQ205783: 6.5 m OD) proved 1.4 m of brown stony clay above 3.25 m of gravel and sand, while at the junction of Cromwell Road and Sutton Court Road (TQ204779: 6 m OD) 6 m of gravel and sand were present beneath 1.5 m of brown sandy clay. A temporary section adjacent to Turnham Green Station (TQ212788: 6 m OD) in 1981 exposed 2.5–3 m of current-bedded gravel and sand beneath 1.5 m of made ground. The lower part of the deposit comprised massive sand matrix-supported medium gravel 1.5 m thick, overlain by interbedded tabular cross-bedded sand, 15–25 cm thick, and narrow fine to

Fig. 38. Sections showing the disposition of deposits in the Hammersmith – Shepherd's Bush area: 1. along the Central Line; 2. along the Hammersmith Sewer line to Barnes; 3. Hammersmith Flyover to Cromwell Road (sources: GLC, IGS).

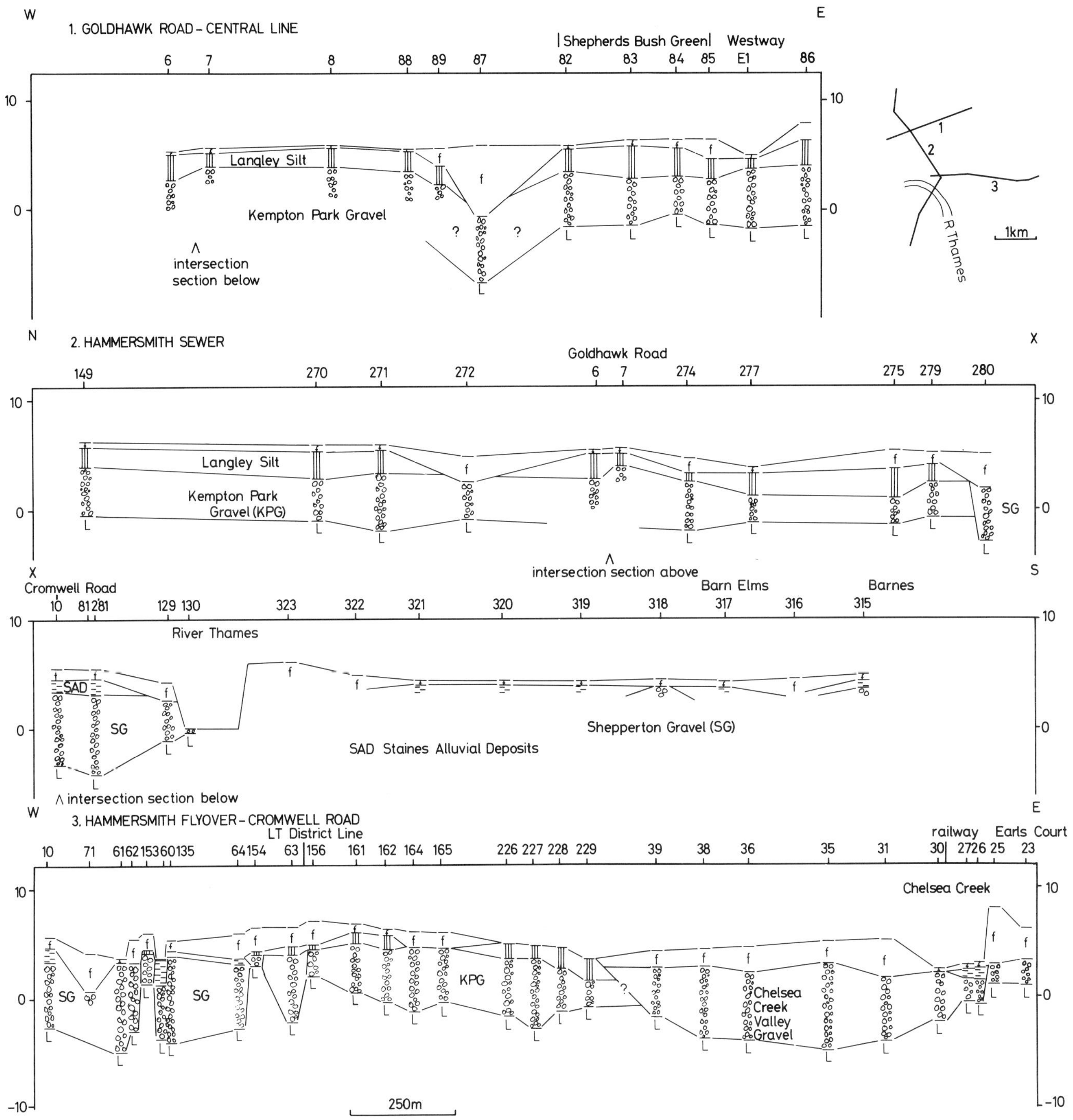

medium gravel bands of the same thickness. Palaeocurrent measurements indicated a flow towards the east and a pebble count sample suggested that the altitudinal correlation was correct. To the south, a break of slope runs parallel to the southern side of Burlington Lane and appears to mark the northern limit of the present floodplain.

Beyond Chiswick, a continuous wide spread of gravel and sand extends as far east as West Kensington, as far south as Parson's Green and as far north as Shepherd's Bush and White City. This spread was noticed by Whitaker (1889) and mapped as 'Upper Floodplain' terrace by Dewey and Bromehead (1921). Although modern exposures are rare in this area, several borehole sections that form a grid provide a three-dimensional view of the deposits. The sections along part of the Central Line, a proposed course of the Piccadilly Line, the Hammersmith Storm Relief Sewer, Hammersmith Flyover – Cromwell Road (Figure 38), and a sewer in Parson's Green (Figure 37) show that, although variable across the area, much of the gravel and sand is altitudinally equivalent to this unit. As much as 5.5–6 m of this material lies on an undulating bedrock surface and is overlain by variable thicknesses of brown to blue sandy clay occasionally containing pebbles. This material, mapped as 'brickearth' by Dewey and Bromehead (1921), is notably more clay-rich and thicker in the northern part of the area, attaining almost 3 m beneath Shepherd's Bush Green (Figure 38). The colour and high clay content suggest that it was derived from local London Clay bedrock by slope processes including solifluction during which local gravel and sand were incorporated. Whitaker (1889, p. 397) came to the same conclusion after noting that at one pit over 10 m of this material passed directly downwards into London Clay.

The relationship of the Kempton Park Gravel in this area to higher and lower units is also shown in other sections (Figure 23). In Figure 29 the base of the unit is separated from the Taplow Gravel by 3.5 m near High Street, Kensington (TQ255798). The deposits of this unit have been dissected by later stream erosion and the infills of tributary valleys, e.g. the Chelsea Creek and the Shepperton Gravel ('Lower Floodplain') of the Thames occur at a lower level. It is, however, extremely difficult to separate precisely these later aggradations by altitude in all cases as can be seen from the sections. Nevertheless, it is clear that at Hammersmith (TQ230783) and Parson's Green (TQ256769), the base of the Shepperton Gravel is about 4.5 to 5 m below that of the Kempton Park member. An anomalously deep isolated hollow fill is present beneath Shepherd's Bush Green (Figure 38: bore-

Fig. 39. Section from Hyde Park to Knightsbridge (where it intersects with Figure 29), constructed using borehole data (sources: GLC, IGS).

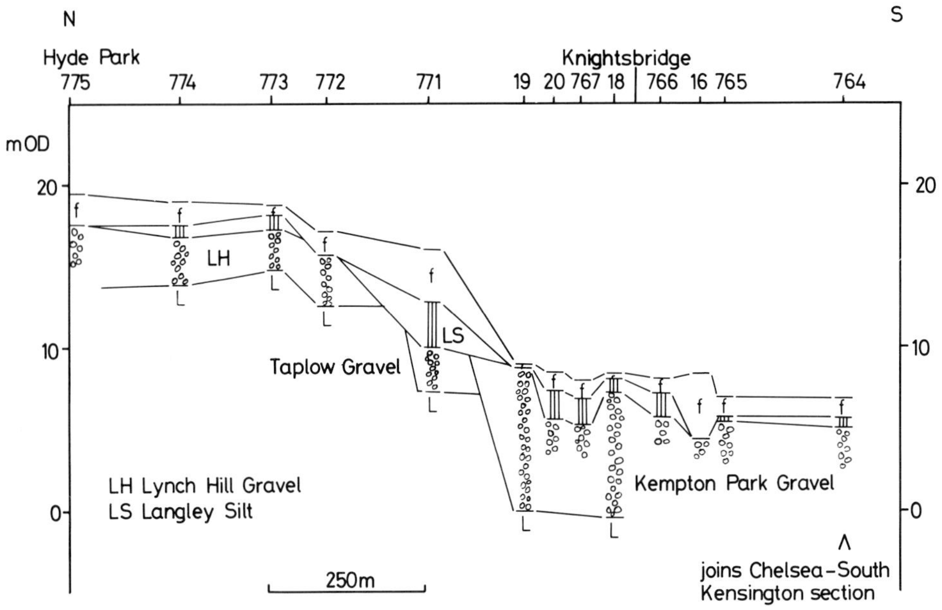

hole 87) where the bedrock surface occurs 5 m below the general level. This hollow may be similar to those described by Berry (1979) from beneath Central London.

The gravel and sand of this unit recurs on the eastern side of the Chelsea Creek beneath West Brompton and underlies Earls Court, South Kensington and Chelsea at the expected height (Figure 36). Boreholes beside the District Line near Gloucester Road Station (TQ261788: 8.4 m OD) proved 3.7 m of pebbly brown clay above 7 m of gravel and sand below 1.2 m of made ground. A nearby temporary section at Cromwell Hospital (TQ256789: 9 m OD) showed 4.5 m of horizontally bedded medium gravel with pebbly sand bands in 1979. The terrace surface feature is well developed in this area and can be followed southwards almost to the river at Cheyne Walk. Large numbers of boreholes at Chelsea College, north of Kings Road (TQ271781: 8 m OD) and beside Fulham Road (TQ268784: 8 m OD) confirm that the same sequence of up to 7 m of gravel and sand is present. The relationships of the deposits to higher and lower units are illustrated by the sections in Figure 29.

The section from Queen's Gate to Chelsea Royal Hospital (Figure 29) demonstrates that the distinct gravel and sand unit abuts the Taplow Gravel beneath the Natural History Museum (TQ266791). It also underlies almost all of Chelsea, but is cut-out by the slope to the modern floodplain aligned parallel to Royal Hospital Road (TQ279781). In spite of later human disturbance the upper surface of the unit appears to fall slightly southwards; deposits of brown silty clay are preserved in places beneath made ground. The section (Figure 39), from Kensington Road to Knightsbridge shows a similar sequence, and clearly demonstrates the relationship of the Kempton Park Gravel not only to the Taplow, but also to the Lynch Hill Gravel.

An important section in these deposits was exposed in late 1980 on the corner of Cromwell Road and Exhibition Road during the construction of the Ismaili Centre (TQ269789). Here a maximum thickness of 8 m of deposits was seen resting on an undulating London Clay surface. The basal sediments comprised poorly sorted stratified medium gravel with a sand matrix. Several lenticular beds of tabular cross-bedded sand were present immediately beneath the overlying deposits. The gravels were overlain throughout the pit by a bed of interstratified silt, silty clay, sand and pebbly sand bands, up to 2.3 m thick. These deposits had a sharply defined base on the gravel, but internally were very variable. In the north section, measured by A. J. Sutcliffe (personal communication), the sediments predominantly consisted of grey silty clay, with some thin blue grey horizons approximately 10 cm thick in the upper half of the bed. The sediments of this bed were oxidised in the upper 25 cm. Throughout the bed, a series of narrow (2–48 cm) sand to sandy gravel bands occurred. Each of these bands had a marked erosional base, but frequently they graded upwards into the overlying finer sediment. In places, the sand was current-bedded and in one place a bed 15 cm thick showed a standing ripple structure. Plant, beetle, Mollusca and Ostracoda remains occurred throughout the fine sediments and have been studied in detail by Coope *et al.*, (unpublished). The overlying gravel and sand was 3.5 m thick. It comprised predominantly horizontally bedded gravel with a few impersistent lenses of tabular cross-bedded sand up to 40 cm thick in the upper 1.5 m of the unit. Current-bedding measurements from the sands in this section gave a flow towards the north-east, the same direction being maintained below, above and within the fine sediments. The pebble lithology of the gravel at this site supports the altitudinal evidence for its correlation with this member (Figure 36).

Numerous boreholes east of Knightsbridge confirm that the gravels and sands continue eastwards. However, they have been removed by the north–south valley of the Westbourne about 100 m east of Sloane Street. The unit is again found beneath Belgrave Square (TQ283794: 9.8 m OD), where 6.3 m of gravel and sand are preserved beneath 2.8 m of made ground, and extends as far north as Hyde Park Corner. It is cut-out to the south again by the modern floodplain. The boundary runs on the northern side of Buckingham Palace Road. To the east, the unit is dissected by the Tybourne Valley, but it recurs in the north-east corner of St James's Park and underlies the area south of the Mall (TQ297800) as far as Whitehall. A transverse section across the deposits along the Jubilee Line (Figure 20) has already been mentioned above and proves gravel and sand at the expected height to be equated with this unit. In this area the Kempton Park Gravel abuts the Trafalgar Square Complex.The boundary between the members can be determined using the dense borehole coverage available (Figure 31). The line runs approximately along the north side of The Mall from St James's Palace (TQ294800), beneath Carlton House Terrace and the northern end of the Admiralty and as far as Northumberland Avenue. The deposits are cut-out approximately 50 m west of Whitehall again by the slope to the modern floodplain.

South of the river few patches of gravel that can be correlated with this unit occur. However, between Mortlake and Barnes well developed terrace-like features can be seen on either side of the Beverley Brook. At its western end the main feature is immediately adjacent to the river, while at the eastern extremity it has a characteristic 'crows-foot' like plan. Boreholes across this feature show that it comprises a degraded gravel and sand remnant up to 6 m thick overlain by possibly soliflucted, clayey gravel in places (Figure 38). Similar deposits underlie Barnes Common (TQ220758). These remnants occur at precisely the altitude expected for correlation with this unit and this strongly suggests that they may represent the confluence of the Beverley Brook at the time.

At several other localities apparent altitudinal equivalents are found. For example, beneath Clapham Junction Station (TQ268754: 6.5 m OD) gravel deposits are found that may be lateral equivalents of this unit beneath degraded 'whale-back' landforms. If so, they would have been deposited at or close to the Wandle confluence. However, the deposits may be tributary fans continuous with those underlying the modern floodplain. At present little else can be said about the relative position and equivalence of these sediments.

Upstream from Kempton Park, altitudinally equivalent deposits have been extensively worked in a quarry at Lower Feltham (TQ108716: 12 m OD). This pit is situated close to the bluff rising to the Taplow Gravel beneath Hanworth (see above). Over 2 m of horizontally bedded gravel with irregularly interbedded lenses of tabular cross-bedded sand up to 10 cm thick was present here. Palaeocurrent measurements indicated a flow towards the east. A pebble count shows that the gravel is of comparable composition to that at the type section (see Appendix). Overlying an erosional surface of the gravel was a bed of light brown very sandy silt, 1 m thick. The basal 20 cm of the silt consisted of alternating laminae 0.5–3 m thick of sandy silt and fine gravel. Above, the silt was massive and calcareous, in places.

Boreholes in the vicinity record a comparable sequence. The deposits are also recorded in the M3 Motorway section (Figure 40) as far east as Shepperton Green (TQ074675) where they are cut-out by the Shepperton Gravel beneath the modern floodplain. In the Staines–Laleham area it is not possible confidently to distinguish between these two units because of the Colne confluence.

East of Windsor the shallow gradient of the Kempton Park Gravel (30–40 cm km^{-1}) appears to bring the deposits to almost the same level as those beneath the modern floodplain. Altitudinal separation of the units is almost impossible except in rare instances.

At Dorney, slightly higher level sediments that could be equated with this unit are present. A borehole at Court Farm (SU933791: 22 m OD) records gravel and sand 5.6 m thick resting on bedrock. However, an extensive spread of gravel occurs on the south side of the Thames at Bray beneath the A308 road. Hare (1947) mapped the surface of this spread at 24–29 m OD (i.e. 2–3 m above the modern floodplain surface) as Taplow. However, as shown above (Section 2.13) it is too low for inclusion in the Taplow Gravel. No exposure was available during the present study, but several boreholes penetrate the deposits. For example, at Water Oakley (SU915777: 25 m OD) the sequence comprises 1.1 m of brown pebbly sand clay overlying 5.1 m of gravel and sand. The deposits fall precisely on the gradient of the Kempton Park member to which they are provisionally assigned.

One final locality should be mentioned here. At Marlow (SU865874), gravel underlying the '3 m terrace' of the Thames contained organic silt according to Shotton and Williams (1971). Analyses by Bell (1968, 1969) showed that the sediment contained both pollen and plant macrofossils as well as beetle remains (G. R. Coope, personal communication). The latter suggested an early Upton Warren Interstadial age, although insufficient material was available to give a finite radiocarbon date (Birm-184: >31 000 BP). The thickness of the gravel and sand that contained the organic sediment is not known, but it is probable that it can be equated with this unit in the same way as that at Bray (Figure 36).

As already stated above, the gradient on the Kempton Park Gravel is 30–40 cm km^{-1} throughout its known area of distribution. If this gradient is extended upstream the deposit would occur at the same or slightly lower level as those beneath the modern floodplain.

The poor exposure of this unit prevented the collection of large numbers of pebble count samples. However, those available indicate that the gravel comprises 86–94% flint (76–90% angular flint, 7–9% rounded flint) 1.5–6% vein quartz, 1–3% quartzite and 1–5.5% Greensand chert, together with some very rare erratic rocks. The content of Greensand chert is markedly higher at the Ismaili Centre site than upstream. This probably reflects incoming of southern tributary material.

The distribution of the brown sandy silt sediment that overlies the fluvial gravel and sand west of London is noteworthy. The material frequently closely resembles the 'brickearth' sediments grouped in the Langley Silt Complex. In section the sediment is

Fig. 40. Section along the M3 Motorway from Virginia Water to Hanworth, constructed using borehole records (source: SERCU).

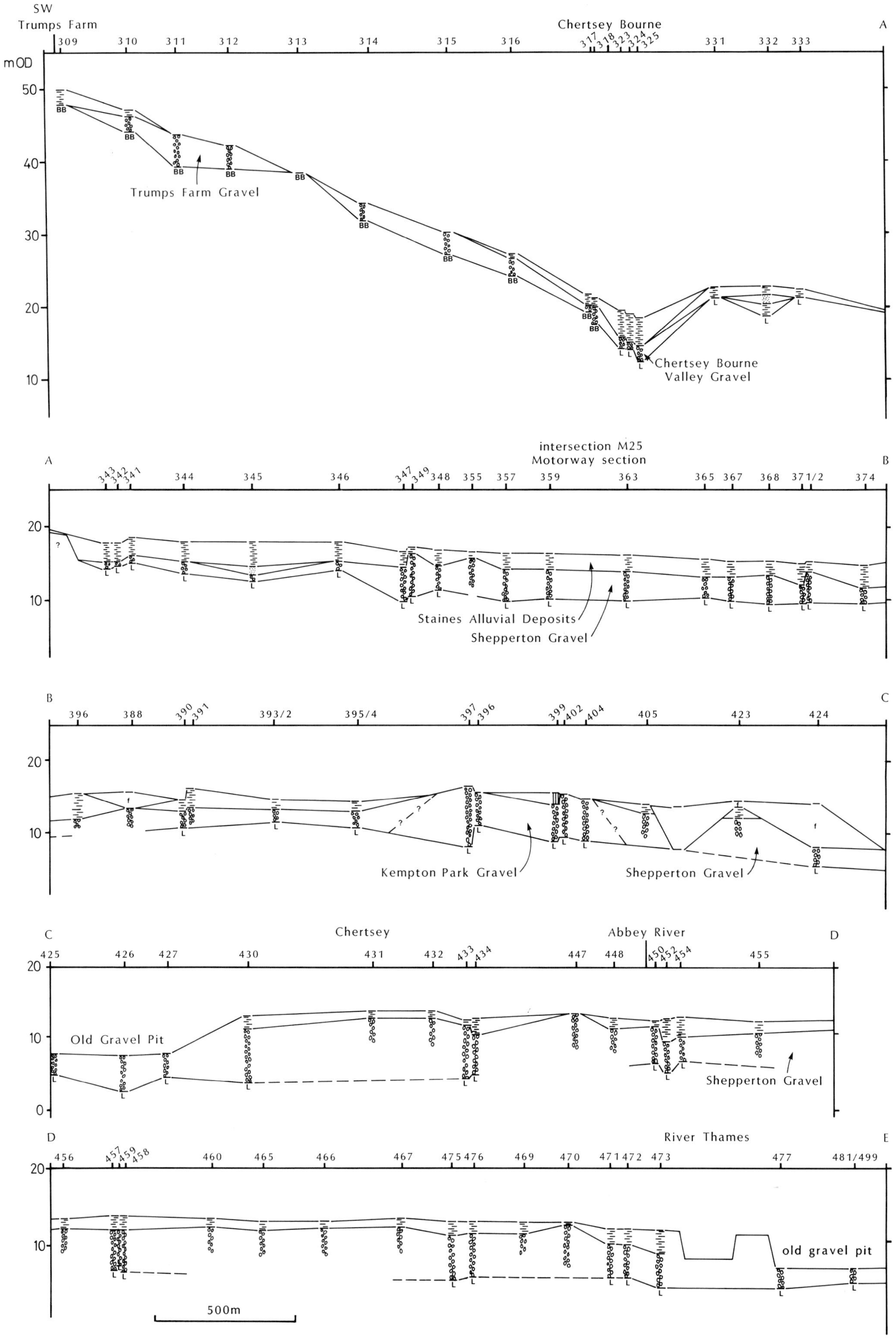

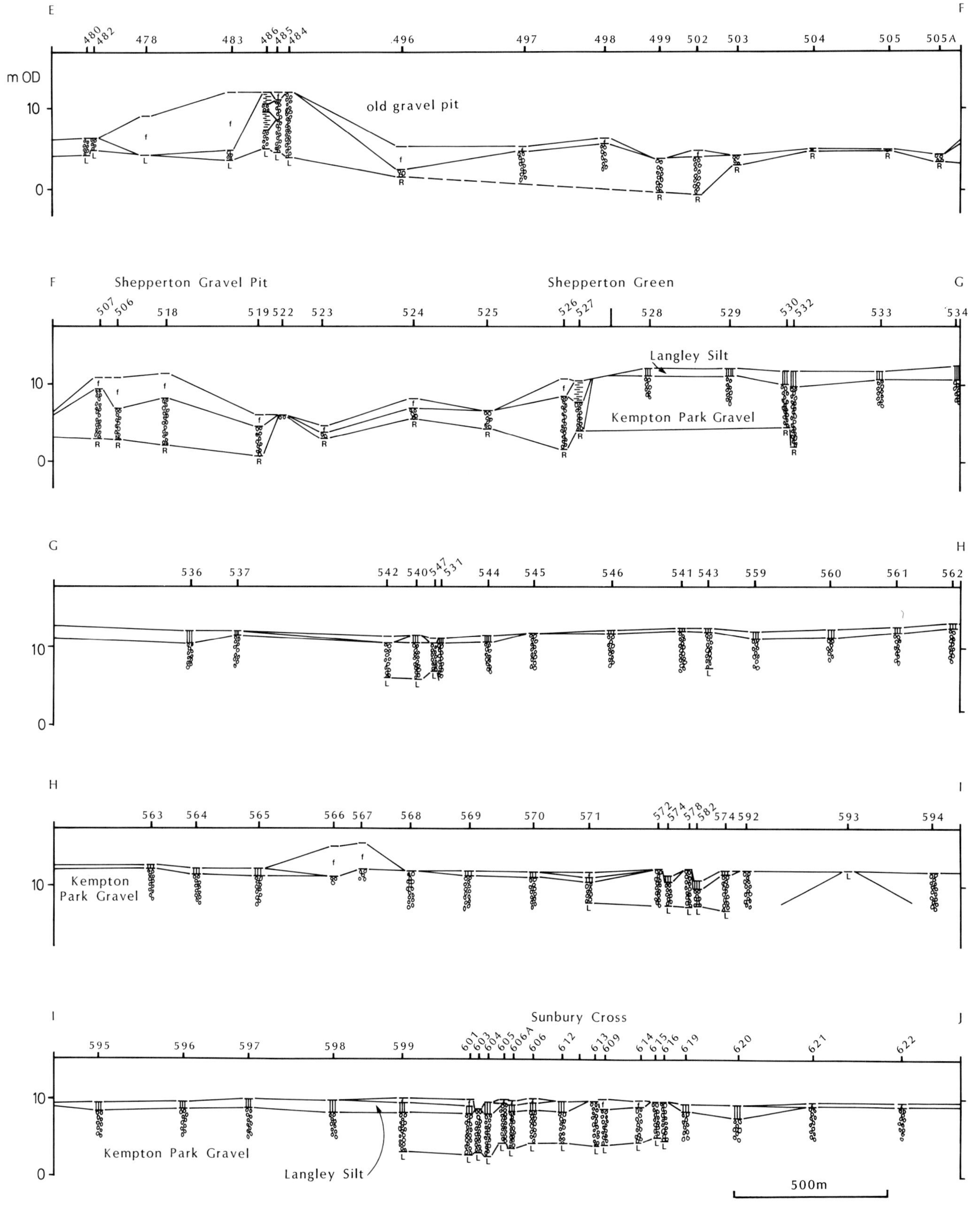

Fig. 40 continued

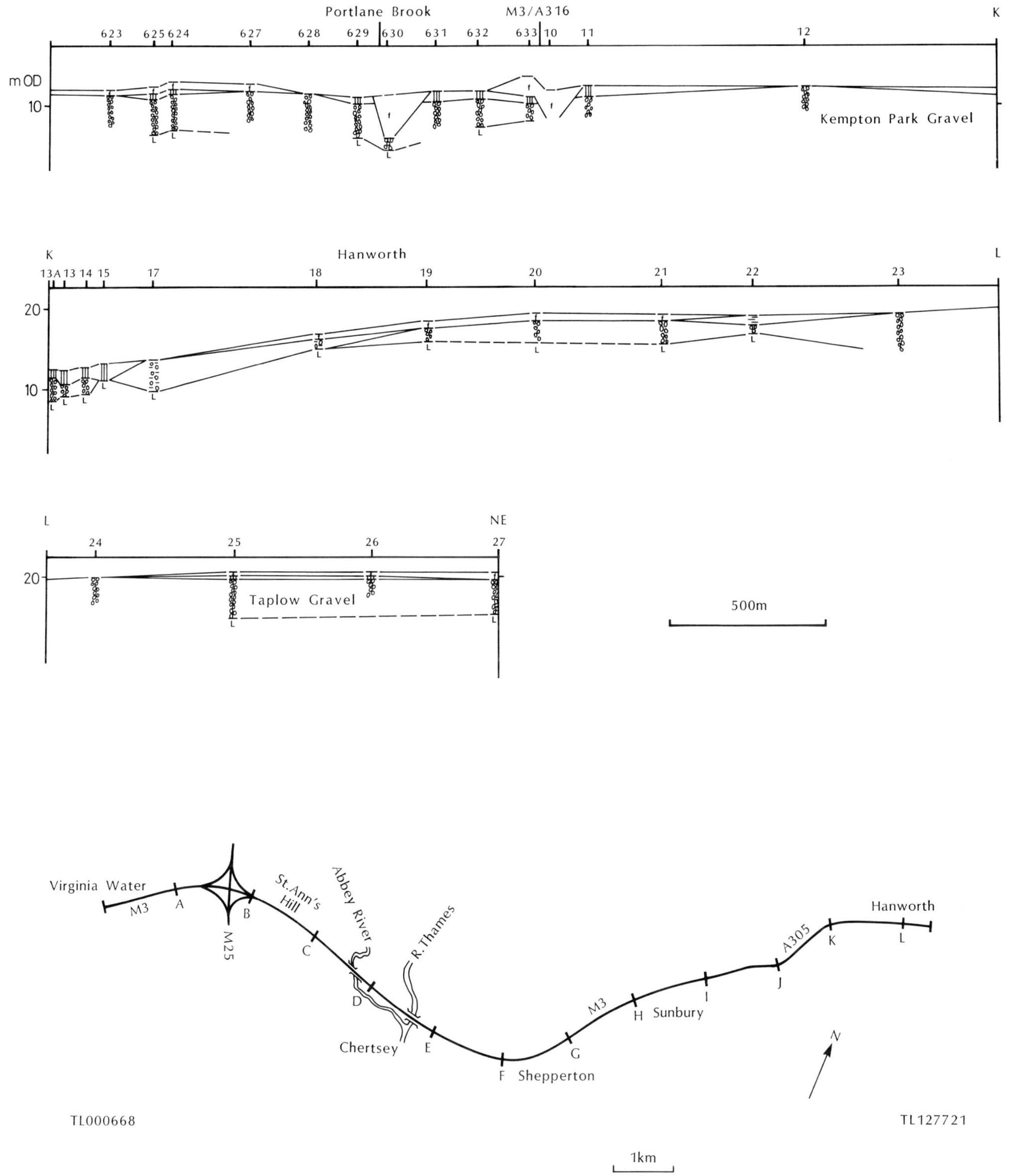

commonly sandy and occasionally shows flow structures. Although some of this material may be reworked London Clay, there are considerable spreads which closely resemble the Langley Silt Complex sediment both in appearance and grain-size distribution. According to the geological map and borehole data available the material is thickest adjacent to the bounding bluff particularly at the slope from the Taplow Gravel against which it is banked. It thins away from this feature.

This distribution strongly suggests that the brown sandy silt is derived from degradation of that sediment overlying the Taplow Gravel, presumably by solifluction. This origin would account for its lateral and vertical distribution and its variable composition resulting from incorporation of gravel, sand and/or bedrock material, including its lateral passage into soliflucted London Clay. This solifluction must therefore post-date the deposition of the Kempton Park Gravel.

Wey Valley (Byfleet Gravel)

The deposits filling the Lower Wey Valley between Byfleet and Weybridge were mapped by Dewey and Bromehead (1915). This south–north aligned southern tributary occupies a wide valley, part of which contains a large expanse of gravel between Byfleet and New Haw. This gravel spread forms the interfluve between the River Wey itself and the Addlestone Bourne.

A section constructed from borehole records for the M25 Motorway from Ongar Place, Addlestone, to Wisley shows the distribution of the deposits (Figure 41). Two major gravel and sand aggradations can be seen in this section. The upper unit comprises gravel and sand resting on a fairly smooth surface of Bagshot Beds at 13.25–14.25 m OD. The gravels are on average 2.5–4 m thick, but locally thicken to 6.5 m in the bedrock hollow near Dartnell Park (TQ055617). They are overlain by brown to mottled brown, green and grey pebbly silty sand, which appears to infill depressions in the surface. It is particularly thick (2.25 m) near the valley side at Wisley. Its mode of distribution and lithological character suggest that it results from local periglacial processes, particularly solifluction.

Exposures in the gravel and sand have not been seen by the writer. However, Dewey and Bromehead (1915, p. 76) saw trench sections 2.5 m deep that exposed deposits consisting almost entirely of sand with seams of flints and occasional chert pebbles. They also noted that mammoth teeth had been found in the diggings.

The gravel and sand of this unit are dissected by the modern

Fig. 41. Section along the M25 Motorway from Ongar Place, Addlestone, to Wisley, constructed using borehole records (source: SCC).

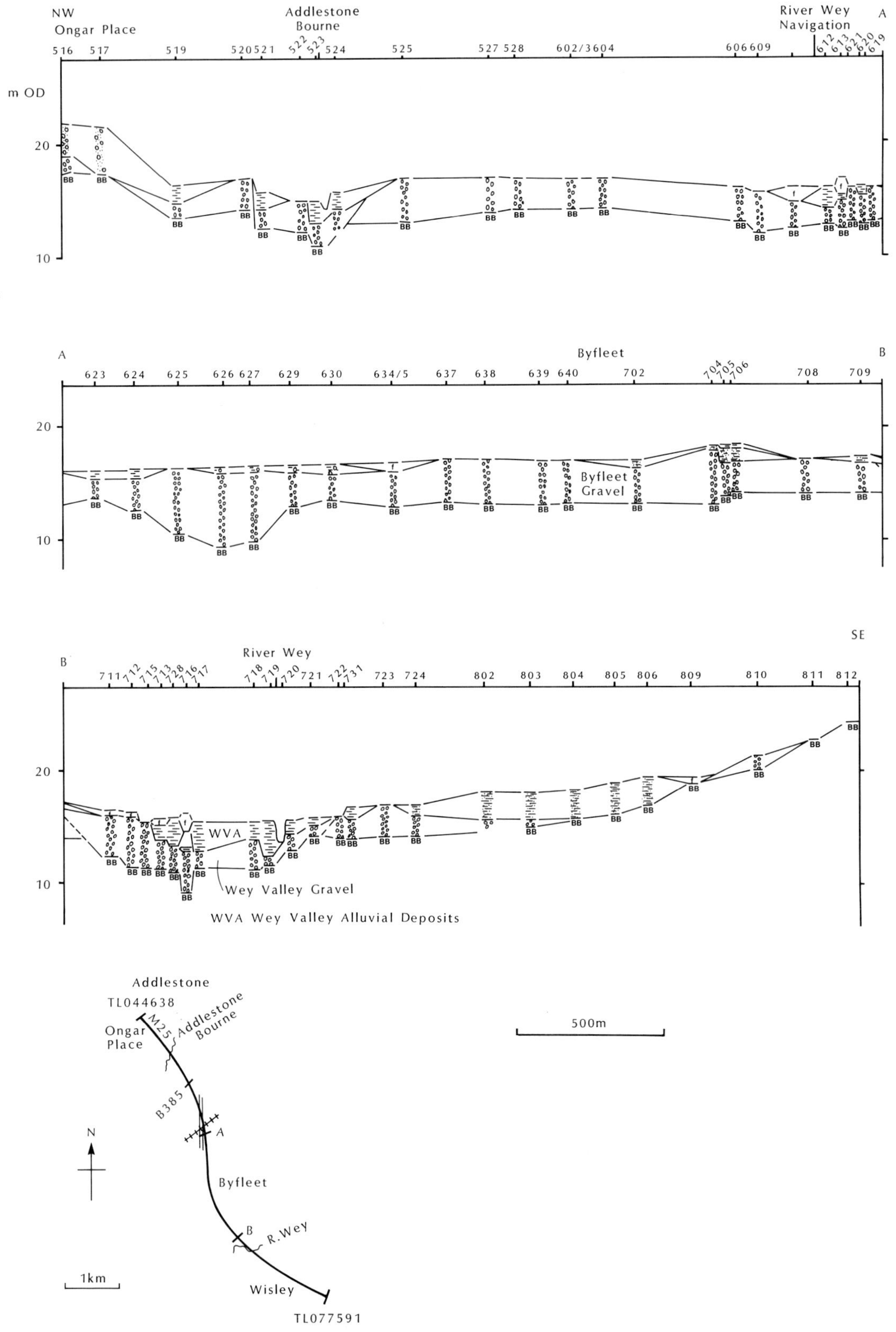

valleys of the Wey and the Addlestone Bourne which are floored by a lower gravel and sand unit (Wisley Gravel: Section 2.19(g) and floodplain fine alluvium.

The upper gravel and sand therefore appears to form a discrete, altitudinally separable aggradational unit, for which the term Byfleet Gravel is proposed (type section: M25: TQ050629–061601). Dewey and Bromehead (1915) assigned this spread to the Taplow-equivalent stage. However, it is too low to represent a Wey-equivalent of the Taplow Gravel. The surface gradient on the gravel unit is approximately 90–100 cm km^{-1} in the section. Downstream projection of the gradient from Byfleet, where the gravel surface is 17.25 m OD, over the 6.25 km to Shepperton gives a surface level of 11 m OD. This precisely intersects with the gravel surface on the Kempton Park Gravel (Figure 36) with which the Byfleet Gravel is tentatively equated.

2.18 Shepperton Gravel ('Lower Floodplain' or Floodplain terrace)

As mentioned above, the term Floodplain gravel was introduced by Bromehead (1912, p. 75) and later subdivided by Dewey and Bromehead (1921) into two units, the lower of which was called the 'Lower Floodplain' gravel. They recognised that its accumulation was separated from the 'Upper Floodplain' gravel (Kempton Park Gravel) by a period of downcutting and erosion during which the 'buried channel' was formed downstream of Brentford. The 'Lower Floodplain' gravel filled this 'channel' and was overlain throughout much of its area of distribution by fine alluvial sediments (discussed below; Section 2.20). King and Oakley (1936) supported these conclusions and suggested that the channel was filled to the level of the present floodplain during the last aggradational stage (their Halling stage; derived from a site in the Medway Valley, Kent). Upstream, where the subdivision was not recognised, Valley gravel (Blake, 1903; White, in Jukes-Brown and White, 1908) or Floodplain gravel (Dewey and Bromehead, 1915; Sherlock and Noble, 1922) was mapped beneath the present floodplain. The age and relationship of the Floodplain gravel to the 'Upper' and 'Lower' subdivisions was discussed only by Dewey and Bromehead (1921). They concluded that the former was equivalent to the Valley or Floodplain gravel upstream. However, as has been shown (Section 2.17) the Kempton Park Gravel can be traced upstream at least as far as Marlow and can be separated from the gravel beneath the

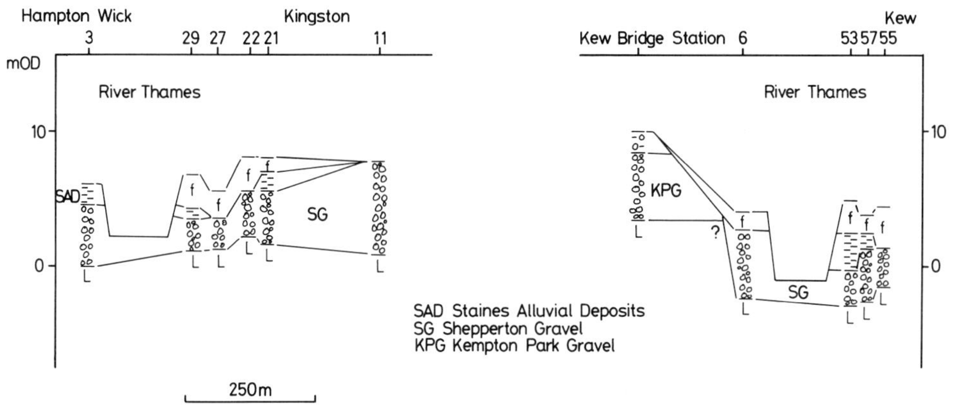

Fig. 42. Sections showing the relationship of deposits to the River Thames at Kingston Bridge (left) and Kew Bridge (right) (source: GLC).

floodplain which must therefore be the equivalent of the 'Lower' subdivision in the London area.

The modern floodplain surface morphology was examined by Hare (1947) to provide 'a yardstick by which the older terraces can be measured'. He noticed that the surface has a local transverse slope of up to 2.4 m that dips to the south on east–west reaches. He also noticed that the present surface shows considerable minor relief, resulting from low gravel hummocks or swells, the intervening hollows being infilled by recent alluvium. The same phenomenon was reported by White (in Jukes-Brown and White, 1908), King and Oakley (1936) and Allen (1978). Hare (1947, p. 324) attributed this morphology to the previously braided habit adopted by the river during gravel aggradation (cf. also White, 1908). He thought that the gravel islands (or eyots) possibly represented fossilised gravel braid bars. The present work supports this theory and shows that these features are present both in the Thames and in tributary valleys. However, Berry (1979) recently suggested that many of the eyots in the present river channel are dissected bodies of fine alluvium and not gravel surface features.

Nevertheless, most authors have not discussed the floodplain gravel in any detail for various reasons. This undoubtedly arises from the difficulty of seeing sections because excavations in the floodplain are invariably flooded. It may also be because the gravels are assumed to be of very recent age. The author, however, takes the view that the Floodplain gravel, being the most recent gravel and sand aggradation, can provide an important analogue for interpretation of older gravel members and therefore these deposits should, as far as possible, be examined in detail (cf. Hare, 1947; Briggs and Gilbertson, 1980).

For this reason and for uniformity a new lithostratigraphical unit is proposed for the 'Lower Floodplain' gravels: the Shepperton Gravel. The lithologically contrasting alluvial deposits that overlie the gravel will be described separately below (Staines Alluvial Complex, Section 2.20). The longitudinal profile along the present floodplain is shown in Figure 44.

The type section for this member is Shepperton quarry (TQ070669: 11 m OD). Here a maximum thickness of 2 m of horizontally bedded medium gravel in a sandy matrix interstratified with occasional lenses of cross-bedded sand 10–20 cm thick occurred in the flooded sections. Palaeocurrent measurements from the sand indicated a flow towards due east, i.e. parallel to the present river course. The composition of the gravel was determined by two pebble counts, the average of which is shown in the Appendix. The base of the gravel and sand was not seen at the site. A maximum of 1.75 m of light brown floodplain silt rested on the gravels with a sharply defined base. Immediately adjacent to the north of the type section is the M3 Motorway, a section along which is shown in Figure 40. Although much of the gravel has been quarried, undisturbed deposits are still present beneath Sheeps Walk (TQ699674), 350 m north of the site. Here the gravel and sand was of variable thickness (on average 5–6 m), rested on bedrock and was overlain by up to 3 m of floodplain alluvial sediment.

As far as St Ann's Hill the Shepperton member forms a continuous spread beneath the modern floodplain and even extends under the River Thames itself at Chertsey. East of the type section, the Shepperton Gravel abuts the Kempton Park Gravel at Shepperton Green. The base of the Shepperton Gravel is at 1.5 m OD here while that of the higher member is 5–6 m OD.

South of the river at Shepperton is the Wey confluence (Section 2.19(*g*)). In this area the floodplain is 2 km wide and the present river meanders over its full width. Downstream the floodplain narrows and a borehole at Apps Court, Walton (TQ109678: 10 m OD) recorded 5.2 m of gravel on London Clay beneath the surface. On the north bank a borehole adjacent to Kempton Park (TQ119694: 11 m OD) proved 4 m of gravel and sand beneath 1.37 m of floodplain silt. The mapped area of alluvial sediments is very narrow between Walton-on-Thames and Chiswick. The confluence of the River Mole with the Thames at East Molesey is marked by a coalescence of the floodplains (Section 2.19(*h*)). North of the river, the distribution of sediments is poorly known in the Hampton Court meander; however, on the basis of the present topography, much of Hampton Court Park should be underlain by deposits of this stage.

The gravels are next recorded at Hampton and Kingston (Figure 42). They also underlie the Ham–Kingston area and reach as much as 8.7 m thick beneath about 1 m of floodplain loam at TQ165726 (8 m OD). Boreholes west of Richmond town centre recorded over 2 m of gravel and sand at the expected height beneath 2 m of organic silty clay and 2.4 m of fill (*c*.TQ176748: 5 m OD). Between Brentford–Isleworth large 'whale-back' gravel landforms (e.g. TQ169776) were identified by Allen (1978) as similar to those recorded by Hare (1947); however, they appear to be formed of eroded remnants of Kempton Park Gravel. At Kew Bridge (Figure 42) the Shepperton Gravel is 5–6 m in thickness on the south side of the river (TQ190778: 4.5 m OD). The relationship of this member to the Kempton Park Gravel can again be seen in the section.

Within the Chiswick meander the gravels have been proven by boreholes at TQ212767 and 213773 (4 m OD). The former encountered 2.7 m of gravel and sand beneath 6.1 m of overlying material, while the latter proved 5.5 m of gravel and sand. Within the Barnes meander the altitudinal equivalent of these deposits is found at TQ220777 (4 m OD) where it is 2.1 m thick and underlies 2.75 m of fill. The Hammersmith Sewer section (Figure 38) demonstrates the relationship of this member to the higher, Kempton Park Gravel in the area. As can be seen the gravels are clearly separable altitudinally, the base of the higher unit occurring at 0.5–1 m OD beneath Hammersmith centre, whereas that of the Shepperton Gravel is at −3 to −4 m OD at Hammersmith Bridge. South of the river the deposits appear to be present beneath the Barn Elms area.

Downstream the gravel can be traced on the north side of the river. Numerous boreholes in the Fulham area show that 6–7 m of gravel and sand are present. A similar thickness is present beneath Kings Road, Fulham (Figure 37) where the Shepperton Gravel again abuts the Kempton Park member. The continuity of this unit beneath the present floodplain and also its merging with the tributary spread of the Chelsea Creek (Section 2.19(*j*)) is well illustrated in Figure 23. Comparable thicknesses of this member are present in the section across Wandsworth and Battersea Bridges (Figure 23). Battersea Festival Gardens is underlain by gravel at the expected height for correlation with this unit. Boreholes on the southern side of the Gardens record as much as 9.5 m of gravel and sand under 1 m of alluvial sediments (TQ280766: 4.5 m OD) and a similar sequence is found along Park Road, Battersea as far as Queenstown junction (TQ287770).

During excavations for gas holders at Battersea Gas Works (TQ287772: 3.9 m OD) up to 7.6 m of gravel and sand were exposed resting on a channelled surface of London Clay (Coombs, 1873). This author observed several sections and described interstratified horizontally bedded gravel and current-bedded sand, the latter filling trough-like channels in the former. A scour channel filled with current-bedded sand was illustrated by Coombs who noted that the maximum dip of the bedding, and therefore the palaeocurrent flow, was towards due east. According to Coombs (1873, pp. 33–4) the gravel comprised predominantly rolled flints, but included fresh Chalk flint, 'bull-head' bed flint, Woolwich pebbles, sandstone pebbles, sarsen, quartzite and white quartz; a composition typical of these deposits. In the north-western corner of the excavation a 1.5 m thick bed of 'bluish-black sandy, muddy clay containing subangular pebbles and frequently intersected with vertical rootlets of plants' was found, 2.4 m above the London Clay surface. This lenticular channel fill was not analysed palaeobotanically and did not contain Mollusca. However, the gravel yielded a number of vertebrate remains.

Immediately adjacent to this site are Battersea Power Station (TQ290775) and the main Gas Works buildings (TQ296773). Extensive borehole investigations on the sites of these buildings have proved the presence of enclosed, often elongate bedrock hollows of anomalous depths (see Berry, 1979, for detailed discussion). These enclosed hollows reach depths of −27.4 m OD beneath the Power Station and −17.6 m OD beneath the Gas Works. Neighbouring features have been proved to a comparable depth. The hollows have very steep and frequently vertical sides and are filled with gravel and sand presumably of the Shepperton member, but often include clasts or masses of bedrock clay at depth. As a general rule, the sediments are occasionally well stratified, but are more commonly disturbed by folding and faulting, presumably by collapse into the hollow. This collapse in many cases has affected the entire thickness of the unit since the overlying fine alluvium is also anomalously thickened above these features (e.g. Battersea Power Station). Adjacent to the hollows the sediments are apparently undisturbed and of the usual order of thickness.

The complex of hollows described by Berry (1979) continues beneath Lambeth and probably in part represents a continuation of those under Vauxhall. The known hollows occur at Clapham Road, South Lambeth (TQ306767), Vauxhall Park (TQ306776) and Vauxhall Grove (TQ304778). They are filled to a maximum depth of 22.6 m with sand, sandy gravel and sandy clay. On the north bank of the river a complex of very similar steep-sided hollows is present beneath the Westminster area at Horseferry Road (TQ297790), Victoria Street (TQ299794), the Ministry of Defence (TQ302800) and Hungerford Bridge beneath the Thames (TQ306801). Depths of down to −27.1 m are again filled with gravel and interbedded sand, although the Victoria Street depression contains a lenticular mass of 'well stratified silty clays, silts and sand strata.' At Whitehall Place (TQ302802), the 'normal' thickness of gravel (5–6 m) is depressed into a shallow hollow, the base of which is at −12 m OD. The resulting depression in the gravel surface is filled with an unusual thickness of fine-grained alluvial sediment.

Where the deposits are apparently undisturbed a 'normal'

Fig. 43. Section showing the disposition of deposits from beneath Buckingham Palace to Southwark (sources: GLC, IGS).

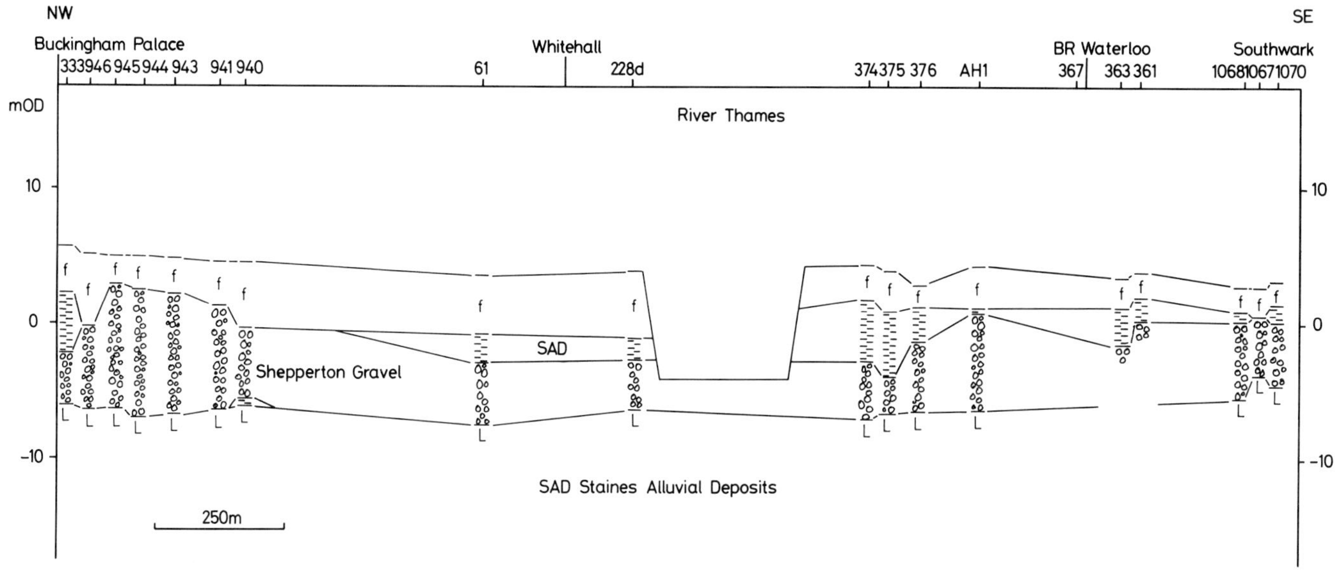

Fig. 44. Long profile section of the Shepperton Gravel and overlying Staines Alluvial Deposits.

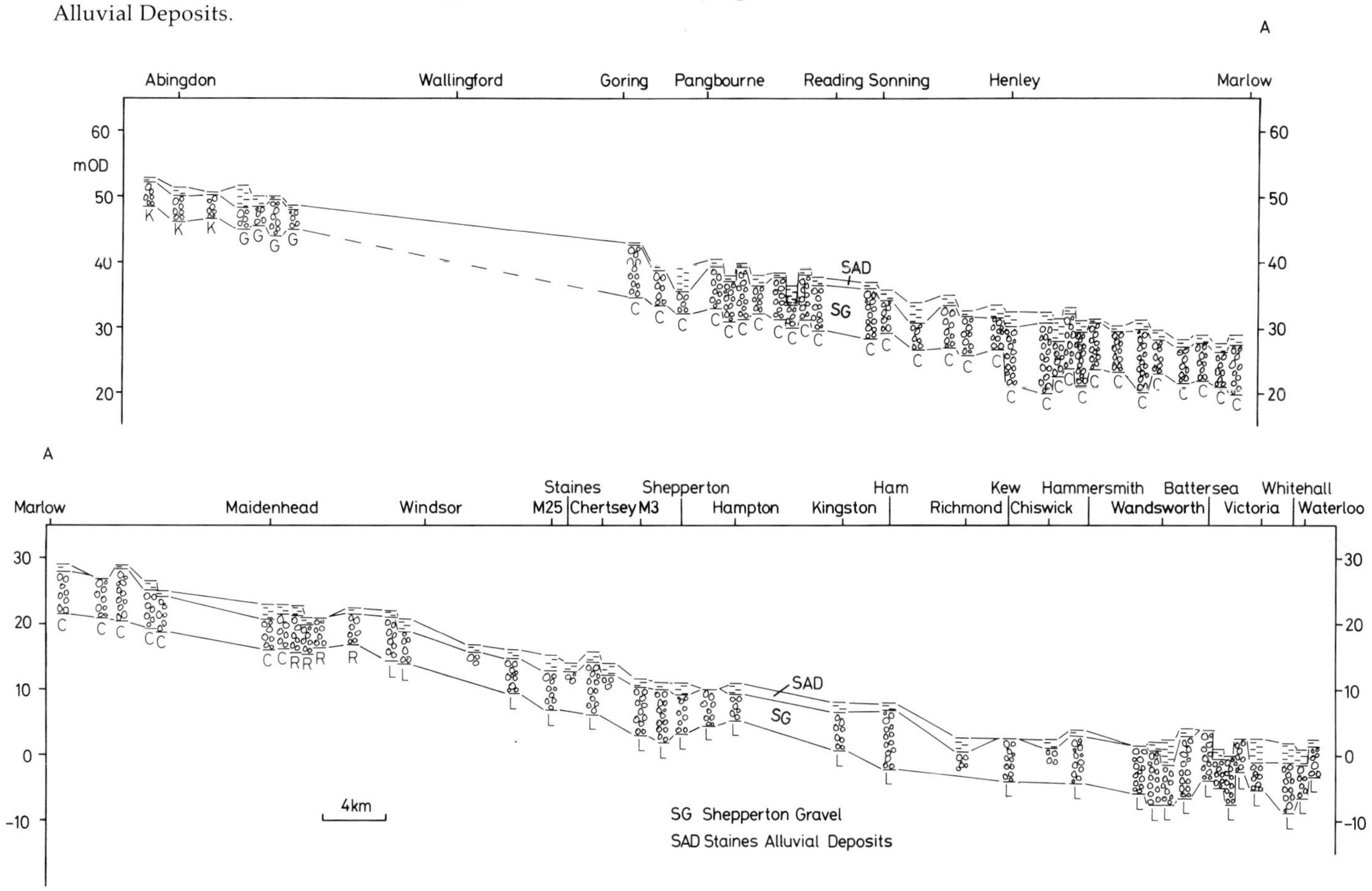

sequence is found. For example, at Kennington (TQ310775: 4.4 m OD) 4.9 m of gravel and sand lie on London Clay and are overlain by 2.45 m of later material. The sequence beneath the Waterloo–Southwark area, where the floodplain is very extensive, is shown in Figures 20 and 43. Here gravel and sand is present beneath the entire area at the expected height for correlation with the Shepperton member (Figure 44). As seen previously the gravel unit continues beneath the river to Westminster near the mouth of the Tybourne, and extends as far north as Buckingham Palace. A borehole in Petty France (bh 940: TQ295795) proved that 45 cm of 'dark grey laminated clayey "peat" and sand with shells and plant fragments' is present under gravel and sand and resting on bedrock. The gravels and sands abut the Kempton Park Gravel on the north side of the valley from Royal Hospital Chelsea (Figure 29) to Belgravia, south of St James's Park and immediately east of Whitehall. Boreholes at Victoria Station (TQ289790: 3.5 m OD) confirm the presence of 2.3 m of gravel and sand beneath 4.5 m of organic sediment and 1.2 m of fill (cf. also Dewey and Bromehead, 1921). In Pimlico numerous boreholes record up to 6 m of gravel and sand beneath 3–4 m of overlying material (e.g. TQ295785: 2.5 m OD). The continuation of these deposits beneath Charing Cross and The Embankment is confirmed by the sections in Figures 22 and 33.

Upstream from the type section, the unit was well exposed at two sites immediately south of the M3 Motorway at Twynersh near Chertsey (TQ032676 and 037676: 14.5 m OD). Both sites showed a very similar sequence comprising approximately 4 m of current-bedded gravel and sand resting on an irregularly channelled London Clay surface and overlain by 1 m of floodplain alluvial clay. The gravels comprised horizontally stratified to massive beds interlain by sand bands up to 25 cm in thickness, occasionally tabular cross-bedded. In one part of the section at the west (Thorpe Road) site a channel-like lens, 2 m deep and 15 m across, filled with gravel and overlain by current-bedded pebbly sand was seen cutting through a gravel bed. An average of pebble counts from the two localities is given in the Appendix.

In the Chertsey–Staines area numerous old flooded gravel workings occur in this unit. For example, at Mixnam's Ferry (TQ042685: 14 m OD) about 1 m of the gravel was exposed above water level; similarly at Hythe (TQ031702: 15 m OD). Boreholes in Staines town centre have recorded a considerable thickness of gravel altitudinally equivalent to this unit. For example, at Staines Pumping Station (TQ041721: 15.7 m OD) 7.9 m of gravel and sand were proved beneath 1.5 m of floodplain alluvium and fill. The

Fig. 45. Section along the M25 Motorway from the M3 intersection to West Drayton, constructed from borehole records (source: SERCU).

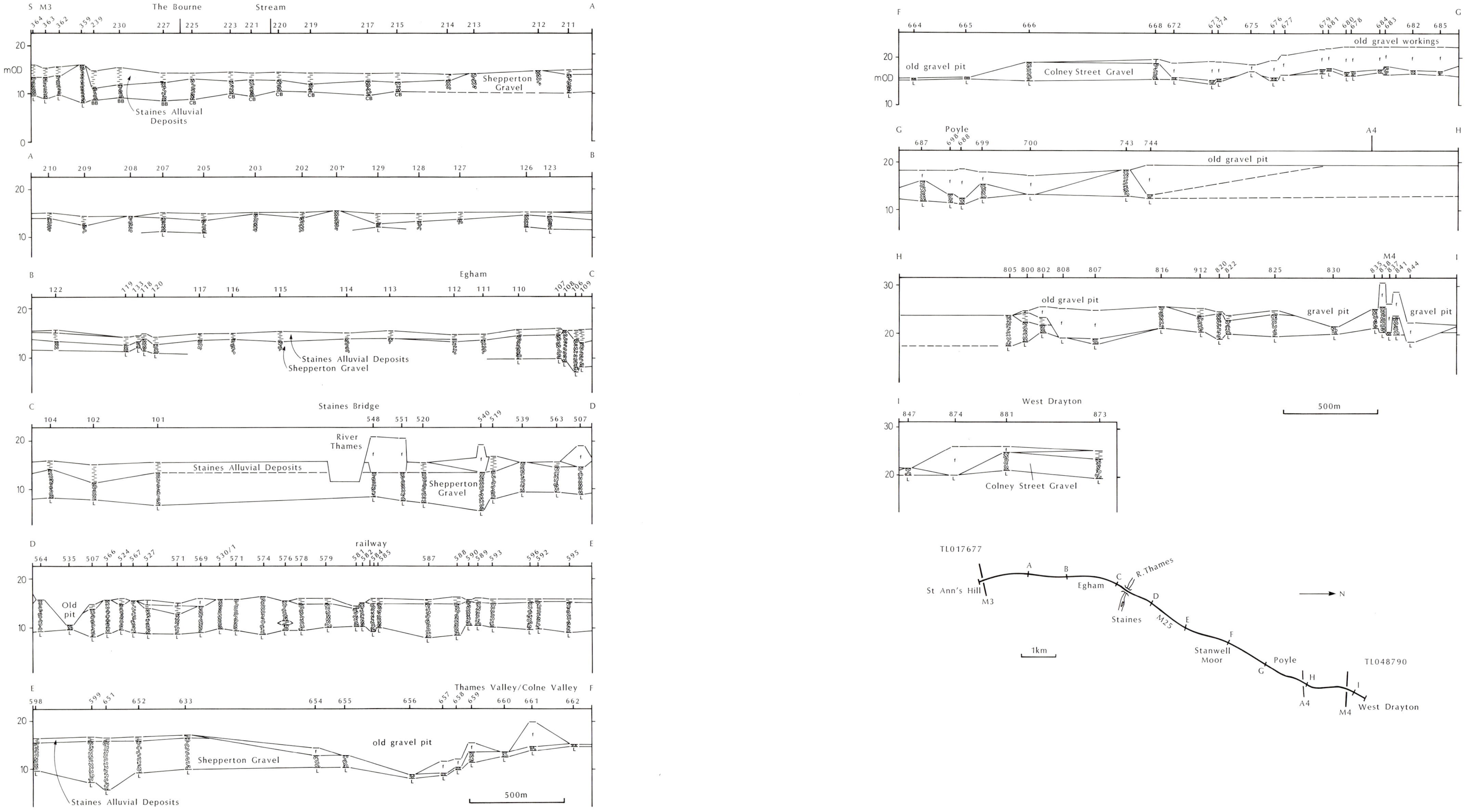

Fig. 46. Section from Datchet pumping station to Ashleys Arch adjacent to the M4 Motorway, constructed from borehole records (source: TWA).

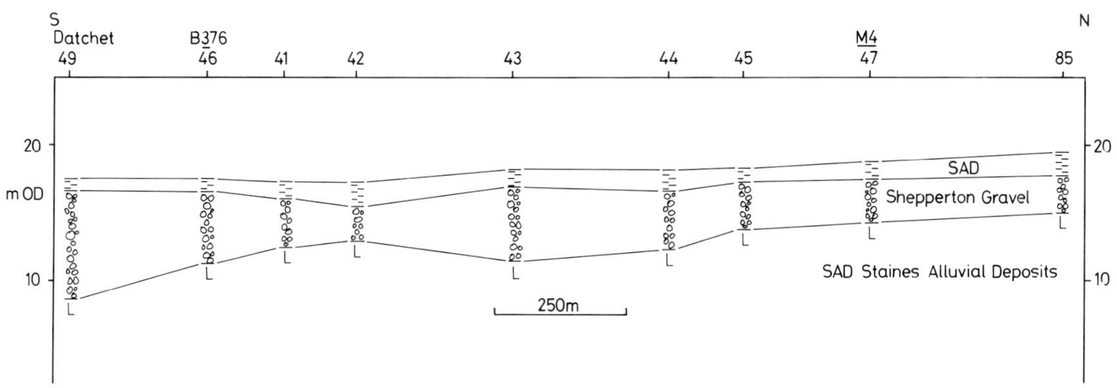

river floodplain in this area is very extensive and, as shown by the geological map (sheet 269), apparently extends from Staines, as far west as Egham. The M25 Motorway section (Figure 45) between the M3 intersection and Staines shows that a continuous thin veneer of gravel and sand underlies the area. It thickens closer to the Thames at Egham itself reaching a maximum of 6.5 m on the south side and 8 m immediately north of the river. This borehole section crosses the floodplain approximately perpendicular to the present river course (M25: Figure 45) and continues along the floodplain of the River Colne. The Colne is at present confluent with the Thames at Staines, but also has a series of distributaries that fan-out to join the Thames at several places. Hare (1947) likened this feature to a delta-cone. In the M25 section, the Colne appeared to have met the Thames near Stanwell Manor (Figure 45, bh 661: TQ033747) to judge from the marked increase in gravel thickness south of this point. It is possible that overdeepening of the bedrock surface may have resulted, to judge by the unusual gravel thickness seen near the Colne confluence. This feature may be analogous to the scour hollows described by Berry (1979) and mentioned above. The equivalent gravel flooring the Colne Valley is discussed below (Section 2.19(*h*)).

In the Wraysbury–Sunnymeads district, there are again numerous gravel excavations. Several pits were working gravel and sand at depths of 3.5–5 m during this study (e.g. TQ011734 and 004750: 17 m OD), but all were flooded and no sections could be seen. A recent borehole at Wraysbury (TQ041721: 21 m OD) proved 6.25 m of gravel and sand, clayey in the upper 1.7 m beneath 0.6 m of fill. A comparable thickness was recorded by Dewey and Bromehead (1915) at Horton (*c*.TQ015757) and Ankerwycke Priory (TQ004730). South of the Thames near Moram Lodge in Home Park, Windsor (*c*.SU983752: 18 m OD), Whitaker (1860) saw about 3.4 m of red gravel and sand, the lower 1 m of which contained bones of an elephant (?mammoth), ox and a deer. The gravels were recorded from several places by Dewey and Bromehead (1915) and are up to 7 m in thickness (e.g. Egham Waterworks: TQ024717). At nearby Albert Bridge (SU985757: 18 m OD) the gravel and sand is 3 m thick, while beside the Southlea Road a series of boreholes showed as much as 4.8 m of gravel and sand in a similar stratigraphical position. A line of boreholes from Datchet pumping station to Ashleys Arch, beside the M4, again confirm altitudinally equivalent gravel and sand thinning northwards beneath floodplain alluvium (Figure 46). West of Datchet this unit is 4.25 m thick (SU987781: 20.5 m OD); north of Datchet, 4.75 m thick (SU993779:

20 m OD); and in Eton it is 3.4 m thick (SU967778: 19 m OD).

Immediately east of Eton the disposition of the deposits is well illustrated by the A332 road section (Figure 25). At the northern end of the section the relationship of the Shepperton and Taplow Gravels can be seen; the difference in their average base levels being a minimum of 11.5 m. The base of the Shepperton Gravel is again gently undulating and the gravels underlie the present river channel, as elsewhere. South of the Thames the deposits extend beneath Windsor immediately west of the Chalk outlier on which the Castle is built. Beside Windsor Bridge where Chalk underlies the deposits of this unit they are 8.5 m thick (SU996772: 20 m OD). Sections exposed at the Infantry Barracks (SU967766: 23 m OD) in 1867 were visited by Whitaker (1889) who saw 5 m of sandy fine gravel, rich in mammalian remains, at a depth of 2.75 m. The remains were mentioned by Dawkins (1880), who wrote 'I found more than one half of these remains belonged to the reindeer, the rest to bisons, horses, wolves and bears.' The gravel thickness at this site has been confirmed by recent borings. A comparable sequence with vertebrate material including reindeer was recorded in Peascod Street (Dewey and Bromehead, 1915, p. 78). At several places in Windsor the flood-plain margin can be confidently mapped using the borehole information available (e.g. beneath the Imperial Service College (SU961764) and St John's House, Clewer (SU954765).

Upstream from Windsor, the gravels continue as shown by Hare (1947, figure 13). Numerous boreholes for the Mineral Assessment Unit survey (Dunkley, 1979) and for water resources, record gravel and sand of fairly uniform thicknesses of 3–6.5 m resting on bedrock and overlain by varying depths of alluvium (Figure 44). However, greater thicknesses have been proved; for example at Tilson Bridge, Dorney (SU942788: 22 m OD), 8.3 m of gravel and sand are present.

Large, partially flooded gravel workings at Bray (SU909786: 21.5 m OD) exposed approximately 1.5 m of interstratified current-bedded sand and lenticular bodies of horizontally bedded medium gravel beneath 1.5 m of floodplain silty clay. The pebble assemblage of the gravel is shown in the Appendix. In one part of the section a laminated bed 70 cm thick, formed of light brown silty sand alternating with white silty clay, filled a channel in the gravels.

The 'classic' Maidenhead area was where Whitaker (1864, 1889) first recognised the existence of a 'lowest terrace' where the gravels protrude through the fine alluvial sediments. Extensive flooded sections were present east of Furze Platt in 1979–80 (SU895828). They showed about 2 m of horizontally bedded, sandy gravel containing lenses up to 75 cm thick of cross-bedded coarse to medium sand. The gravel lithology is presented in the Appendix. Palaeocurrent measurements from the sands indicated a local flow towards east-south-east.

Northwards from Maidenhead to Bourne End the present river flows at the bottom of a steep slope up to the Black Park Gravel Terrace at Cliveden. The valley is floored by the continuous spread of gravel and sand equated altitudinally with the Shepperton member. The bend in the river at Maidenhead almost coincides with the Reading Beds – Chalk bedrock interface. The Chalk apparently confined the valley, which is only 2–2.25 km wide in this reach, probably because it is less easily mechanically eroded. Again several boreholes record gravel and sand beneath fine alluvial sediments here. To judge from the data available the deposits are thinner on the western than on the eastern side of the valley (3.9 m thick at SU889837 compared with 5.7 m thick at SU903854), although this may only be an artefact of the limited information.

At Bourne End, the Thames is confluent with the River Wye and during deposition of the Shepperton Gravel a contemporary equivalent was laid down in this valley (Section 2.19(*e*)). Boreholes at the mouth of the Wye record anomalous thicknesses of 7.2 and 8 m of gravel and sand (SU897862: 25 m OD and 888869: 29 m OD, respectively), reminiscent of the overdeepened scour hollows at tributary confluences described by Berry (1979). However, at Bourne End Mill (SU898871: 29 m OD) and Wharton Grove (SU901866: 31.5 m OD) more 'normal' gravel thicknesses of 6.1 m and 5.6 m, respectively, are present.

Continuing upstream the gravel can be followed from Marlow to Henley. Here, as before, the deposits rest on Chalk bedrock and therefore the valley is narrow and steep sided. Extensive gravel workings east of Marlow (SU865875) were almost completely flooded during this study, so no sections could be seen, but according to workmen the deposits were about 6 m thick here. This is confirmed by boreholes in the area, for example at Westthorpe House (SU867873: 27 m OD), where a depth of 5.5 m was found. Comparable depths were recorded in IGS borings.

Close to the south side of the valley at Temple Farm (*c*.SU840836) Sherlock and Noble (1922, p. 45) saw a 3 m section 'of chalky sand with pebbles of chalk, flint and sandstone, and large almost unworn flints'. Very similar deposits were reported by Whitaker (1889, p. 388) from near Rosehill (SU808839: 30 m OD) and comprised 'shattered flints and small pieces of chalk in a sandy matrix, together with lenticular masses of sand', up to 1.8 m thick, overlying buff clayey sand, above sandy gravel, 'of subangular flints, pebbles of flint, of quartz and of quartz-rock and pieces of chalk'; not bottomed at 2.2 m. This description suggests that the deposits represent material, derived from the valley side by the action of frost and solifluction, overlying fluviatile gravel and sand. More typical fluviatile deposits occur in the centre of the valley, based on the borehole data, for example at Low Grounds (SU834848) and Frogmore Farm (SU810834), which record up to 5.8 m of gravel and sand. At Hurley (*c*.SU825835: 28 m OD) a pit yielded an extensive mammalian fauna, from a depth of 5.5 m, in the mid nineteenth century, but it has since been dispersed (Whitaker, 1889, p. 389).

The spread of altitudinally equivalent gravel and sand continues upstream to Aston, Mill End and then southwards to Henley. On the north side in this area beneath the Henley–Medmenham road (A4155) are fan-like accumulations of chalk-rich flint gravel, called 'the younger coombe deposits' by the Geological Survey. These gravels merge with those of the Shepperton member and were originally described by White (1904) and more recently by Sealy and Sealy (1956). A section in this material at Oaken Grove (SU767856: 38 m OD) exposed 5 m of very coarse flint gravel (with many pebbles very large and unabraded) with a few pebbles of other rock types including vein quartz (derived from higher ground) in a chalk putty-like matrix decalcified in the upper 1 m to form a brown stony clay (cf. also Whitaker, 1889; White, 1904). White (1908) described two further sections in the neighbouring fan at Mill End and Harpsden Court. The fans, the largest of which is 1 km across at Greenlands (SU775858) appear to be derived from the series of narrow dry valleys that dissect the northern valley side above the river. A nearby borehole at New Cottages, Fawley Court (SU763848: 34 m OD), recorded 10.1 m of chalky gravel and sand, clayey in the upper 3 m. In comparison, a borehole in the valley centre proved 5.7 m of gravel and sand beneath 3.6 m of fine alluvium (SU768843: 31.5 m OD). At Mill End and Henley small tributary gravel spreads flooring the Hambleden and Assendon valleys join the Shepperton Gravel (Section 2.19(*d*)).

At Henley, the floodplain is exceptionally narrow reaching a maximum of only 750 m. The town centre is built on another of the fan-like accumulations associated with a small east-north-east-trending dry valley, a stream course from the latter having dissected the fan. Coarse flint, possibly locally derived from washing of the 'coombe deposits' or from bedrock, forms the basal bed of the Shepperton Gravel beneath the town, as shown by a boring at the Brewery (SU762828). South of the town the gravel and sand is of variable thickness. A borehole at Henley Lock (SU772815: 32.5 m OD) penetrated 9.1 m of gravel and sand beneath 2.2 m of alluvium, while at Lower Bolney Farm (SU772806: 33.5 m OD) only 4.7 m of gravels were present.

In the Wargrave–Sonning area the Shepperton Gravel is apparently buried everywhere by later fine sediment. Boreholes in the Thames floodplain show that an altitudinally continuous gravel spread of variable thickness rests on the Chalk. Several boreholes prove gravel thicknesses of up to 7.3 m near Shiplake (e.g. SU766777: 34 m OD). On the eastern valley side at Wargrave, further chalk-rich gravel, 4.5 m thick, is present at SU784779 (36 m OD). Between Twyford and Sonning the Loddon Valley and underlying gravels join the Thames (Section 2.19(*c*)). Upstream from the confluence a borehole at Sonning Bridge (SU754758: 36 m OD) proved an anomalous depth of 11.2 m of gravel and sand beneath 3 m of fill. Two partially flooded gravel pits occurred in the immediate area: the first at Sonning (SU745754: 36 m OD) and the second, upvalley, south of Lower Caversham (SU734750: 37 m OD). Both sites showed an exactly similar sequence and can be described together. They exposed 2.5 m of horizontally bedded alternating medium gravel and coarse sand beds, each approximately 10–15 cm thick. No current-bedding could be seen at either locality. At both places the deposits were overlain by 1–1.5 m of floodplain silts. A borehole close to the Lower Caversham locality showed the gravel to be 7.7 m thick here (SU735749: 37 m OD).

North of Reading town centre the spread continues beneath the floodplain fine alluvium. Here the southern valley side is closely followed by the Western main railway line. A section across the valley is shown in Figure 34. Here the Shepperton Gravel can be seen underlying the valley floor and abutting the Reading Town Gravel at the railway station.

From Reading to Purley the gravel and sand is continuously present beneath the modern floodplain (Figure 44). In Cow Lane, Reading (SU703742: 38 m OD), it is 6.8 m thick; north of Tilehurst (SU683753: 38.5 m OD) it is 6.4 m thick; and at Purley (SU661767: 40 m OD) it is 7.8 m thick. The River Pang joins the Thames at Pangbourne and the deposits that floor the tributary valley merge with the Shepperton Gravel between Pangbourne and Purley (Section 2.19(*a*) and Figure 47).

Beyond Pangbourne and Whitchurch the valley narrows markedly to as little as 310 m. The gravel and sand, still continu-

Fig. 47. Section across the Thames Valley and along the tributary River Pang Valley near Pangbourne, constructed from borehole records (source: IGS).

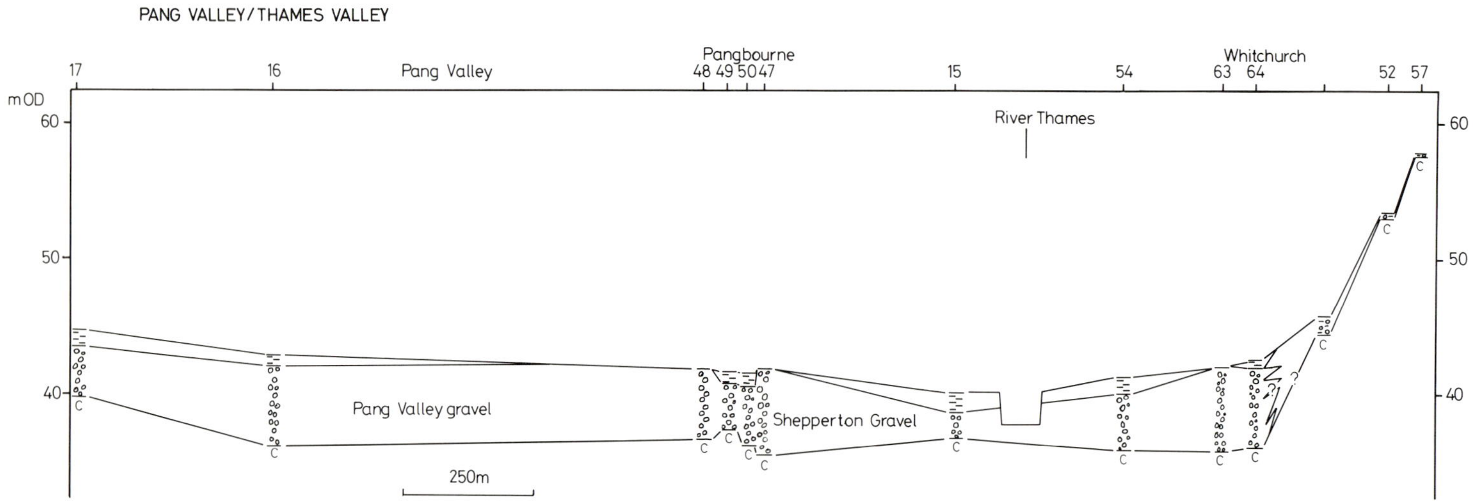

ously present lying on Chalk bedrock, is 5.1 m thick at Basildon (SU615786: 39.5 m OD) and 7.9 m thick at Gatehampton Farm, Goring (SU604797: 43 m OD).

The pebble lithology of the Shepperton Gravel (see Appendix) is based on a limited number of samples (ten). However, they show strong consistency and comprise 92–95.5% total flint (3–5% rounded flint and 87.5–92% angular flint), 1.5–3.5% vein quartz, 1–3% quartzite and 0–1% Greensand chert, together with some minor constituents. The downstream increase in Greensand chert and vein quartz mirrors the incorporation of detritus from tributary and older gravel sources.

The downstream gradient of the Shepperton Gravel in the region is 35–40 cm km^{-1} and is very similar to that obtained by Sealy and Sealy (1956) for the floodplain surface. The figure also closely parallels those for the older gravel aggradations in the Thames Valley.

2.19 Tributary valley floodplain gravels

Several tributaries join the River Thames along its course from Goring to Central London. In most cases these tributary valleys are floored by gravel and sand aggradations which can be altitudinally correlated with, and in many cases can be shown to grade into, the Shepperton Gravel (Section 2.18). The implications of these correlations are very important for relating the evolution of tributary valleys to the Thames and vice versa. The tributaries, for which sufficient information was available during this study, will be dealt with in order downstream from Goring.

(*a*) *River Pang*

Since the work of Hawkins (1926) the Sulham Gap, a wide valley between Theale and Pangbourne, has been thought to represent the recently abandoned valley of the Kennet. This view has persisted and was supported by Sealy and Sealy (1956) and Thomas (1961). It follows therefore that the Pang has apparently had a rather short history as an independent Thames tributary stream.

The borehole section (Figure 47) from Whitchurch into the Pang Valley suggests that at least part of the gravel and sand aggradation beneath the valley is equivalent to the Shepperton Gravel. Boreholes in the Pang Valley (Squirrell, 1976) indicated that a substantial thickness of 7–8 m of gravel and sand underlies the valley upstream of Tidmarsh. According to Squirrell (1976), the gravel comprises angular to rounded flint with occasional

Fig. 48. Long profile sections of the Kennet (above) and Loddon (below) floodplain deposits showing their relationship to the Thames's Shepperton Gravel and Staines Alluvial Deposits (borehole source: IGS).

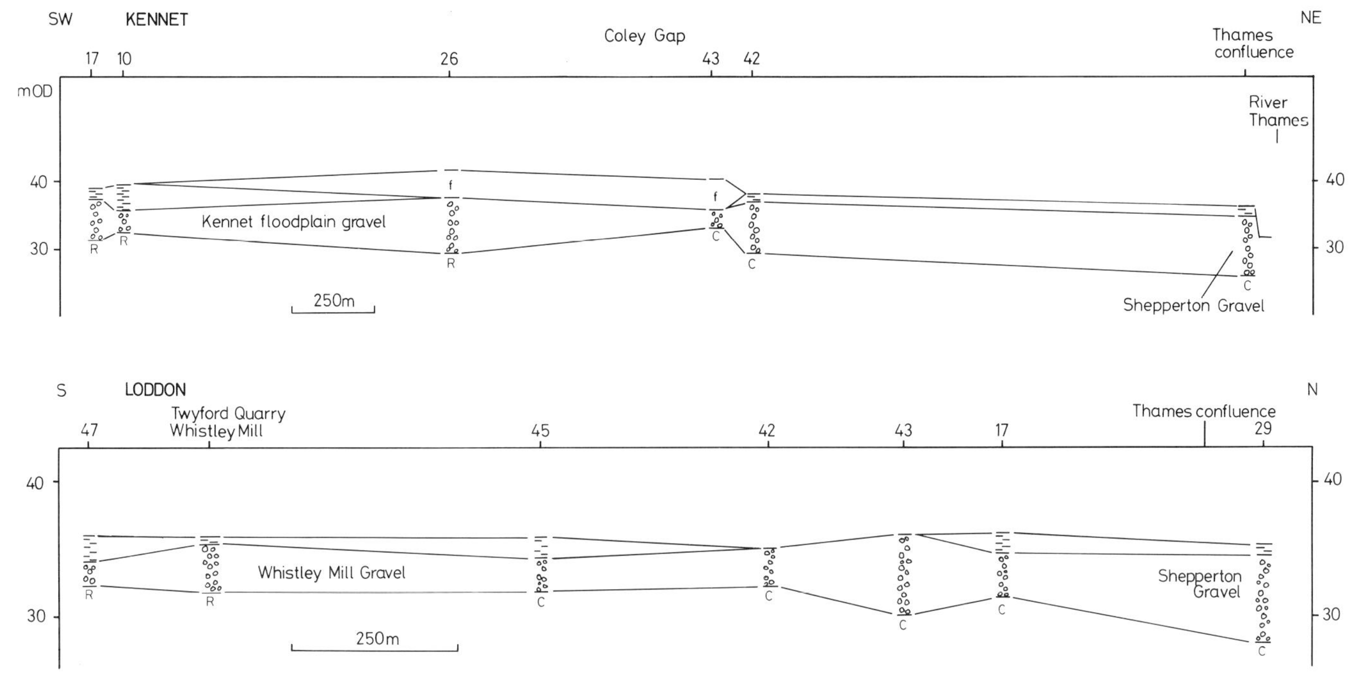

quartz pebbles, and chalk clasts in the basal part. A lithostratigraphical re-examination of the Sulham Gap sedimentary fill would clearly be advantageous.

The Pang therefore appears to have followed its present course at least during Shepperton Gravel times.

(*b*) *River Kennet*

Substantial deposits of gravel and sand underlie the present Kennet Valley, south of Reading. This wide expanse of gravel was thought by Hawkins (1926) to have resulted from constriction of the stream through the gorge-like Coley Gap, downstream in Reading itself. Thomas (1961) supported the suggestion that the Coley Gap was of recent formation and included the wide gravel spread immediately upstream in his Kennet floodplain terrace. The gravel underlying the floodplain has recently been described by Cheetham (1975, 1980) and was included by him in the deposits beneath the Beenham Grange Terrace. The latter is apparently only recognisable as an altitudinally separable morphological entity west of the area under discussion.

The section across the Coley Gap in central Reading (Figure 34) has already been mentioned. Gravel and sand 7.3 m thick underlies alluvial organic sediment in the Gap. A longitudinal profile from Fobney Meadow through the Gap to the Thames at Kennet Mouth (Figure 48) demonstrates that a continuous gravel spread is present that grades laterally into the Shepperton Gravel of the Thames Valley. Borehole descriptions indicate that the Kennet floodplain gravel comprises predominantly subangular to rounded flint with a few quartz and quartzite pebbles (Squirrell, 1978). The Coley Gap therefore functioned during Shepperton Gravel times.

(*c*) *River Loddon*

As with the other streams in the Reading area, the evolution of the Loddon was considered by Hawkins (1926), Sealy and Sealy (1956) and Thomas (1961). The Loddon Valley gravel and sand deposits were examined by Whitaker (1889), Blake (1903) and Walder (1967), and have recently been re-evaluated by Clarke and Dixon (1981). These last authors suggested that the gravel aggradation underlying the present floodplain surface (their terrace 1) could be equated with that in the Thames Valley on a morphometric basis. The confluence of the Loddon with the Thames has been investigated during this study.

To examine the nature of the confluence a borehole section has been constructed from the Thames Valley, south of Bath Road

(Section 2.18 and Figure 44) to Whistley Park (Figure 48). The section demonstrates that a continuous gravel spread exists beneath the Loddon Valley throughout the length of the section and grades into the Thames's Shepperton Gravel. As shown by Clarke and Dixon (1981), the Loddon gravels continue upstream. The deposits were investigated in a large gravel pit at Whistley Mill, Twyford (SU788745: 36 m OD). Here they comprised a sequence of well stratified gravel and sand, 3.75 m thick, lying on an irregularly channelled surface of Reading Beds clay and overlain with a sharply defined base by 0.5–1 m of alluvial clayey silt. The medium to fine gravel was present as narrow lenticular or elongated beds up to 70 cm thick that frequently exhibited strong pebble imbrication indicating a palaeocurrent towards the north-east. The gravel beds alternated with medium sand, present as channel-like fills, or more persistent tabular cross-bedded bands up to 50 cm thick. Palaeocurrent measurements from the sand indicated a flow towards the north-east.

The lithology of the gravel is shown in the Appendix, and comprises only flint (a high proportion of which is of 'Tertiary type'), Greensand chert and, rarely, quartz pebbles. The counts, however, show a slightly greater total flint frequency than in the higher, Woodley Gravel. This results from the correspondingly lower content of Greensand chert present. While this may simply be an artefact of the counts, it is to be expected because the Loddon's tributary, the Blackwater, no longer drains the Lower Greensand area as it did in the past (Clarke and Dixon, 1981; Gibbard, 1982). The proportion of Greensand chert will therefore diminish during successive aggradational phases; all Greensand material present must have been derived by reworking of older deposits. Conversely, contemporaneous incorporation of recycled and newly eroded local flint will also have occurred. The separation of low-level individual Loddon gravel aggradations may therefore be possible using this criterion of progressive Greensand chert dilution.

Meanwhile, however, since it is possible to recognise the unit as distinct both altitudinally and lithologically from higher members, it is suggested that the exposure described above be used as the type section for the gravel and sand underlying the present Loddon floodplain and resting on bedrock. The term Whistley Mill Gravel is proposed for this member.

(*d*) *Henley area*

Several dry valley chalk-rich gravel fills grade into the Shepperton Gravel in the Henley area, for example the Harpsden Valley (White, 1904, 1908). The area is underlain by Chalk bedrock. However, there are more substantial valley fills in the Assendon and Hambleden valleys. The former has not been studied in detail, but was observed by White (1908, p. 98) as comprising 'loamy flint gravel' of unstated depth at the Sewage Works, Lower Assendon (SU742850: 53 m OD).

The Hambleden Valley is aligned just east of south. It is at present drained by the small Hamble Brook and joins the Thames at Mill End. The valley is underlain by 'mainly coarse to very coarse angular to subangular flint with some chalk pebbles and rounded quartz' (Squirrell, 1978; Whitaker, 1889). The section along the valley from Hambleden Lock to Flint Hall, Upper Hambleden (Figure 49) illustrates the apparent gradation of the gravel fill into the Thames Shepperton Gravel at Mill End. Although the nature of the deposits is based on only three boreholes, they indicate a consistent gradient of 800 cm km^{-1}. At Mill End the gravel fill appears to dissect one of the 'chalky gravel' fans mapped by White (1904, 1908) and therefore must post-date the feature.

(*e*) *River Wye*

To the author's knowledge the sediments beneath the steep-sided Wye Valley cut into Chalk bedrock have never been studied apart from isolated observations by Whitaker (1889), White (1908) and Sherlock and Noble (1922).

The River Wye joins the Thames at Bourne End, where a fan of material deposited at the stream's mouth can be recognised at 29–30 m OD, i.e. 2.5–3.5 m above the general Thames floodplain level in the immediate area (Hare, 1947).

The deposits within the Wye Valley are shown in Figure 49. The longitudinal profile extends 4.8 km along the valley from Bourne End to Loudwater. The deposits comprise gravel and sand overlain by alluvial light grey silty clay. Boreholes described by Dunkley (1979) show the gravel consists of subangular to rounded flint and chalk clasts together with some quartz and quartzite pebbles. The latter were presumably derived from local older Thames gravels since they are the only major source of such material within the Wye catchment. The section shows that the gravel and sand deposit continuously present beneath the valley bottom is of variable thickness, rests on bedrock and grades into the Shepperton Gravel of the Thames. The gradient of the Wye Valley gravel is approximately 450 cm km^{-1}.

The transverse section (Figure 6) at Loudwater shows the gravel member beneath organic floodplain deposits and in

Fig. 49. Long profile sections of the deposits underlying the Hambleden Valley (above) and Wye Valley (below), showing their relationship to the Thames's Shepperton Gravel and Staines Alluvial Deposits (borehole sources: IGS, TWA, ERCU).

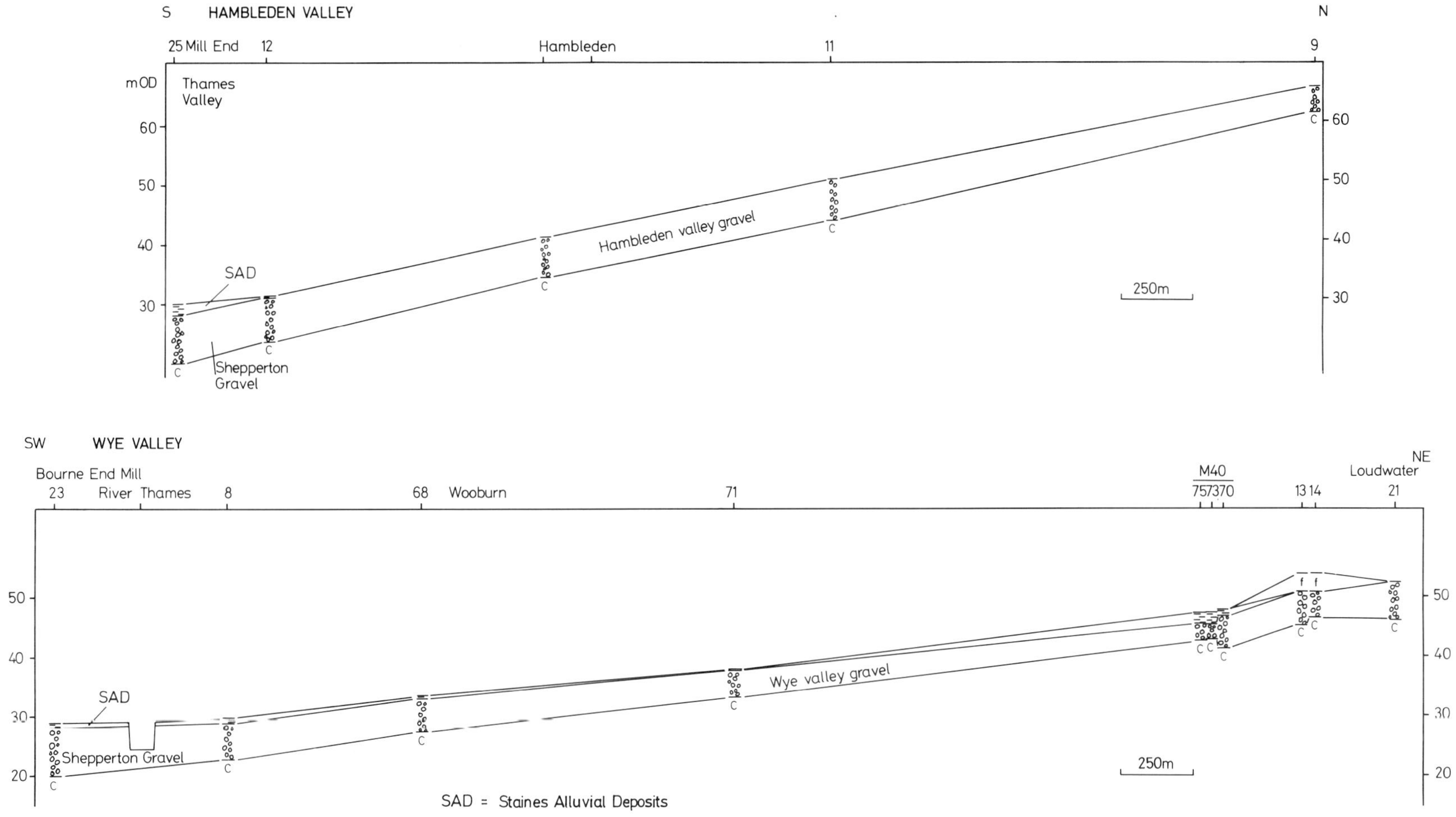

borehole 63/1 beneath tufa or 'rubble' chalk (hard chalk blocks and flints in a putty-like chalk matrix; 'coombe rock' of some authors). The valley sides comprise a veneer of brown silty clay with angular and rounded flints overlying weathered or 'rubble' chalk of varying, often indeterminate depths. An interesting additional feature is the sequence present only in borehole 65/1, which recorded 2.4 m of brown fine silt above 4.75 m of brown pebbly clay and 3.5 m of weathered chalk with bedrock beneath. Such a sequence is similar to those found elsewhere in bedrock solution hollow fills of which this may be an example (see Sections 2.7, 2.10 and 2.17).

(*f*) *River Colne*

Previous research on the deposits filling the Lower Colne Valley has been very restricted. Whitaker (1889, p. 393) noted that sections at Horton and Colnbrook exposed gravel, but gave no details. Geological mapping by Dewey and Bromehead (1915) and Sherlock and Noble (1922) was followed by geomorphological mapping by Hare (1947). He showed that the Colne and Thames floodplains merge near Staines where a 'delta-like' fan of slightly raised ground is present. Hare (1947) attributed this fan to the spreading-out of the Colne distributaries (Colne Brook, Wraysbury River, Colne, Yiewsley River, etc.) Lacaille (1961a, 1963) described the gravel at two localities in connection with discoveries of Mesolithic artefacts on its surface. In the Upper Colne Valley in Hertfordshire, Godwin (1964) and Gibbard (1974, 1977) showed that organic deposits occurred in the gravel and sand underlying the modern floodplain. Recently similar plant-bearing channel fill sediments were found in the gravels flooring the Lower Colne Valley in workings at West Drayton and Colnbrook (Coope, 1982; Gibbard and Hall, 1982). The organic deposits have all yielded fossils indicative of cold climates and have been radiocarbon-dated as follows:

Colney Heath	13560 ± 210 BP (Q-385)
Colney Street	14320 ± 210 BP (Q-1120)
West Drayton	11230 ± 120 BP (Q-2020)
Colnbrook	13405 ± 170 BP (Q-2021)

The results indicated that the organic deposits, and therefore their containing gravels, accumulated during the Late Devensian substage (Gibbard and Hall, 1982).

The gravel and sand beneath the floodplain of the Colne in Hertfordshire was assigned to a lithostratigraphical member, the Colney Street Gravel (Gibbard, 1974, 1977). A continuous spread

Fig. 50. Long profile section along the Lower Colne Valley of the Colney Street Gravel and floodplain alluvial sediments. The relationship of these aggradations to those of the Thames is also illustrated.

Table 3. *Pebble counts (as percentages) from the Colney Street Gravel in the Lower Colne Valley derived by examination of sieve grades 33, 16 and 8 mm. Grid references for sample sites are given in the text.*

	Flint (angular)	Flint (rounded)	Flint (total)	Vein quartz	Quartzite	Greensand chert	Palaeozoic chert	*Rhaxella* chert	Total number of pebbles
Colney Street Gravel									
Colnbrook	80.00	13.56	93.56	5.98	0.5	—	—	—	435
West Drayton	79.97	12.19	92.16	7.40	0.25	0.15	0.15	—	689/702
Stanwell Moor	88.16	5.15	93.31	3.60	2.09	0.68	0.16	0.16	718

of gravel and sand underlying post-glacial floodplain deposits is present throughout the length of the Colne Valley. In view of this, together with the similarity of their sedimentary structures, altitude, etc., Gibbard and Hall (1982) suggested that the gravel and sand deposit flooring the Colne Valley in West London should be the lateral continuation of the Hertfordshire Colney Street member. The relationship of this unit to the Thames sequence has not been demonstrated previously.

The Lower Colne Valley is aligned north–south and is cut through older Thames terrace sediments resting on Lower Tertiary, and above Denham, Chalk bedrock. The valley has been heavily exploited for gravel resources and many sections above Iver are completely flooded. However, the extensive borehole programme for the M25 Motorway has made the construction of a continuous borehole section possible. The section (Figure 50) extends from the Colne confluence at Staines, 8 km along the valley to Thorney near West Drayton and thence, after a 1.5 km gap, 2.75 km from Iver to Mansfield House, near Uxbridge (see also Figures 11 and 45).

In spite of gaps created by gravel removal, the section shows that gravel and sand continuously underlie the modern floodplain of the Colne from Stanwell Moor as far as Thorney. At the former place, they merge with the Thames Shepperton Gravel. The gravel and sand are of variable thickness, being usually 3.0–4.0 m, but locally reaching as much as 5.5 m. They rest on a gently undulating London Clay surface and are overlain by generally less than 1 m, but occasionally as much as 2.5 m, of floodplain organic and fine alluvial sediments.

Sections across the Colne floodplain in this area have been previously published by Gibbard and Hall (1982, figures 4 and 5). The second of these, along the M4 Motorway, is reproduced here (Figure 17). This section intersects with the longitudinal profile (Figure 50) and demonstrates the relationship of the Colne Valley gravel to the probable Thames–Colne equivalent Taplow Gravel beneath Harmondsworth (Section 2.13). The valley gravel forms a distinct aggradational unit, the base of which is at about 18 m OD on average in this section, but greater depths are occasionally present (bh 960) and may represent overdeepening by scour or bedrock solution.

Exposures in the Colne Valley deposits have been examined at several sites. The sequences of stratified sand and gravel at West Drayton (TQ053792: 29 m OD) and Colnbrook (TQ026776: 22 m OD) have been described in connection with their contained organic sediments by Gibbard and Hall (1982). Pebble counts from these sites are given in Table 3.

Downstream at Stanwell Moor (TQ046748: 20 m OD) the sediments were 3.5 m thick and comprised interbedded stratified gravel and sand. The medium to fine gravel was predominantly supported in a silty sand matrix and occurred as elongate beds, up to 70 cm thick, and occasionally more massive horizontally bedded units as much as 1.5–2 m in thickness. Alternating beds of tabular cross-bedded medium to coarse sand up to 50 cm thick were present. In several places narrow light grey silt bands often interlaminated with sand occurred. None were fossiliferous. Palaeocurrent measurements from the sand indicated a flow towards the south-south-east. A pebble count from the gravel is given in Table 3.

Beside the M4 at Lambourne Farm (TQ043785: 23 m OD) large, partially flooded workings exposed 3 m of poorly stratified medium gravel in a sandy matrix overlain by 1.8 m of floodplain alluvial clay.

At Iver the gravel and sand unit is again present beneath the floodplain as far as Mansfield House (Figure 50) and can be equated with that at Thorney, West Drayton and further south with which it is altitudinally continuous. Close to the proposed M25, sections in the Iver Bridge quarry (TQ043813: 27.5 m OD) showed 2 m of interbedded sandy gravel bands up to 70 cm thick and lenticular sand beds beneath alluvial silts. Gravel in the same stratigraphical position was present at the now infilled Sandstone (TQ044836) and Willowbank (TQ054855) sites (Lacaille, 1963) and is present beneath Ivy House roundabout (TQ044858: 36 m OD) in boreholes for the M40 (Figure 30). Lacaille (1961b) also noted the occurrence of floodplain gravel underlying alluvium at Dewe's pit, Harefield (TQ054884). Stratified gravel and sand 2 m thick below 50 cm of alluvial clay was found at Harefield (TQ045894: 38.5 m OD). Further upstream (7 km) at Batchworth, Rickmansworth (TQ063940: 47 m OD) several boreholes prove as much as 5.8 m of gravel and sand on bedrock Chalk, overlain by up to 1.2 m of post-glacial organic deposits.

The altitudinal and stratigraphical correlation of all the sites mentioned in the Colne Valley is shown in Figure 50. The gravel unit, demonstrably continuous with the Thames Shepperton member, has a downstream gradient of 175–180 cm km^{-1}.

The lithology of the Colne Valley gravel (?Colney Street Gravel) listed in Table 3, comprises 92–93.5% total flint, predominantly the angular type (80–88%), but with consistent amounts of rounded flint (5–13.5%). The content of vein quartz and quartzite is more variable, being 3.5–7.5% for the former and 0.25–2% for the latter. The content of all the lithologies presumably results from reworking of older gravels, particularly those of the Thames, together with the incorporation of local bedrock. The similarity of catchment source rocks to those in the Thames Valley results in the gravels being lithologically very like those of the Thames itself. A possible means of separating the gravels of the two rivers is the frequency of Greensand chert. This rock is virtually absent from Colne gravel, except adjacent to outcrops of Thames gravels, as is the case at Stanwell.

Assuming that the correlations presented here and in Section 2.11 are correct, the suggestion by Wooldridge and Kirkaldy (1936) that a knick point may be present in the Colne Valley at 35 m OD would seem to be unfounded. Equally, Hare's (1947) proposal that the Colne flows on 'Taplow terrace' (Colne-equivalent) upstream of this point, is also contradicted by the modern evidence.

North of Mansfield House (TQ041837) two smaller west-bank tributaries are confluent with the Colne. They are the Alder Bourne immediately to the north, and at Denham (3 km further north) the River Misbourne. The floodplains of these two streams grade into that of the Colne without any discernible break. Both valleys are incised into the Denham Village Gravel remnant on the western side of the Colne Valley (Section 2.13). Both valleys are crossed by the M25 and transverse sections of their infill (Figure 11) show that they are both underlain by gravel and sand. A longitudinal profile along the Alder Bourne Valley (Figure 51) demonstrates that it is floored by generally about 3.75 m of coarse to fine gravel of angular to rounded flint with some quartz and quartzite pebbles and sand (Squirrell, 1974) beneath post-glacial alluvium. However, an anomalous 12.5 m deep infilled hollow is present near Rush Green (TQ027852). The gravel and sand beneath the Alder Bourne Valley grade into those beneath the River Colne floodplain at Mansfield House.

Likewise, the Misbourne Valley is underlain by gravel of similar lithology (Whitaker, 1889; Squirrell, 1974) and sand, but reaching greater thicknesses. The transverse section (Figure 11) shows that the Chalk bedrock surface beneath the deposits is very irregular and results in gravel thicknesses of up to 11 m. The longitudinal profile along the Misbourne Valley (Figure 51) shows that gravel and sand is continuously present beneath the floodplain and merges with that beneath the Colne at Denham. It also shows that these unusual gravel thicknesses occupy isolated deep hollows, as in the Alder Bourne Valley. Bedrock solution

Fig. 51. Long profile sections along the Misbourne (above) and Alder Bourne (lower) valleys showing their relationship to the deposits beneath the Lower Colne Valley (borehole sources: IGS, ERCU).

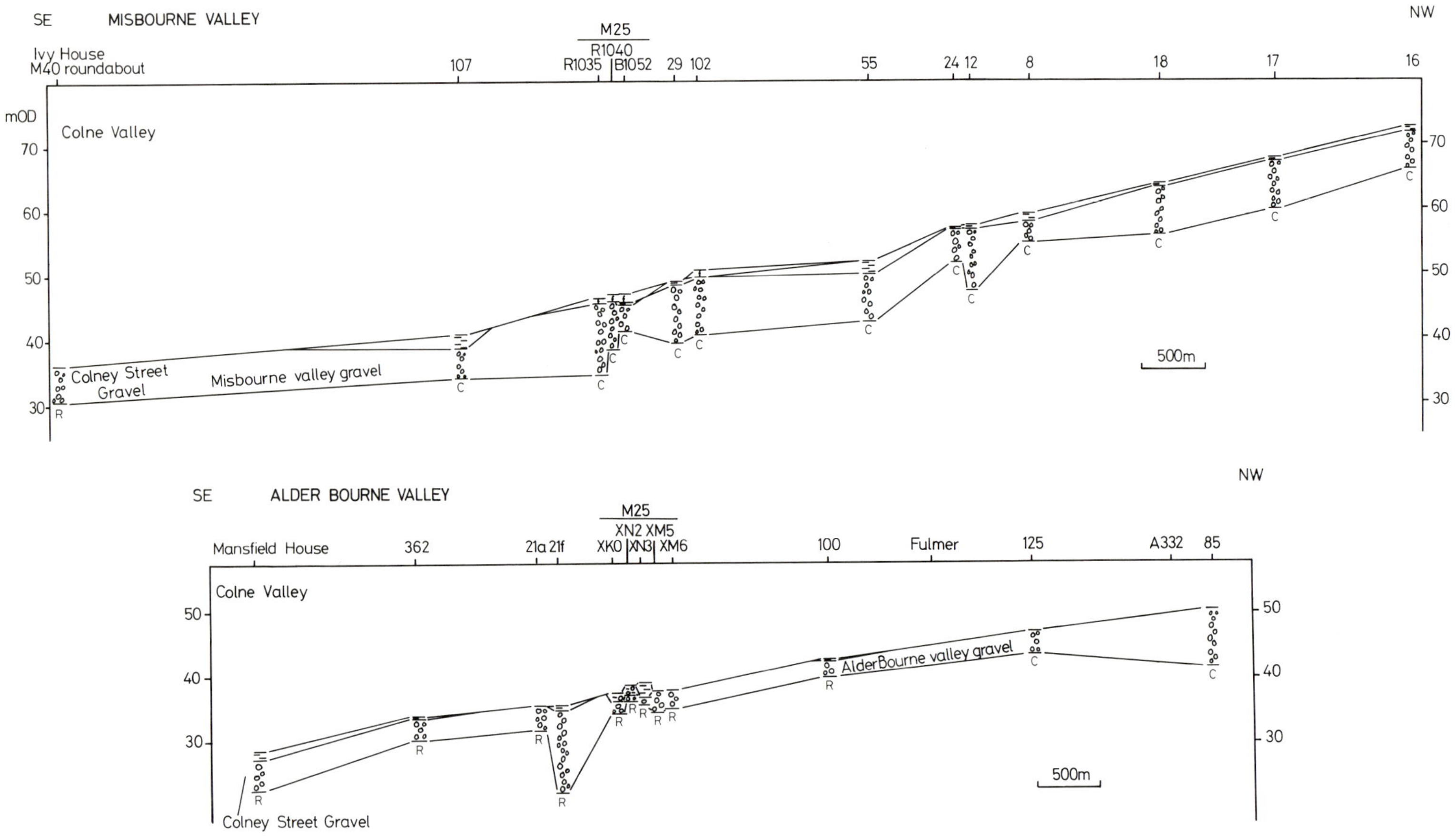

may be responsible for the overdeepening of the bedrock surface. If these hollows were formed during aggradation of deposits this would explain the extreme thickness of gravel and sand present in them. These hollows closely resemble those found beneath several of the members discussed in other sections (e.g. Sections 2.7, 2.10 and 2.17). They are also similar to those described by Berry (1979) and attributed to scour. The extreme depth (10.75 m) of brown silt overlying bedrock chalk (bh 1008/1) on the southern side of the Misbourne Valley (Figure 11) appears to be an isolated solution 'pipe' infill. This has been confirmed by field observation. Both the infill and the apparent form are comparable with those described from other parts of the area. However, the similarity of form and infill does not imply that all these features in the region are of the same age. On the contrary, they may be expected to have formed at many different times.

(g) *River Wey*

The River Wey meets the Thames at Weybridge (TQ074657), the northern part of its valley being also drained by a small stream, the Addlestone Bourne. The lower Wey Valley is aligned in a south–north direction between Wisley and the Thames Valley and is excavated into Bagshot and Bracklesham Beds. The deposits filling the valley have never been studied in detail, but were mapped by Dewey and Bromehead (1915).

In order to examine the deposits' distribution a section was constructed along the M25 Motorway from Ongar Place, Addlestone, to Telegraph Hill, Wisley (Figure 41). The section crosses both the Addlestone Bourne and River Wey valleys. The deposits appear to consist of two major gravel and sand accumulations, the upper unit called here the Byfleet Gravel is discussed above (Section 2.17). At present, the lower unit is separable from the upper, which it abuts, only by altitude and stratigraphical position, i.e. it underlies modern floodplain sediments. The lower unit fills a valley 625 m wide at Wisley excavated into the higher, Byfleet Gravel and bedrock. Although of irregular thickness, the gravel and sand appears to be 2.5–3 m thick on average. As yet there are no lithological studies of these deposits available. However, for ease of reference, the term Wisley Gravel is proposed for this lowest gravel unit of the Wey Valley (type section: M25: TQ061601–067970).

The downstream gradient of the present River Wey is 95–105 cm km^{-1}. Projection of this gradient the 6.25 km downstream from Wisley to Shepperton in the Thames Valley gives a height of 4.95 m (base) to 8 m OD (top) for the gravel at the confluence. This

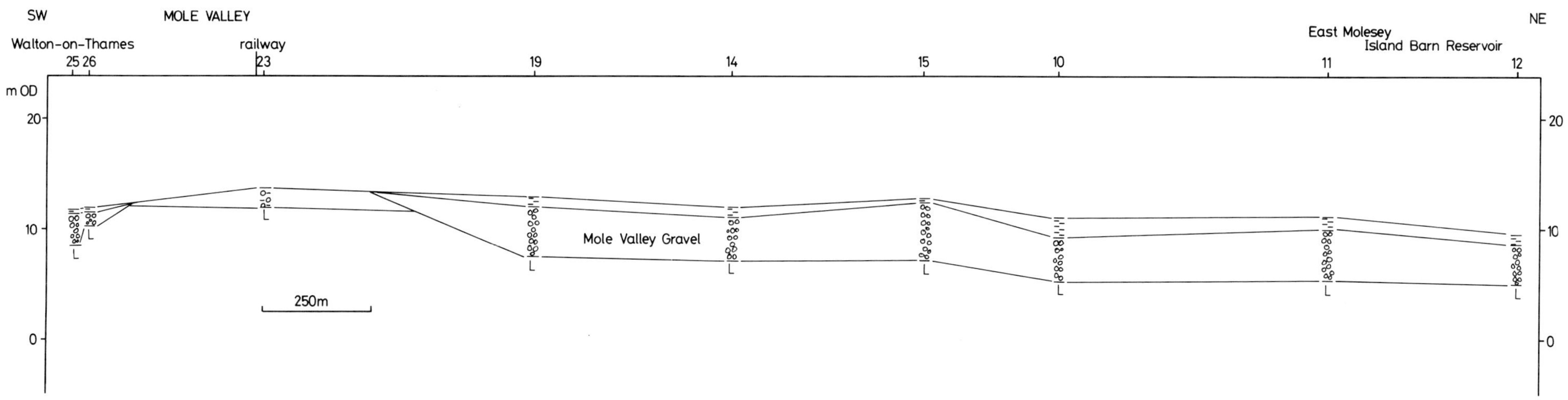

Fig. 52. Section showing the deposits beneath the Mole Valley near East Molesey, constructed from borehole records (sources: IGS, TWA).

falls precisely within the height range of the Shepperton Gravel in the area (Figure 44). Gravel in an exactly comparable position underlies the Addlestone Bourne Valley in the section and may also be the lateral equivalent of the Shepperton Gravel in the same way.

(h) River Mole

The River Mole is confluent with the Thames opposite Hampton Court Palace (TQ156682). The wide floodplain of this stream was described by Dewey and Bromehead (1921, p. 66) while the terrace morphology and deposits of the stream further up valley were studied by Bull and Wooldridge (1925) and Bull *et al.* (1934).

A section constructed from drillings in the Mole floodplain between Walton-on-Thames and Island Barn, East Molesey, is shown in Figure 52. A continuous spread of gravel and sand up to 5 m thick overlies London Clay and is covered by fine-grained alluvium beneath the floodplain in this area. The gravel accumulation has a downstream gradient of 115–120 cm km^{-1}. If this gradient is projected the 2 km downstream from Island Barn to the Thames, a height range of approximately 2.7 m (base) to 5.8 m OD (top) is obtained. This range coincides precisely with that of the Shepperton Gravel at Hampton Court (Figure 44). As in the other tributary valleys described, the gravel and sand flooring the Mole (Mole Valley Gravel) appears to represent a Mole-equivalent of the Shepperton Gravel.

(i) River Brent

The floodplain of the River Brent has been greatly modified by canalisation and dock construction at its confluence with the Thames. The borehole section across the Brent Valley along the M4 Motorway (Figure 17) shows that the valley is excavated into the Thames Lynch Hill and Kempton Park Gravels and London Clay. The valley is underlain by up to 3 m of gravel and sand resting on bedrock; floodplain organic and alluvial silts cover the gravel and sand. The gravel and sand occurs between 5 m (base) and 7.75 m OD (top). The gradient on the present Brent floodplain surface is 400 cm km^{-1}. Projection of the gradient from the M4 at Boston Manor to the Thames confluence at Brentford (TQ177772) gives a height range of −1 m (base) to +1.75 m OD (top), which falls precisely into the height range of the Shepperton Gravel (Figure 44). The Brent Valley Gravel may therefore be the lateral equivalent of the Thames Shepperton member. According to Lobley (1887), the Brent Valley Gravel comprises 93.7% flint, 3.65% quartz and 2.5% granitic rocks.

(j) River Wandle

The River Wandle joins the Thames at Wandsworth (TQ255753), where the ground has been greatly modified by dock basin and warehouse construction. The history of the River Wandle has recently been reviewed by Peake (1971, 1982) based on research upstream around Mitcham and Croydon.

The river has a well defined floodplain underlain by modern alluvium along much of its length. The deposits beneath have been described in detail by Peake (1971) from Mitcham from where she described gravel and sand containing lenses of organic sediment that have been radiocarbon-dated to the Late Devensian substage. Boreholes in the Wandle floodplain at Wandsworth (Figure 24) demonstrate that the floodplain fine alluvium and organic sediment is underlain by gravel and sand up to 2 m in thickness. As with the Brent, these gravels can be shown to grade into the nearby Thames Shepperton Gravel (Figure 44) as suggested by Peake (1971).

(k) Chelsea Creek

The borehole section along the proposed West Cross – Westway route from Battersea to Wandsworth follows the north bank of the Thames (Figure 23), and crosses the confluence of the Chelsea Creek (TQ265770). This area of wharfs and docks called Chelsea Basin is underlain in part by Thames Shepperton Gravel and overlying floodplain organic deposits. However, the north–south section along the old Chelsea Creek Valley shows that it is underlain by a shallow gravel and sand accumulation, in spite of much modification by man. The gravel and sand is overlain by fine alluvial sediment and appears to grade into the Shepperton Gravel as in other tributary valleys. The width of the unit is very difficult to determine in transverse sections, since the gravel appears to fill a valley excavated into Thames gravels beneath West Kensington and Baron's Court (Figure 38). Where the valley can be recognised the gravels appear to be generally 2.5–3 m thick but locally reach 5 m in thickness. Further upstream the Chelsea Creek Valley cannot be reliably identified.

2.20 Staines Alluvial Deposits

Alluvium is the term that has been applied to the predominantly fine-grained deposits of silt, mud, clay, peat, tufa, marl, sand and gravel that underlie the marsh and water meadow land fringing the rivers in southern Britain. These low-lying areas are liable to flooding when rivers overflow their banks, although human interference (locks, embankments, weirs, etc.) has reduced the frequency of such events.

The modern floodplain of the River Thames is variable, although frequently narrow in width as has already been mentioned above (Section 2.18). Whitaker (1889) in describing the Thames alluvial sequence immediately upstream from London noted that it is generally restricted to narrow, short strips; the river often flowing between gravel banks. In the Maidenhead area, he noticed that in many places gravel 'islands', completely surrounded by alluvium, protrude through the marsh ground. These islands have subsequently been recognised throughout the Middle Thames Valley and also in tributary valleys, e.g. the Colne, by many workers including White (1908), Dewey and Bromehead (1915, 1921), King and Oakley (1936), Hare (1947) and Lacaille (1961a) (see discussion, Section 2.19). The alluvial deposits appear to infill depressions or channels between these 'islands', but almost everywhere they rest on gravel and sand continuous with that forming the gravel shoals (Shepperton Gravel).

Internally the alluvial sediments show a considerable variability both vertically and horizontally. They frequently contain abundant plant fossils of all types from fruits to wood. Fossil Mollusca from both terrestrial and freshwater environments abound at many localities in all sediment types, but are particularly abundant in tufa and marl. Vertebrate remains have often been collected and include a very wide range of species such as elk, reindeer, roe deer, red deer, fallow deer, ox, horse, beaver and boar, and recent domesticated animals (Dewey and Bromehead, 1921, p. 65).

Archaeological remains are abundant from the floodplain alluvium. The earliest are of Mesolithic age and have been found in considerable numbers from the modern river channel throughout the area (Lawrence, 1930; Wymer, 1977a). Both authors have suggested that this proves the Thames has followed its present course since that time. Mesolithic artefacts have also been found in the basal part of the alluvial sediments and on the surface of the gravels beneath (Lacaille, 1961a, 1963). Later discoveries of Neolithic, Bronze and Iron Age, Roman and post-Roman material are also represented both in the sediments and the modern Thames channel (Dewey and Bromehead, 1921; Lawrence, 1930; Wymer, 1977a).

Work on the alluvial sequence both in the Thames and in its tributary valleys has been very limited. The most important contributions are those by Kennard and Woodward (1906), Cooper (1907, 1913) and Dewey and Bromehead (1915, 1921). More recent studies by Lacaille (1963) in the Colne, and the outstanding work by Holyoak (1980) in the Kennet and Preece and Robinson (1982) in the Thames Valley, illustrate the potential in a study of this sequence.

Although the alluvial sediments of the Thames drainage system have not been studied in detail by the writer, the deposits have frequently been observed, and recorded in borehole sections. For the sake of continuity and uniformity, therefore, generalised descriptions of the sediments together with their mode of occurrence are presented here. Everywhere they have been examined they form a distinct aggradational unit which will be referred to here as the Staines Alluvial Deposits. All available evidence indicates that these sediments are exclusively of Flandrian (post-glacial) age and that accumulation would be continuing today if the rivers were not controlled.

The type section for this member is near Mixnam's Ferry (TQ042685: 14 m OD), 3 km south of Staines. It was recently described by Preece and Robinson (1982). Here a channel 2.85 m deep, in the underlying Shepperton Gravel, was filled with fossiliferous sediment. The sequence was as follows:

	Ground surface
0–70 cm	brown silty clay, mottled in places;
70–120 cm	grey, well sorted, current-bedded sand;
120–165 cm	dark grey organic-rich silts with wood fragments, hazel (*Corylus*) nuts, etc.;
165–205 cm	shelly sand with several distinct organic seams up to 3 cm thick;
205–210 cm	gravel;
210–265 cm	shelly silts and sands, some plant debris;
265–275 cm	orange gravel and sand;
275–285 cm	dark grey organic-rich silts with wood fragments.
	Orange gravel and sand

The base of the channel was a sharply defined non-sequence. The five prominent irregular seams of plant debris, each about 3 cm thick, in the silts below 165 cm suggest periodic flooding. A substantial increase in flow velocity was indicated by the current-bedded sand bed (70-120 cm) that rested on an erosional surface of the silts below. Above the sand, the clay (0–70 cm) rich in swamp and river-side Mollusca was thought to represent overbank deposition.

Detailed investigations of the faunal remains by Preece and Robinson (1982) indicated that the silts and sands represent an accumulation formed under shallow-water, fully temperate conditions. Two samples of wood from the channel sediments were radiocarbon-dated as follows: 135–145 cm: 665 ± 55 BP (Q-2043) and 275–285 cm: 8360 ± 100 BP (Q-2042). The channel fill is therefore of early Flandrian age.

The alluvium occupies a wide area between Staines and Chertsey drained by several small streams. Several other fossiliferous alluvial deposits have been reported from the area. Exposures close to the type section within the Penton Hook meander (TQ043692) were described by Cooper (1922) and more recently by Howard (1952). The latter author examined a channel 1.8 m deep and 23.3 m wide in the Shepperton Gravel filled with light grey silt rich in Mollusca, near the base of which were bones and Iron Age sherds. Beneath was a bed of shelly peat. A smaller channel about 15 m wide had cut through the grey silts and been filled with 'a darker clay deposit' (Howard, 1952, p. 262). At nearby Hythe (TQ032703: 13 m OD) large flooded sections exposed over 1.5 m of alluvial sediments in the south-east corner of the pit. Although the base was not visible, the lowest sediments (of unknown thickness) comprised black detritus mud rich in mollusc shells. This was overlain conformably by grey silty marl, about 1 m in thickness, with abundant plant fragments and Mollusca. The calcareous content of the sediment diminished upwards as the deposit graded into shelly grey silty clay with occasional fine (2–3 cm) black, plant-rich bands. The upper 70 cm comprised a mottled brown to grey silty clay that lacked fossils, but rested conformably on the silty clay beneath. Lateral variation of the deposits was apparent with individual horizons lensing out, as the underlying gravel rose closer to the surface. Where the gravel was only 30–50 cm from the present surface it was covered only by brown clay.

The neighbouring sections along the M25 Motorway from Staines to Thorpe (Figures 11 and 45) illustrate the variation in thickness of the deposits. The gravel surface highs are covered by a thin veneer of brown silty clay often with a few stones, (e.g. bh 115), while in the deepest hollows as much as 2 m of sediment, frequently including wood peat or detritus mud, occur (e.g. bh 120). This section also extends into Chertsey Bourne Valley where similar deposits now underlie the floodplain.

A closely comparable sequence is present in the M3 section which crosses the Thames floodplain at Chertsey (Figure 40). In this area the alluvial deposits clearly thicken towards the present river channels to reach a maximum of 3 m near the Thames, but also near the Abbey River which occupies an almost-filled channel. In deeper hollows in the gravel surface dark grey shelly silts and clays are present. In places, these are overlain by peat and in turn capped by the widespread brown silty clay with occasional stones that again covers all but the highest gravel surface shoals. The two quarry sections at Twynersh, Chertsey (TQ032676 and 037676: 14.5 m OD) exposed up to 1 m of shelly brown clayey silt resting directly on the undulating gravel surface. However, nearby, the centre of Chertsey is built on a gravel island (Dewey and Bromehead, 1915).

Downstream of Chertsey the floodplain narrows and the alluvial spread is about 750 m wide immediately east of the Wey confluence. At Shepperton quarry a maximum of 1.75 m of light brown silt rested with a sharply defined base on the type Shepperton Gravel. A borehole at Kenyon Court, near Kempton Park (TQ119694: 11 m OD) proved the same thickness of brown silty clay with plant remains. At this locality the alluvial spread is restricted to a width of about 100–150 m as far as Hampton Court. At Hampton (TQ135693: 10 m OD), Whitaker (1889) recorded a sequence of dark sandy clay, rich in shells and plant fossils and up to 1.2 m thick, resting on sandy gravel. The clay was overlain by green sandy clay with freshwater molluscs and brown clay up to 1 m thick. A Mesolithic artefact was also found from the lowest organic sediment at this site (Wymer, 1977a). Downstream of Hampton the floodplain widens considerably at the Mole confluence where the Mole and the smaller Ember stream join the Thames.

In the Hampton–Kingston section (Figure 42) the alluvial sediments comprise 90 cm of grey silty clay with plant remains (bh 24). They continue to occupy only a very narrow strip each side of the river downstream as far as Chiswick. The alluvial sequence beneath Kingston has been summarised by Penn and Rolls (1981). Beneath Ham Fields (TQ165726: 8 m OD) about 1 m of sandy loam occurs and has yielded Mesolithic material (Lacaille, 1936). Lawrence (1930) has also described microliths from Eel Pie Island in the Thames between Twickenham and Ham (TQ164731).

The deposits are well developed at Kew Bridge (Figure 42). Here a maximum thickness of 2.75 m of grey to brown silty clay with shells, plant remains and wood are present (bh 53). According to Lacaille (1961a) Kew Bridge has been a major site for Mesolithic discoveries almost all dredged from the modern river channel. In the Chiswick meander, boreholes at TQ212767 and 213773 (4 m OD) show the deposits comprise 2.4 m of organic silty clay at the former and 2.75 m of yellow clay at the latter.

In the Hammersmith Sewer section (Figure 38) there is some difficulty in demonstrating conclusively the relationship of the alluvial deposits to the older sediment using the borehole records alone. The most probable situation is shown in the diagram, but this may need to be modified when exposures become available. The alluvial deposits were described as brown sandy clay in the records. Dewey and Bromehead (1921, p. 66) remarked on the considerable extent of the alluvial deposits in the Barn Elms area (TQ230772). However, within the Fulham meander the alluvium is poorly represented, the river flowing between gravel banks. The alluvium recurs at Waltham Green (Figure 37), where it comprises 70 cm of brown clay with some pebbles, overlying 60 cm of dark brown organic silty clay resting on gravel and sand.

The variability in thickness of the alluvial sequence and its relationship to the modern channel can best be seen in the West Cross Route sections (Figure 23). Near the river channel the alluvial sediments are thickest reaching a maximum of about 4.5 m and rest on an undulating gravel surface with an amplitude of at least 3 m. Immediately north of Wandsworth Bridge they consist of 3 m of dark grey silty clay with shells and plant remains resting on gravel and overlain by 1.5 m of grey sandy clay with occasional stones. A similar thickening of the deposits is present at Chelsea Basin where Chelsea Creek meets the Thames (Figure 23). Here again they comprise grey to black organic silty clay, almost 4 m thick. The sediments clearly extend beneath both banks of the river, but are certainly best-developed near the mouth of the tributary. A comparable situation is seen at the mouth of the Wandle (Figure 24).

Mention has already been made of the extensive complex of gravel and sand-filled hollows in the bedrock surface in the Lambeth–Battersea area (Section 2.18). South of Battersea Park a topographic low filled with fine-grained organic alluvium (e.g. at TQ287762), that appears to represent the former course of the River Falcon from Balham, trends towards the deep hollows at Battersea Power Station and Gas Works (TQ290775–296773) studied by Berry (1979). The hollows are overlain by as much as 5 m of 'fine grained silty clays and sands with thin clay partings' (Berry, 1979). Close by, boreholes for a gas holder (TQ293779: 2 m OD) proved 3.1 m of grey organic clay, resting on sand and gravel, and including a bed of black peat (1.4 m thick) 0.45 m above the base.

On the north side of the river, alluvial sediments are well developed from Chelsea Bridge beneath the Pimlico and Westminster areas associated with the mouths of the Westbourne and Tybourne streams. The gravel island, formerly called Thorney Island, now the site of the Palace of Westminster, is surrounded by alluvium (Dewey and Bromehead, 1921). A former channel aligned along Victoria Street rejoining the present river immediately north of Westminster Bridge is now filled by these deposits. Numerous boreholes are available from this area; for example at Victoria Station (TQ289790) up to 4.5 m of organic clay and peat are present (cf. also Dewey and Bromehead, 1921), while beneath Pimlico (TQ295786) only 3 m of similar sediments occur.

Beneath Whitehall Place (TQ302802) another of the hollows is filled with 4–5 m of well stratified, fine-grained alluvium, the base of which is at −7 to −8 m OD (Berry, 1979). The transverse sections across the Thames at Westminster (Figures 20 and 43) demonstrate the continuity of these sediments on both banks of the river. They are well represented under the Waterloo – South Bank – north Southwark area and reach a maximum of 4.9 m beside County Hall (TQ308799). However, they again thin southwards away from the modern channel of the river. At all localities the modern alluvium rests on the Shepperton Gravels. The gravels again have the complex surface topography of channels and shoals with an amplitude of about 3 m.

West of Staines, Kennard and Woodward (1906) described sections 'a mile west of [the town] on the tow path to Old Windsor' (*c.*TQ016721), which exposed 60–90 cm of calcareous loam rich in shells and plant remains, overlain by 30 cm of shell marl and 90 cm of brown clayey silt. The section they described was capped by 75 cm of coarse gravel. However, to judge from the sequences observed elsewhere and summarised here the last-mentioned material was probably made ground. This section occurred only 400 m upstream from the M25 road bridge. The section along this motorway (Figure 41), has already been discussed in part above, however between Egham and Stanwell a wide spread of alluvial sediments can be found. At Glanty near Egham (bh 102), 3.8 m of brownish-yellow to grey sandy clay and silty clay rests on the Shepperton Gravel, while immediately north of the river 3.0 m of brown to grey sandy clay is present. The deposits thin markedly northwards. As elsewhere the gravel surface topography is preserved beneath the alluvial sediments, so that the latter are restricted to depressions or old channels. Where the deposits are very thin or absent, for example around Wraysbury, the upper 1–2 m of the gravels and sands often have a clay-rich matrix. The M25 section (Figure 11) shows that similar deposits continue beneath the River Colne floodplain where they have not been removed for gravel exploitation (cf. Lacaille, 1961b).

At Home Park, Windsor (*c.*SU983752) 1.5 m of 'sandy loam' equivalent to this unit was observed by Whitaker (1860). Boreholes for the nearby Albert Bridge (SU985757) proved 0.4 m of peat resting on gravel, succeeded by 0.4 m of blue clay with shells and 0.75 m of brown silty clay. The section from the M4 to Datchet (Figure 46) shows that contrary to the geological map (sheet 269) the gravels are covered by alluvial deposits which again comprise a basal peat, 0.75 m thick, above which is mottled brown to grey clay 1.7 m thick in this area.

North of Windsor the alluvial aggradation is well illustrated by the A332 road section (Figure 25). Here the sediments again fill lows in the gravel surface. They are thickest (3.5 m) nearest the present channel and comprise a basal 2 m of grey silty clay with plant material above which are 1.5 m of grey brown sandy clay with occasional pebbles (bh 193). A possible channel (bh 203) is infilled by 3.4 m of brown sandy clay. The same type of sediment seems to form a mantle throughout the section. It is interesting to note the presence of gravel 'islands', so often mentioned, that lack an alluvial cover. The section also crosses Windsor Eyot which is bounded on the south by the Thames and the north by a small stream. It is worth noting that the eyot is not a gravel surface feature but simply an island of alluvium (cf. Berry, 1979). Beneath Windsor itself the alluvium normally rests on gravel for example at the Imperial Service College (TQ961764). It is generally a brown to grey sandy clay about 60 cm thick. However, at the extreme southern corner of Victoria Barracks (TQ967764), 1.2 m of alluvial clay rests on Reading Beds.

Between Bray and Eton the floodplain is about 3.5 km wide, the alluvial deposits being present as isolated channel fills; the intervening gravel ridges have either a thin sandy clay veneer or lack any cover (Figure 25). The large numbers of boreholes available in the Dorney–Bray area support these general statements and will not be elaborated further here. Many sections in the alluvium of the area have also been observed by Dewey and Bromehead (1915) and the author. The former authors particularly noted those at Boveney Lock (SU945778) and Amerden House on the north bank (SU911797). The latter section in the river bank exposed brown sandy loam that yielded bones of horse, ox, sheep, pig and dog, together with many Mollusca and part of a bronze sword-scabbard (now in the British Museum).

Extensive gravel workings at Bray (SU909786) exposed sediments equivalent to this unit resting on Shepperton Gravel. They consisted of a basal light grey sandy silt with pebbles (10 cm thick) above which was a black silty clay (10 cm), capped by brown clayey silt. As elsewhere, the upper surface of the gravel undulated and was apparently unmodified by periglacial activity or channelling before deposition of the silt. Nearby at Braywick (SU894793) Mesolithic artefacts were discovered in the brown loam during construction of the M4 Motorway.

Upstream from Bray the floodplain is of generally uniform width (1.5–2.5 km) and continues as above. Sections in the large working quarry at Furze Platt (SU895828) confirmed that no alluvial cover was present in the sections. In the channels on either side of the gravel 'islands' in this reach, up to 2.0 m of yellow brown to grey silty clay are present, e.g. at Widbrook Common, Cookham (SU889837). Cookham village is built on a gravel island. A comparable situation is found in the Marlow – Bourne End district as far as Hurley. Boreholes within the alluvial channels in this area normally penetrate brown mottled grey sandy clay 1–2 m in thickness resting on gravel. However, close to the river at Hurley (SU815839) a borehole penetrated 2.0 m of alluvium comprising brown silt at the surface becoming blue–grey silt and clay with shell fragments at depth. Near Medmenham, on the north bank, the floodplain remains about 1.5 km wide. Here the alluvium is grey brown silty clay with organic layers, e.g. at Manor House (SU803842) where it is 1.4 m thick (cf. White, 1908). White (1908) also saw comparable material beneath the meadows south of Greenlands (SU774854) and north of Aston (SU785845). Stratified clayey silt with pebbly beds is present at Hambleden Lock (SU782852).

From Henley to Wargrave the deposits are restricted to a narrow spread near the river and infills of depressions in the underlying gravel, e.g. at Lower Bolney Farm (SU772806). At Wargrave itself (*c.*SU783788), White (in Kennard and Woodward, 1906) found loam with molluscan shell-rich seams 60 cm below the meadow surface. Here as elsewhere gravel underlies the loam, as shown by local boreholes (e.g. SU780790). Beyond Wargrave, the gravels are apparently everywhere buried by alluvial clayey silts varying in thickness from a few centimetres to as much as 2 m. At the Warren, Shiplake (SU766777), 3 m of brown to blue grey silt with some peat is present.

The continuation of the alluvium into the Loddon Valley can be seen in Figure 48. At the Whistley Mill, Twyford quarry (SU788745) the sediment comprised poorly bedded clayey silt with pebbles, 0.5–1 m thick, grey at the base becoming brown in the upper 40 cm.

In the Thames Valley at Sonning (SU745754) and Caversham (SU734750), gravel workings exposed the same type of sequence. At both sites the undulating gravel surface from which the alluvium has been removed clearly showed a series of channels and bars. This surface was previously buried beneath 0.5 m to 1.5 m of brown partially decalcified alluvial silt. 'Brownish to yellowish loam' was also described from the area north of Sonning by Blake (1903). A borehole at nearby Long Ait on the south bank of the Thames (SU760767) proved as much as 3.1 m of alluvial sediments that comprised 1.1 m of light brown to greenish grey and black silty clay overlain by 1.7 m of light brown silty clay, both horizons rich in Mollusca.

The distribution of alluvium at Reading is shown in Figure 34. In this section borehole 6 recorded only 0.2 m of brown sandy clay, while borehole 49 penetrated 1.35 m of blue grey clay with shells. In both this and Figure 48, the Kennet alluvium flooring the Coley Gap is present and comprises 1.2 m of black organic silt.

Between Reading and Pangbourne the alluvial sediments cover almost the whole floodplain, but are only of significant thickness (2–3.5 m) in depressions between highs in the gravel beneath, the latter covered by about 1 m of 'dark brown silty sandy clay with scattered flints', for example at Home Farm, Purley (SU661769). In contrast, at Mapledurham (SU686747) a channel fill comprises 0.4 m of 'dark grey-brown "peat" with shells', overlain by 2.0 m of grey-green to mottled brown clayey silt with scattered pebbles and 0.3 m of brown silty clay.

The section across the Thames at Whitchurch and Pangbourne (Figure 47) demonstrates the relationship of the alluvial deposits to the underlying gravels in the area. It also indicates that the alluvial sediments continue into the Pang Valley, where they comprise brown silty clay and peat (bh 49,50). North-west of Pangbourne the river enters the 500–750 m wide stretch of valley leading to Goring and Wallingford. Here the alluvial deposits continue to floor the valley, but are restricted to the ground immediately adjacent to the modern channel. For example, at Basildon Park (SU622774) 1.1 m of dark brown organic mud with scattered shells underlies 2.6 m of silty clay, green to blue green at the base becoming mottled brown and grey and ultimately dark brown in colour. As downstream the higher gravel areas are capped by only a thin bed, about 0.5 m of brown sandy clay, e.g. at Lower Basildon (SU675786: SU67NW6).

As has already been suggested the tributary valleys contain alluvial aggradations whose mode of occurrence is exactly similar to that of the main river valley.

The deposits in the Kennet Valley have been thoroughly studied by Holyoak (1980) and show a broadly similar sequence to those described from the Thames. As will be evident from the discussion of tributary valley floodplain gravel spreads (Section 2.19), alluvial deposits continuous or probably continuous with those in the Thames Valley are present in the Pang, Loddon, Wey, Mole, Brent, Wandle (Peake, 1982) and Colne valleys. The last-mentioned should be regarded as the most important, after the Kennet, since the deposits have been studied in the greatest detail.

The Colne alluvium has been summarised by Lacaille (1961a, 1963). In the second publication he described at length the stratigraphy and contained artefacts of two sites, the first, Sandstone pit, north of the Uxbridge–Slough road at Iver (TQ044836: 30 m OD) and the second Boyer's pit, Denham (TQ054855). Both sites are now infilled. The sequence at both showed the main characteristics of the Colne alluvial sediments. At Sandstone pit, the deposits comprised:

0–0.76 m	Made ground;
0.76–1.00 m	peaty clay;
1.00–1.45 m	grey calcareous mud with mollusc shells;
1.45–2.30 m	peat (compressed, clayey and decayed; containing pieces of tree (mainly birch, hazel and pine) and also a tine of red deer);
	gravel and sand with artefacts on top surface.

The Mesolithic artefacts described by Lacaille (1963) were found in the lower part of the peat and on the gravel surface immediately beneath. Both pollen and molluscan analyses were also undertaken at this site. The former indicated a late Boreal age for the peat.

At Boyer's pit, the sequence was similar, with the alluvial deposits being banked against a gravel high. Here brown clayey silt covered a bed of compressed peat of variable thickness. Beneath the peat a lenticular bed of 'shell-marl and *Chara* mud' occurred which was present only in a channel adjacent to the gravel ridge. A rich molluscan fauna was obtained by Kennard from the marl, but was originally described by Howe and Skeats (1903) when it was first exposed. Mesolithic artefacts were again found beneath the peat on the gravel surface at this pit. At additional sites at Dewe's pit, Harefield (TQ054844) and near Dewe's Farm (TQ055879) peat seems to have underlain much of the surface, being particularly thick at the former site (Lacaille 1961a, p. 118).

Because of extensive gravel exploitation the section along the M25 from the Thames confluence (Figure 11) does not demonstrate the continuity of the floodplain alluvium mentioned by both Dewey and Bromehead (1915) and Lacaille (1961a). In the borehole section, the alluvial sediments are not continually present south of the Harmondsworth area. Isolated patches are present, for example, beneath the Airport at Perry Oaks (TQ052747) near the River Colne. Here 1.7 m of dark brown organic silty clay (bh 693) is present. Beside the Wraysbury River at Harmondsworth (bh 800) 2.2 m of grey mottled brown silty organic clay with shells near the base was proven.

Alluvial deposits have been observed by the writer at several localities in the Colne Valley. At Colnbrook (TQ026776), the gravel was overlain by 0.2 m of brown silty clay (Gibbard and Hall, 1982). Several sections in old workings near the M4 Motorway exposed these sediments; for example, immediately west of the Staines–Yiewsley railway line (TQ046783) 10–15 cm of chalky shell marl rested on gravel and was overlain by 40–60 cm of peat rich in plant remains. In sections at West Drayton (TQ053792) 20 cm of black detritus mud lay on gravel and was covered by mottled grey to brown silty clay.

The irregularity of the distribution of these sediments across the Colne Valley is shown in Figure 17. In this section the greatest thickness of alluvium is concentrated on the western side of the valley. However, from descriptions by Lacaille (1963) it is unlikely that this situation typifies the whole Rickmansworth–Stanwell reach.

The M25 section along the Colne Valley (Figures 11 and 50) proves as much as 2.6 m of brown sandy silty clay with up to 2.6 m of brown sandy silty clay with organic inclusions immediately west of Thorney (TQ042794). These deposits are also intermittently recorded between Iver and Mansfield House. Near the former the section intersects a partially infilled channel of the Colnbrook. The deposits here comprised:

0–2.10 m	Brown silty clay with plant remains;
2.10–2.70 m	grey silty clay with Mollusca;
2.70–3.00 m	black 'peaty' clay;
	gravel and sand.

Adjacent to Mansfield House another complex of channel fill deposits was encountered adjacent to the valley side. The deposits consisted of as much as 2.3 m of grey silty clay with shells and plant remains, becoming grey in the upper 0.5 m.

The borehole and exposure data therefore confirm Lacaille's (1961a, 1963) observations from Iver, Denham and Harefield, and also descriptions from the Upper Colne Valley in Hertfordshire (e.g. Gibbard, 1974).

3

Analytical comparison of the pebble lithological composition of the gravel members

In the descriptions presented above the pebble lithology of each gravel member and the origins of the material have been discussed. However, if the lithostratigraphical scheme is valid it is necessary to demonstrate as far as possible that the lithological assemblages are sufficiently characteristic to identify the individual members. Such a comparison also provides insight into both local and regional catchment changes that may perhaps be present, together with other possible temporal patterns of lithological change.

In order to express and compare multivariate lithological compositional data from the gravel members (see Appendix) as simply as possible, the analyses were compared numerically using the multivariate statistical techniques Canonical Variates Analysis (CANVAR) and Principal Components Analysis (PCA). Of the techniques available for the comparison of groups of samples consisting of an assemblage of variables, the most useful are those that ordinate all the samples together. Ordination procedures such as CANVAR and PCA seek to represent in a low-dimensional space the similarities defined in a mathematical sense between sample group means or the individual samples. In both techniques, group means or samples of similar composition, irrespective of which group they are derived from, will, if the ordination is an accurate representation of the data, be positioned together in the ordination (Blackith and Reyment, 1971; Davis, 1973; Gordon and Birks, 1974). The geometrical disposition of group means and samples will thus allow the detection of clusters of similar lithological composition that may originate from the same lithostratigraphical unit and of groups of dissimilar lithological composition that may originate from different units.

3.1 Canonical Variates Analysis (CANVAR)

This technique calculates mean compositions for predetermined groups and uses these to generate transformed axes, the first in the direction of the greatest variability of the group means, the second axis perpendicular to the first and inclined in the direction of the next greatest variability and so on. In this analysis individual samples are then plotted on the axes thus generated so that the compositional variability and validity of the predetermined groups can be assessed (Blackith and Reyment, 1971).

To examine the compositional variability of the gravel members defined in the field, a CANVAR of chord distance using square root transformation and co-variance matrix was used. The samples of the Westland Green and Stoke Row Gravels were not included in the main calculations, but entered as unknowns because the variability of these units could not be assessed. Samples from the Kempton Park and Shepperton Gravels were combined as the Floodplain gravel because of the small number of samples available from each unit.

In this ordination (Figure 53) the first canonical variate axis explains 63.89% of the variance. It shows high positive scores (>0.4) for samples rich in angular flint, and high negative scores (<−8.8) for those rich in rounded flint. The second canonical variate axis explains 15.88% of the variance and has high positive scores (>2.9) for samples rich in rounded flint and high negative scores (<−1.4) for samples with high quartz, quartzite and erratic rock frequencies.

The plots obtained indicate that the predetermined members can be separated using their mean compositions (Figure 53a). When all the samples are plotted (Figure 53b) there is a conspicuous overlap between member group distributions. The plots indicate that the gravel members fall broadly into a three-fold division. The main points arising from the analysis are as follows:

1. There is a generalised trend indicated by the line joining the plotted means (Figure 53a) from gravel rich in rounded flint with low quartz and erratic rock frequencies, to those with high quartz and erratic contents and progressively reduced rounded flint, to those which almost totally comprise angular flint.

2. The Nettlebed Gravel plots distinctly separately from the lower members.

3. The members and their samples rich in quartz and other erratic rock types (Stoke Row Gravel, Westland Green Gravel, Satwell Gravel, Beaconsfield Gravel, Gerrards Cross Gravel) also

Fig. 53. Plots of Canonical Variates Analysis of the Middle Thames Valley pebble count data: a. group means only with trend shown by line; b. all samples. For explanation see text. The Kempton Park and Shepperton Gravel samples have been grouped as Floodplain gravels because of the relatively small number of samples.

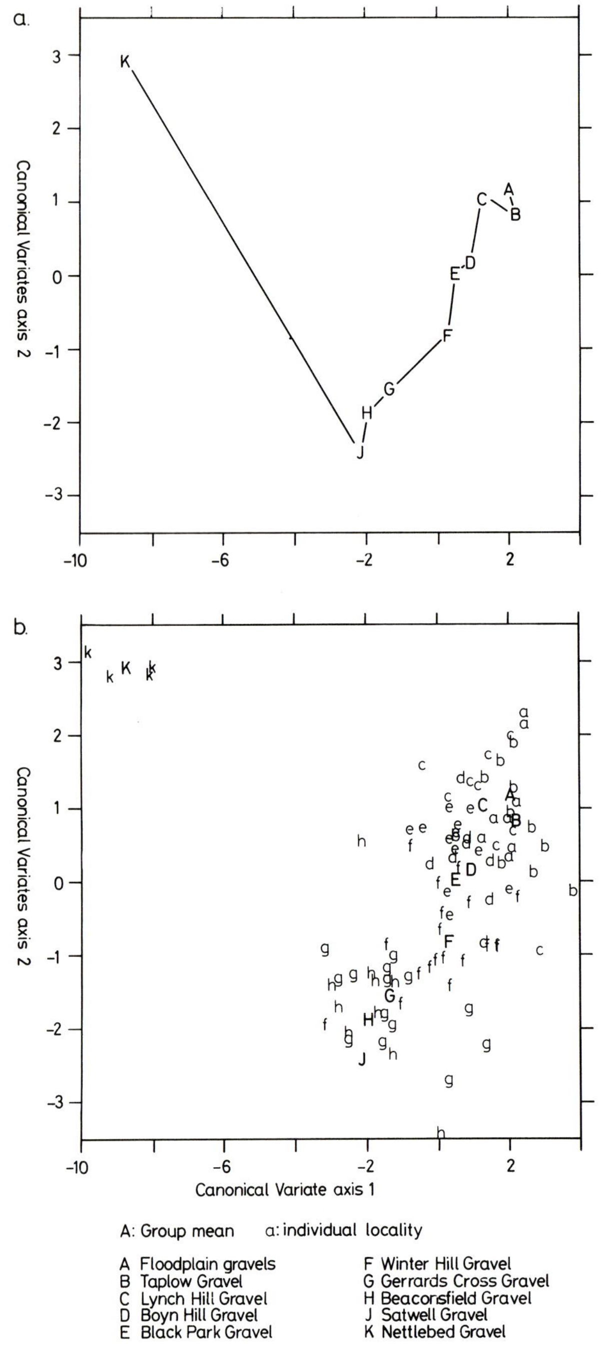

Fig. 54. Principal Components Analysis of the Middle Thames pebble count data. For explanation see text.

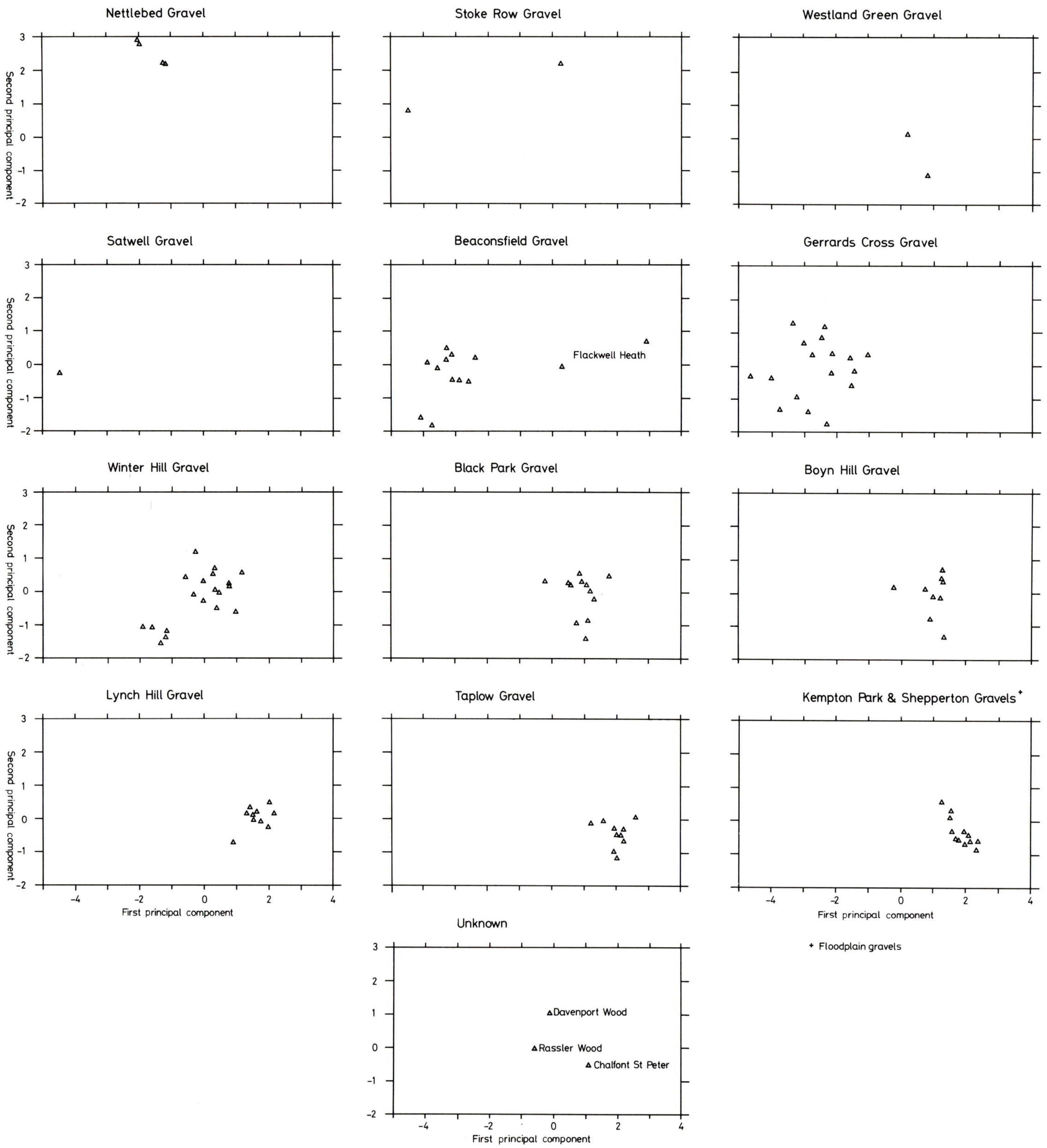

plot as a distinct association. An underlying trend towards decreasing content of rounded flint and dilution of the quartz and erratic rocks by addition of angular flint at successive stages is apparent from the alignment of means towards the right of the diagram.

4. The member means and their samples rich in angular flint form the third association and all plot close together at the far right. That minor differences are indeed present is, however, highlighted by their areal separation. The gradual diminution of the quartz and quartzite frequency is borne out by the trend shown in Figure 53a (see discussion below).

5. The transitional composition of the Winter Hill Gravel is clearly indicated, both by the mean and the individual samples that bridge the gap between the quartz and erratic-rich, and flint-rich associations.

3.2 Principal Components Analysis

In this analysis, a conventional R-mode PCA of chord distance using a co-variance matrix with square root transformation was used to compare samples on the basis of all lithological types present. The results of the analysis are shown in Figure 54. Although all the samples could be positioned on a single plot, for the sake of clarity samples have been separated into plots for each member. The positions of individual samples, as represented by their component scores, are plotted on the first and second principal component axes. These axes remain constant for all members.

In this ordination the first principal component accounts for 70.63% of the total variance. It effectively distinguishes between samples containing significant amounts of angular flint (high positive scores >0.5) and to a lesser extent Greensand chert, and those containing rounded flint, quartz, quartzite and minor erratic rocks (high negative scores). The second principal component represents a further 12.55% of the variance and has high positive loadings for rounded flint (>0.84) and Greensand chert, and negative loadings for angular flint, quartz, quartzite and erratic materials. The patterns on the third and remaining principal components were not interpretable, except possibly as arising from local variability.

The numerical results of the PCA show a generally consistent grouping of samples. It is clearly possible to discern several patterns of variability and sample grouping. They may be summarised as follows:

1. The Nettlebed Gravel samples rich in rounded flint and relatively low frequencies of quartz and quartzite plot completely separately from all other samples, as in the CANVAR plots.

2. There is again a marked separation of the samples rich in quartz, quartzite and rounded flint from those rich in angular flint and poor in quartz and erratic lithologies. Indeed the entire assemblage, with the exception of the Nettlebed Gravel can be divided into these two associations.

3. Within the quartz and erratic-rich group, apart from the relatively wide range in composition, no doubt arising from cross- and down-valley variations, there is a slight trend towards progressive fall in total quartz and erratic rock frequency from the Stoke Row to the Gerrards Cross stages. This was again also reflected in the CANVAR plots. The pebble lithological assemblage of these gravels has also been studied in the western part of the area by Green and McGregor (1978) and McGregor and Green (1978). However, these authors did not find this trend; instead, they found the maximum enrichment in quartz and erratic rocks to be in the Gerrards Cross Gravel. This contradiction may arise from the different grain-size range used by these authors (33–11.2 mm). They also found no evidence for down-valley effects in their results. While this is to some extent true for the Beaconsfield Gravel, there is certainly a down-valley trend towards increasing enrichment of angular flint accompanied by a decreasing content of quartz and erratic rocks downstream in the Gerrards Cross Gravel. The limited size of the sample area investigated by Green and McGregor (1978) may account for their not detecting this effect.

4. As mentioned above, the Winter Hill Gravel samples form an intermediate grouping between those rich in quartz and those rich in angular flint. This assemblage is characteristic since it is only present within the Winter Hill member. Again there is an enrichment in angular flint and a proportional decrease in quartz and erratic rocks downstream. This trend is continued downstream in the western part of the Vale of St Albans in the lateral correlative of the Winter Hill Gravel, the Westmill Gravel (Gibbard, 1974, 1977 and Section 2.8). There are, however, several Winter Hill samples with compositions similar to those of the preceding Gerrards Cross Gravel, all of which were obtained from the Caversham Channel area between Reading and Henley. In this area the river was restricted to a narrow course through the Chalk. Reworking of older gravel must have been inevitable

here and will have locally enriched the Winter Hill Gravel with quartz and erratic pebbles. Such local enrichment will have undoubtedly given rise to cross-valley as well as down-valley compositional variation.

5. At each of the Beaconsfield, Gerrards Cross and Winter Hill stages the gravels repeatedly show a distinct separation into two sample groups. This separation consistently reflects the two main areas of outcrop, i.e. the Caversham Channel (Reading–Henley) area and the Bourne End – Watford area, the latter being richer in angular flint and poorer in exotic lithologies than the former. The general absence of gravel remnants of these members between Marlow and Henley presumably accounts for this feature. However, the River Kennet is known to have been confluent with the Thames at Henley during these stages and will have certainly introduced additional flint (Gibbard, 1982). It is interesting to note that this regionality is not found in the compositions of the lower gravel members.

6. The post-Winter Hill members are all rich in angular flint. Down-valley changes appear to be quite small except where tributaries enter the valley and therefore on the basis of total assemblage there is little gross variation in composition. However, a close inspection reveals a progressive trend from the Black Park to the Shepperton Gravel of decreasing quartz, quartzite and erratic content accompanied by a proportional increase in angular flint, again recorded by CANVAR. Indeed at each successive stage the gravel assemblage seems to become increasingly uniform.

Using both techniques, CANVAR and PCA, additional unknown samples can be incorporated to test their affinities to particular groups or compositions. An example is shown in Figure 55 in which samples from the type locality of the 'Rassler Terrace' of Sealy and Sealy (1956) from Rassler and Davenport Woods were plotted. The similar positioning of the Rassler sample to those of the Winter Hill Gravel indicates that part of the spread is equivalent to that unit, in contrast to that from Davenport Wood. The latter, as suggested by the field evidence, does not represent part of a major gravel spread, but rather older gravel reworked by slope processes (Section 2.8).

In conclusion, it would appear that the member groups can be used lithostratigraphically since they show effective separation. The overlap of individual samples within members is, however, a problem and underlines the fact that pebble counts cannot be used in isolation for lithostratigraphic correlation, but must be closely related to other lithological parameters, sedimentary structures and altitude. The need to sample undisturbed sediments below the level of cryoturbation or solifluction material has been stressed elsewhere (Gibbard, 1982) and should not be underestimated. Surface or near-surface samples frequently have a pebble assemblage that differs significantly from that in an underlying stratified gravel, as exemplified in the 'Rassler Terrace' example cited above. This may arise from preferential weathering of lithologies or from incorporation of particular lithologies (usually angular flint in this area) during the solifluction process; the latter process was also noted by McGregor and Green (1978).

Finally, Green and McGregor (1978) have shown that the high-level gravels in the eastern part of the area contain a decreasing proportion of rounded pebbles with decreasing altitude. They have used this change as a stratigraphical marker and have attributed the decrease to the increased denudation of Chalk bedrock, the latter contributing angular flint to the gravels. This interpretation is only partially true since a major source of rounded flint in the study area is the Reading Beds. Many of the high-level gravel remnants rest on Reading Beds rather than Chalk, the latter being first extensively exposed beneath the Beaconsfield Gravel throughout much of the area. It therefore follows that an increase in angular flint derived from denudation of Chalk must be expected and is indeed found at the Beaconsfield stage (cf. McGregor and Green, 1978). The progressive increase in angular flint at the expense of rounded flint thus reflects the increased incision at each subsequent stage and the progressively greater altitudinal distance from the source of rounded flint pebbles. For this reason it would appear unwise to interpret increased angular flint as representing increased frost or other denudational processes.

4

Sedimentary structures and depositional environment of the gravel units

All the units of waterlain, sorted gravel and sand repeatedly show suites of sedimentary structures that indicate deposition under broadly similar fluviatile conditions throughout the area studied. It is therefore possible, in general terms, to discuss the sedimentary sequences of deposits together, with the exception of those of the Winter Hill Upper Gravel, which were considered in Section 2.8.

Previous investigations into the sedimentary structures of the gravel units have been presented in association with descriptions of fossiliferous fine sediment channel fills in the gravels. For various reasons, these have concluded that the sediments originated in a cold climate braided river depositional environment. However, detailed facies assemblage studies of the type undertaken by Corner (1975) in Norway or Bryant (1983a,b) in England have not as yet been presented for the Middle Thames region. On the basis of the sequences examined during this study and noted above, a generalised facies description for the gravel members is presented. The sediments can be assigned to the braided river environment on the basis of facies models developed by Miall (1977, 1978). Individual gravel members may show minor variations from the scheme. In addition, the facies descriptions can only be applied to those units that are exposed at large sections where the vertical and lateral relationships of facies can be unequivocally observed.

4.1 Descriptive summary

In sections aligned parallel to the valley axis, the sedimentary bodies appeared to be elongated in the downstream direction. This phenomenon has frequently been observed from gravel bed rivers (Miall, 1977; Bryant, 1983b) and apparently results from downstream migration of the sediment body.

The most commonly occurring facies is massive or horizontally crudely-bedded gravel (facies Gm). The gravel clasts show a great size range, but are most frequently of medium gravel (2–30 cm). Larger clasts occur and may reach as much as 50 cm in diameter, but these are rare. The gravels are normally massive, but often show poorly developed horizontal bedding. Individual clasts are only rarely imbricated. This may arise because of the relative paucity of blade and rod-shaped pebbles, the irregular shape of angular flint and high sphericity of vein quartz and quartzite rocks (cf. Bluck, 1967). The gravel is generally matrix-supported, but clast-supported gravel is common, particularly in the coarser particle sizes. The interstices are filled with silt, sand and granules. Secondary clay may also be present. Individual gravel units are typically up to about 90 cm thick but are frequently superimposed to produce units of greater thickness. Laterally the gravel units may be very persistent and can be traced across quarry faces for over 25 m. They usually rest on an erosional base and are often 'channelled' into the beds beneath.

Broad, shallow channels are often found excavated into the gravel facies. The channels are filled by interbedded tabular cross-bedded gravel and sand (facies Gp and Sp). The channels range up to 15 m in width and 1.5 m in depth, but are often much smaller. Individual foresets comprise a complex of pebbly sand, sand and gravel, the grain size showing considerable variation. The bases of such units are normally erosional. The coarsest clasts frequently form a gravel lag at the base. Decreasing upward dip in foreset angle is often observed. Reactivation surfaces (cf. Collinson, 1970) are common. There is a tendency towards upward fining channel fills, but this trend is seldom completed, since the sequences are often truncated by a reactivation surface and overlain by additional sediments recording the input of fast-flowing water.

Trough cross-bedded pebbly sand (facies Gt and St) occurs at all levels in the deposits normally filling channel-like features cutting across earlier sediments and alternating with gravel facies Gm. The channels are of similar proportions to those mentioned above in cross-section, but individual units, often elongated downstream for 20–30 m, may be as much as 2–3 m in thickness. The base of these deposits is again erosional.

Of the sand facies present only two types are well represented.

The most common is the scour–fill sand (facies Ss). This consists of scours, often asymmetric in form, ranging in depth from a few centimetres to about 50 cm and up to about 1.5 m in width. The lenses of sediment often occur within or resting upon bar facies gravel (facies Gm) and comprise pebbly sand. The bedding of the sand is frequently parallel to that of the scour surface. The only other sand facies present appears to be horizontally bedded, often massive sand (facies Sh). These beds are normally of very local extent and up to 20 cm in thickness.

Fine-grained sediments occur very infrequently. Only facies Fl is represented by beds of silty clay, clayey silt and organic sediment. The latter is apparently restricted to the lower gravel members. These deposits appear to fill channels or floodplain depressions which may have scoured into underlying sediments or remained as unfilled hollows. They occasionally represent the completion of a fining upward sedimentary sequence beginning as facies Sr (ripple-bedded medium to fine sand) and terminating in facies Fl. The sediments are often grey to brown in colour and contain interbedded sand and even gravel bands. Organic remains, where present, almost invariably include both autochthonous aquatic floral and faunal fossils and also allochthonous material washed in during flood events from vegetated stable tracts and neighbouring slopes (Rust, 1972).

A further sedimentary facies occurs in the sediments particularly near the contemporary valley side. This is the lenticular wedge or tongue-shaped pebbly clay of mud flow or solifluction origin. These have been described above (Section 2.7). 'Rubble chalk' (see also Section 2.7 for description) can also be found in a similar relationship to the fluvial gravels near valley sides developed in Chalk bedrock areas, where the gravels have not been post-depositionally decalcified.

Large-scale planar erosion surfaces are a common feature of the gravel members, as noted by Bryant (1983b) in the Upper Thames Valley. They almost invariably underlie medium to coarse gravel accumulations (facies Gm and Gt) and in any one exposure may be present at various levels. They may be very persistent and are often traceable almost continuously across a quarry face. Where ice wedge casts have been found they often occur beneath such surfaces and indeed may be truncated by them. It is interesting to note that these surfaces are particularly well developed where the contemporary valley was comparatively rather wide, for example downstream of Maidenhead. The features are much less common in narrow valley stretches, e.g. between Maidenhead and Henley.

Micro-involution zones, generally 10–20 cm thick, immediately beneath shallow channels and reactivation surfaces were observed at several sites particularly within the Taplow Gravel (Section 2.13). They have been interpreted as a type of 'arctic structure' soil formed by subaerial freeze–thaw processes.

4.2 Interpretation

Several authors have stressed the difficulty of distinguishing between meandering (high sinuosity) and braided (low sinuosity) rivers (Miall, 1977). Recent models for coarse grained alluvial sequences (Miall, 1977, 1978; Bluck, 1979; Rust, 1972) have provided a means for discriminating and interpreting the sequences described above.

The main mass of gravels (facies Gm) appear to have arisen from accumulation and migration of low-amplitude longitudinal bars. The gravel was probably laid down in horizontal beds, but stratification is often obscured by accumulation of finer particles in open interstices. Deposition of such units only occurs at peak discharges as shown by Hein and Walker (1977). Interbedded sand lenses probably resulted from flow in secondary channels over bar surfaces during low-energy waterflow.

Shallow sand and gravel filled channels have been observed in many modern braided rivers (for summary see Miall, 1977). Minor channels are thought to originate from scour associated with local eddies. Major channels are formed by lateral channel migration at high water stage or bar dissection during falling water conditions (Williams and Rust, 1969; Collinson, 1970; Miall, 1977). These channels, particularly the larger features, show multiple infills of facies Gt, St and Ss.

Facies St and Sh result from bedform migration. The migration of large-scale ripples or dunes gives rise to trough cross-bedded sand either singly or in groups and corresponds to facies St (Miall, 1977). Sedimentation by sand waves, of similar proportions to dunes but with straighter crests produces planar tabular cross-bedding (facies Sp). Such large bedforms are produced during upper regime flow for which there is ample supporting experimental evidence (Allen, 1970; Friend and Moody-Stuart, 1972). Horizontal bedding (facies Sh) may be formed under lower flow regime in shallow water (Harms *et al.*, 1975) or during the flood stage (Harms and Fahnestock, 1965). In the latter case the bed is formed by streaming of particles across the surface under upper flow regime conditions.

4.2. Interpretation

During low-water periods only small-scale bedforms are produced and these are generally restricted to the infilling of minor channels and hollows on bar surfaces by ripple-laminated sand (facies Sr) or in shallow water by horizontally bedded sand (facies Sh). Sand wedges may develop during falling water stages at bar margins. Such wedges normally show planar cross-bedding. Silt, clay and organic material (facies Fl) accumulates in abandoned or partially-abandoned channels in standing or trickling water especially in topographically higher areas of the floodplain (Williams and Rust, 1969; Rust, 1972). In spite of their apparently low preservation potential (cf. Cant, 1976) these beds are occasionally found.

Braided rivers commonly only occupy part of the valley bottom at a particular time and large areas of the floodplain may become temporarily stable and may even be colonised by vegetation (Williams and Rust, 1969; Rust, 1972; Miall, 1977). During exposure these surfaces are subjected to subaerial weathering and ice wedge polygons and cryoturbation may develop in favourable localities, particularly on higher level stable tracts (Washburn, 1968; Bryant, 1983b). Renewed channel migration will degrade such a surface, terminate subaerial weathering and result in talik development leading to the degradation of ground ice phenomena. Infill and truncation of ice wedge casts developed in the surface, together with the truncation and burial of soliflucted material and abandoned channel fills may therefore be expected at this time. The extensive planar erosional surfaces have been attributed by Bryant (1983b) to truncation of fluvially inactive areas, raised slightly above the active areas of the river, by large-scale channel migration. The association of truncated infilled ice wedge casts and other periglacial phenomena, channel fills and earlier sediments in such surfaces supports this interpretation (Chapter 5).

The assemblage of sedimentary facies and structures present in the Thames gravel members broadly comprises a sequence of interposed channel fills and superimposed longitudinal bars. This assemblage most closely approximates to the Scott-type depositional model proposed by Miall (1977, 1978). However, the greater abundance of trough cross-bedded sand channel fills and association of some units of fine sediment shows some affinities with the Donjek model of the same author. The facies observed in this study closely resemble those described by Bryant (1983b) from Stanton Harcourt (Upper Thames).

Taken together all these factors indicate that the deposits accumulated in a braided river environment. The consistent association with cold climate indicators such as ice wedge casts, solifluction deposits and cryoturbation structures all suggest aggradation under a predominantly periglacial regime. This is strongly supported by the facies present which have been shown to typify nival-flood dominated rivers in the modern Arctic (Bryant, 1983a).

5

Solution of Chalk bedrock

The dissolution of carbonate rocks by carbon dioxide dissolved in groundwater is a well established phenomenon. Because carbon dioxide solubility is inversely proportional to temperature, the lower the temperature the greater the solution effect on limestones. The Chalk is known to have undergone considerable solution in southern Britain and Northern France (for summary, Jones, 1981). In the study area, Hare (1947) reported abundant 'swallets' or dolines developed along the floors of dry valleys near Burnham. Pipes and surface cavities have been repeatedly observed along the line of the M40 Motorway particularly where a thin covering of younger deposits occurs (Higginbottom and Fookes, 1971). Recently, the author recognised an extremely large doline-like structure near Denham (Gibbard, Bryant and Hall, in prep.) which he attributed to solution of the underlying bedrock Chalk.

During this study pipes or subsidence hollows have been found penetrating bedrock immediately beneath almost all gravel members. These features are normally restricted to the Chalk, but also where a thin veneer of Tertiary clays occurs above it. Although they are generally circular in cross-section, elongate hollows have also been observed. These features are variable in size from a few centimetres to several metres in both width and depth. They often have steeply sloping or vertical sides and are filled with a variety of sediment. On the basis of their time of formation and infilling, the features can be classified into two groups with respect to the gravel members: syndepositional and post-depositional.

Evidence for contemporaneous fluvial gravel and sand deposition and subsidence of sections of the underlying bedrock floor has frequently been found. A typical example was described above from Westwood quarry (Section 2.7). These features comprise graben-like structures into which gravel and sand has collapsed. This is indicated by the chaotic bedding of the fill. In some instances large structures such as groups of high-angle normal faults offsetting but otherwise not disturbing depositional bedding have been found. Deposits overlying such fills or faulted zones are often undisturbed and therefore indicate that subsidence had been completed before accumulation of the later deposits. Similar features have been described by McGregor and Green (1983). Larger examples of these features are the anomalously deep hollows beneath the floodplain gravels of the Misbourne and Alder Bourne (Figure 51). These features reach a maximum depth of 12.5 m (M25: bh 21F: TQ027852), i.e. 8.75 m deeper than the average sediment depth.

More common than the former type is the pinnacled Chalk surface beneath gravel and sand accumulations. This surface form is also frequently present beneath mass-flow deposits and even weathered regolith. The amplitude of these near vertically-sided features (pipes) is very variable, but in some cases, for example at Highland's Farm (Black Park Gravel) or Chalfont Lane (Gerrards Cross Gravel), they may reach 4–5 m in depth. In many cases the deposits are highly disturbed by complete collapse into the structures. This is expressed by oversteepened bedding, faulting and in extreme cases by hollows in the gravel and sand unit infilled by younger sediments, e.g. at Remenham (Lynch Hill Gravel). In the latter case the sides of the sedimentary fill resemble steeply dipping or vertical faults cutting across the gravel and sand strata. The younger fill may have various origins, but could include material washed into the hollow by surface flow, blown in by wind, soliflucted from nearby, or collapsed from the sides. These post-depositional structures broadly increase in intensity and 'maturity' of development with height above the modern river and therefore with presumed antiquity.

An extreme example of this type of collapse feature is present at Slade Oak Lane, Denham (TQ018897) where a maximum of 37.5 m of redeposited Reading Beds and fluvial gravel and sand fill a funnel-like doline depression in Chalk (see Section 2.10). This feature has been described by Gibbard *et al.* (in prep.) in connection with interglacial organic sediments intercolated in its filling.

All the structures appear to result from subsidence caused by Chalk bedrock solution beneath a superficial sedimentary cover. This cover may comprise gravels and sands, Tertiary clays or even Clay-with-flints (Hare, 1947; Catt and Hodgson, 1976; Sperling *et al.*, 1977).

Modern opinion on the contribution of chemical weathering in cold climate environments is divided. French (1976) following Bird (1967) concluded that in general solutional effects are weakly developed in Arctic regions because the presence of permafrost prevents water from penetrating into bedrock limestones. However, Washburn (1968) stressed that chemical weathering may be locally important, particularly where water movement is able to occur. This is particularly true where subpermafrost taliks exist. During the cold stages in Britain, permafrost may not have been continuous at all times. Indeed, in transitional periods, during permafrost disintegration for example, bedrock solution could have been very severe.

It is probable that the syndepositional features resulted from contemporaneous solution at and beneath the stream bed. The warming effect of the stream water results in an unfrozen zone or talik (Washburn, 1972) beneath the stream. In this zone the migration and interchange of ground and river water would be expected. Conditions would therefore be ideal for removal of calcium carbonate, particularly along fissures and joints. Local springs would further encourage solution and may have been responsible for formation of some of the larger features. Subsidence might be expected to continue until the permafrost table was reached. However, when the active river channels migrate to another part of the valley and the warming effect of the river water is removed, permafrost would develop in the deposits beneath the stable tract. This has frequently been observed in modern Arctic environments. This would halt further solution temporarily or permanently. Hollows formed syndepositionally will be progressively infilled by gravel and sand or, if isolated from the main flow, by silts and clays (cf. Williams and Rust, 1969; Rust, 1972; Washburn, 1972; Bryant, 1983b).

The similarity of the form and sedimentary fill of these features to the infilled pipes described by Edmunds (1931), Higginbottom and Fookes (1971) and Berry (1979) from Central London is remarkable (see also Section 2.18). These drift-filled depressions are funnel-shaped or oval in plan and vary from 20 to 150 m wide. They are steep sided and become less regular and more narrow downwards. Their maximum depth below the average gravel floor level is about 25 m. These structures were originally interpreted as having formed by solutional collapse of Chalk beneath the Tertiary rocks by Edmunds (1931). However, they are not developed in Chalk, but in London Clay and Reading Beds. Because of the impermeability of these rocks and the presence of the pipes close to tributary stream confluences with the Thames, they were thought (by Zeuner (1959) and Berry (1979)) to have resulted from fluvial scour. However, while some of these features may be scour hollows particularly those at confluences, the form and depth of the majority negates this interpretation. Hutchinson (1980) suggested that they may have formed as feeder channels for open-system pingos. Relatively little is known about the internal structure of pingos, but at present they occur in permeable valley spreads where they are spring fed.

Suitable conditions may have prevailed during deposition of the gravel spreads, but the pingo interpretation does not explain the large-scale collapse of both the Pleistocene and Tertiary sediments in deep hollows in apparently impermeable bedrock. While open-system (or East Greenland type) pingos are fed from an aquifer at depth beneath permafrost (Embleton and King, 1975), all the evidence to support the pingo hypothesis is circumstantial, since no pingo structure, for example low ramparts or rims (Embleton and King, 1975), has been observed in sediments of the River Thames.

Hutchinson (1980) noted that the hollows occur mainly 'at the feather edge of the London Clay as it thins against the inlier of the Lower London Tertiaries and Chalk produced by the Deptford pericline' (anticline). All the depressions lie close to or within the area of artesian springs (Lucas, 1887; Hutchinson, 1980) many of which flowed during recent times. Under periglacial conditions the Chalk would have been an intensively confined subpermafrost aquifer. Water under high hydrostatic pressure is an even more efficient solvent of calcium carbonate than normal cold water and therefore solution of the Chalk would have been intensified at this time. Fissures and joints are almost certainly present in the overlying Tertiary rocks as a result of tensional stress during the anticlinal folding that formed the Deptford structure. These joints would have provided channels that could have permitted the high-pressure water to escape to the surface and form the funnel-like depressions. Periodic collapse caused by removal of the Chalk in solution, and possibly of the other lithologies by flowing water, would enlarge the holes still further. Fluvial activity at the surface would certainly have enlarged and elongated the openings of the hollows by scour. At the same time gravel and sand together with eroded bedrock clasts would be expected to have fallen into the hollows thus formed. All the observed characteristics of the depressions and their fills would therefore be explained by this hypothesis without the need to

UNIVERSITY COLLEGE LIBRARY CARDIFF

invoke pingo development. Collapse of some of these hollows has continued into the Flandrian, e.g. Southwark (Berry, 1979). This continued subsidence during an interglacial stage can only be attributed to solution of bedrock Chalk beneath the site.

Finally, it would seem likely that the continued flow of artesian water might be expected to have resulted in naled or icing formation rather than pingo development. Pingos require restricted groundwater flow (French, 1975) since too large a rate would prevent freezing of the pingo, whereas icings can form with quite considerable groundwater flows. Icings are commonly formed in periglacial river valley situations (Church, 1972; Washburn, 1972; Embleton and King, 1975).

As regards post-depositional collapse, this is closely similar to that arising from doline formation. When permafrost is present no solution is possible except within the active zone. However, when the permafrost melts or is only discontinuously present, migrating water is able to attack the Chalk. The soluble calcium carbonate becomes dissolved in the groundwater and the Chalk is rapidly removed probably along joints or fractures in the rock. Percolation of water from rain and snow melt will leach-out carbonates above the ground-water table. This removal of Chalk will give rise to the post-depositional collapse and decalcification of gravels and sands seen in the field.

6

Vertebrate faunal assemblages

On occasions over the last 250 years vertebrate remains have been recovered and recorded from the deposits of the River Thames. Many of these remains have long since been destroyed or lost. However, since the mid nineteenth century, collections from various sources have been conserved and stored in museums or private collections. Such collections are normally dominated by large vertebrate remains which were easily recognised during excavation, smaller remains have often been overlooked. By far the largest numbers of fossils were found during the latter half of the last century by local collectors when excavations were undertaken by hand. Since the advent of machine working, a decline in the frequency of finds has taken place.

A synthesis of individual vertebrate records from the Middle Thames has never been discussed. While the assemblages from the interglacial deposits at Trafalgar Square (Franks *et al.*, 1958; Stuart, 1976) and Brentford (Trimmer, 1813; Zeuner, 1959) have been considered, those from the gravel members have received relatively little attention. For this reason the following discussion will concentrate on the vertebrate fossils from the gravels. The fossil assemblages are summarised in Table 4, which includes the assemblages from the interglacial sites for comparison.

6.1 Taphonomy and preservation

Most of the finds of vertebrate fossils have been from gravel member sediments which accumulated in braided river channels formed predominantly under cold climates, as shown above. The vertebrate remains, generally in a disarticulated form, have been recovered in various states of preservation. Specimens may be rolled or broken, although decalcification appears to be a common feature of material from older gravel units.

The taphonomy of vertebrate remains in river sediments has recently been considered by Hanson (1980) and reviewed by Stuart (1982) to which reference should be made for detailed discussion. Both, however, stressed the need to consider the various stages through which skeletal parts pass before they become fossilised in sediment. An attempt will be made here to consider these stages relevant to fossils from the Thames gravels. In the Middle Thames region, partial or complete mammal skeletons are exceedingly rare, but have been found on two occasions in fine sediments. Complete skeletons of mammoth *Mammuthus primigenius* were recovered from Lynch Hill Gravel deposits at Southall (Brown, 1889) and from sediments beneath Endsleigh Gardens (Hicks, 1892). These animals must therefore have died at or close to their depositional sites.

However, as noted by Stuart (1982, p. 65), disarticulated vertebrate remains are much more common. These have been transported to the depositional site by the river and/or periglacial slope processes. The abundance of terrestrial vertebrates in the sediments suggests that animals came to the river to drink and to feed on the riverside vegetation. The corpses of animals that died on the river floodplain could have been attacked by scavengers and the skeletons left in a disaggregated form. The floating cadavers could also have been carried by the stream until they became waterlogged and sank, disaggregated by the continual buffeting or until they came to rest on an obstruction such as a gravel braid. Winter freezing may have tended to preserve corpses and therefore aided longer downstream transport before the flesh decayed. Related to this are remains that entered the river in soliflucted material from the banks or valley slopes. Bones and teeth that remained on the ground surface will have undergone slow erosion, decalcification and breakage due to surface processes, although in modern Arctic regions skeletal parts may remain virtually unaltered for many years. If, however, they were incorporated by cryoturbation into regolith, particularly carbonate-rich material, then the chances of preservation of the vertebrate remains would have been greatly increased. Solifluction of the regolith could then have carried the material to the river where it would have become incorporated in the river floodplain deposits.

The role of rivers in concentrating detrital remains from throughout their catchment suggests that vertebrate material found in river sediments could give an over-representation of the abundance of numbers of individuals present at any particular time. This possible over-representation could arise not only from the interaction of river and slope processes, but also from animals visiting rivers to drink and feed. Furthermore, in areas lacking shrub or tree vegetation cover, smaller river valleys may also provide refuge from wind. These factors suggest that it might therefore be difficult to obtain a true estimate of numbers of vertebrate individuals from river sediment accumulations.

The problem of reworking of previously deposited vertebrate remains must be considered. Erosion of the river bed and banks during channel migration or slope denudation may result in the incorporation of previously fossilised material not only from bedrock, but also from older Pleistocene sediments.

Nevertheless, once the bones, teeth or antlers enter the river they become sedimentary clasts subject to the same processes as all other sedimentary particles in the fluvial system. Most of the vertebrate material found in coarse river sediment is relatively large and this selection must result from sedimentary sorting. Smaller remains (e.g. vole teeth) may therefore be expected to occur in coarse sands. While large vertebrate remains have been found in almost every sedimentary facies of braided rivers, they are particularly frequent in gravel braid (facies Gm) and channel scour and fill (facies Gt) deposits. In the former, large vertebrate remains transported predominantly by high velocity flow during flood events will be deposited on braid bars together with gravel-sized clasts. They become incorporated into the sediment by burial during subsequent aggradation. In scoured channels cut into bar facies gravel or often at the scoured bedrock or underlying sediment surface, concentrations of large vertebrate remains and artefacts and large clasts are frequently found juxtaposed. The coincidence of the occurrence of these components appears to result from a 'lag' effect following erosion of underlying sediment, or a river bank. In this situation, the finer transportable material is winnowed away by swift currents and eddies resulting in clasts being progressively lowered and marshalled into concentrations by rolling along the channel bottom. This accounts for the occurrence of unrolled and apparently fresh specimens mixed with heavily rolled or broken bones or artefacts; a commonly observed phenomenon in some studies of fluvial sequences. Subsequent rapid burial ensures the preservation of a proportion of the fresh forms along with the degraded specimens.

Experiments using elongated bones in rivers suggest that they are transported and deposited parallel to the current flow (Hanson, 1980). This observation supports the hypothesis that bones are often imbricated parallel to the clasts in gravel beds.

6.2 Faunal assemblages

The faunal list in Table 4 was compiled from various sources most of which are mentioned in the description of members in Chapter 2. The remaining records are from the Geological Museum and the British Museum (Natural History) collections. For comparative purposes the faunas from Swanscombe (Sutcliffe, 1964; Stuart, 1982), Trafalgar Square (A. J. Sutcliffe, personal communication) and Brentford (Trimmer, 1813; Zeuner, 1959) have been included. The faunas from the sites of unknown age at Endsleigh Gardens (Hicks, 1892) and Redlands, Reading (Poulton, 1880) are also shown and will be considered below. It should be stressed that the identifications of vertebrate fossils are based on those used in the original publications or lists and therefore may include some minor errors. The nomenclature has been modified to present-day classification following Stuart (1982). In addition the identification of *Bison priscus* and *Bos primigenius* is confused, many of the older records referring simply to 'ox'. The separation of the remains of these species is very difficult and therefore where there was any doubt the records were assigned to *Bos/Bison* except where there has been modern revision.

The number and diversity of preserved remains appears in general to be proportional to age; more remains are consistently recorded from younger deposits. This is likely to be a consequence of the increasing effect of carbonate solution with the increasing antiquity of the deposits. It undoubtedly explains the absence of vertebrate material from older, higher units.

The oldest gravel member that has yielded vertebrate material is the Boyn Hill Gravel from which only one tooth of horse *Equus ferus* and a few bone fragments were found by Treacher (in Wymer, 1968) at Toot's Farm quarry. The remains were apparently in a decayed state.

A more varied fauna has been found in the Lynch Hill Gravel. The most frequent finds were those of mammoth *Mammuthus primigenius*, extinct bison *Bison priscus* (*Bos/Bison* sp.) and *E. ferus*, while woolly rhinoceros *Coelodonta antiquitatis* and red deer *Cervus elaphus* have also been found. The most prolific site was Grovelands (Stevens, 1882; Shrubsole, 1890). According to Wymer (1968), this site also produced one molar of straight tusked elephant *Palaeoloxodon antiquus*. Abundant remains were also discovered in the Law Courts excavations in the Strand (Hudleston and Price, 1873), the most common being those of bison.

Fewer species have been recovered from the Taplow Gravel member. The most famous vertebrate find is the distinctive skull of musk ox *Ovibos moschatus* from the type section, which was associated with remains of *M. primigenius* and *C. antiquitatis* (Owen, 1856; Prestwich, 1856; Oakley, 1937). Fossils of musk ox are extremely rare in Britain and may indicate that it was present only during the most severe climatic phases (Stuart, 1982). Today this animal is confined to Arctic tundra in Alaska and Canada, where it avoids deep snow and lives on grasses, sedges, rushes, dwarf shrubs and willows (Banfield, 1974). *M. primigenius* has been recorded at two other Taplow Gravel sites, Donkin Hill (Shrubsole, 1890; Wymer, 1968) and Ferns, Marlow (Treacher, 1916).

The Reading Town Gravel has yielded limited faunal material. Remains of *M. primigenius* were found at three localities within the gravel spread (Shrubsole, 1890). The Kennet Mouth pit has provided both teeth and tusks of mammoth together with a tooth of *E. ferus* (Stevens, 1882). The discoveries of several hippopotamus *Hippopotamus amphibius* teeth and some broken pieces of bone at Kensington Road, Reading (Shrubsole and Whitaker, 1902) are from this unit. The association of hippopotamus, mammoth and horse suggests that the fauna is 'mixed' and includes elements associated with temperate woodland stages (i.e. hippopotamus) and regional grassland conditions (cf. Stuart, 1982).

By far the richest fossil vertebrate assemblages have been found in the Kempton Park Gravel, which is of Middle Devensian age (Gibbard *et al.*, 1982; also above). The earliest finds from this member are those from Kew Bridge (Morris, 1850), which included *M. primigenius*, *Bos*/*Bison* sp., *C. elaphus*, reindeer *Rangifer tarandus*, *C. antiquitatis*, *H. amphibius* and polar bear *Thalarctos maritimus*. The last-mentioned was identified by Kurtén (1964), but could perhaps be brown bear *Ursus arctos* (A. J. Stuart, personal communication). Lane-Fox (1872) found a similar fauna nearby at Brown's Orchard, Acton, and at Turnham Green Road, Chiswick. These sites yielded extinct rhinoceros *Dicerorhinus hemitoechus*, *E. ferus*, *H. amphibius*, *Bos*/*Bison* sp., fallow deer *Dama dama*, *C. elaphus*, *R. tarandus*, *Ursus* sp. and *M. primigenius*.

In 1891, Leeson discovered a skull of saiga antelope *Saiga tartarica* from gravel at Twickenham and soon afterwards Leeson and Laffan (1894) found a vertebrate assemblage associated with organic sediments at Strawberry Hill, Twickenham. The remains were found both in and immediately above the organic deposit, the latter presumably derived from the deposit. They were dominated by bison bones although *C. elaphus*, *R. tarandus*, wild boar *Sus scrofa* and wolf *Canis lupus* were also found.

During the late 1950s and early 1960s many large vertebrate remains were collected by J. W. Simmons from Willment's quarry at Isleworth. The organic silty clay that occurs in the gravel and sand contained abundant fossil flora and fauna (see above) and was radiocarbon-dated to the Upton Warren Interstadial (Coope and Angus, 1975; Kerney *et al.*, 1982). The vertebrate fossils comprised *M. primigenius*, *C. antiquitatis*, *R. tarandus* and *B. priscus*. The remains of reindeer and bison were found within the organic deposit and were therefore contemporary with the accumulation. The remaining material was from the overlying gravel. According to Stuart (1982) the numerous shed reindeer antlers were mostly from males indicating the presence of herds during the winter months but this does not exclude their presence in summer. Small mammal finds from the site include northern vole *Microtus oeconomus* and tundra vole *M. gregalis* (Sutcliffe and Kowalski, 1976).

Isolated finds of mammoth and bison remains have been recovered from several excavations in the Earls Court and South Kensington areas. A single limb bone of *B. priscus* was found during the Ismaili Centre excavations and pollen analysis of adhering sediment showed a spectrum typical of the Upton Warren Interstadial at the site (Coope *et al.*, unpublished).

The vertebrate communities of the Middle Devensian were recently discussed by Rackham (1978) in connection with faunas from gravels in the Witham Valley, Lincolnshire. He considered that these faunas are 'mixed' in that they contain both steppe and tundra elements; the former comprising bison and reindeer and the latter mammoth and woolly rhinoceros, these representing geographically and climatically distinct populations. There is some support for Rackham's hypothesis from the Isleworth sections. Here large numbers of bison seem to be present during the Upton Warren Interstadial, represented by the organic deposit, and are apparently markedly reduced in number during the cold stadial, represented by the gravels.

Apart from the faunas typical of cold, treeless phases, the remains of animals normally associated with temperate woodland have also been found at the sites mentioned above in the Kempton Park unit. They include *H. amphibius*, *Bos primigenius*, *D. dama*, roe deer *Capreolus capreolus* and *D. hemitoechus*. In addition, a

6. Vertebrate faunal assemblages

Table 4. *Vertebrate faunas of the Middle Thames gravel aggradations. Finds: + occasional, ++ common.*

Species	Aggradational unit: Swanscombe Lower Gravel	Swanscombe Middle Gravel	Boyn Hill	Lynch Hill	Taplow	Trafalgar Square	Brentford	Endsleigh Gardens	Redlands Sands	Reading Town	Kempton Park	Shepperton
Hominidae (Hominids)												
Homo sp.		+										
Leporidae (Hares and rabbits)												
Lepus sp.	+	+										
Cricetidae (Voles and lemmings)												
Arvicola sp.								+				
Microtus agrestis		+										
Microtus gregalis											+	
Microtus oeconomus											+	
Clethrionomys sp.		+										
Lemmus sp.		+										
Canidae (Dogs)												
Canis lupus	+	+									+	+
Ursidae (Bears)												
Ursus spelaeus	+										+	
Thalarctos maritimus											+	
Mustelidae (Mustelids)												
Martes cf. *martes*	+											
Hyaenida (Hyaenas)												
Crocuta crocuta							+					
Felidae (Cats)												
Panthera leo	+											
Proboscidea (Elephants)												
Palaeoloxodon antiquus	++			+		+	+				+	
Mammuthus primigenius				++	+			++	++	++	++	+
Equidae (Horses)												
Equus ferus	++	++	+	++				+	+	+	+	+
Rhinocerotidae (Rhinoceroses)												
Coelodonta antiquitatis				+	+				+		+	+
Dicerorhinus hemitoechus	+										+	
Dicerorhinus kirchbergensis	++	+										
Suidae (Pigs)												
Sus cf. *scrofa*	+										+	

badly rolled tooth of *P. antiquus* has been collected from this member at Hammersmith. Almost all these species are present in the Ipswichian interglacial sediments at Trafalgar Square and Brentford (Table 4) which are remnants of a more continuous spread that must have existed at the time. One is led to conclude that the hippopotamus and other 'temperate mammal' remains must have been reworked from these interglacial deposits.

The fauna of the Shepperton Gravel is very poorly known. This is probably because excavations in this unit are nearly always completely flooded. The list in Table 4 was compiled from five sites: Battersea Gas Works, Hurley and three sites in Windsor at Peascod Street, Home Park and the Infantry Barracks. The first included remains of *M. primigenius*, *C. antiquitatis*, *E. ferus*, *B. priscus*, *C. elaphus* and *R. tarandus* (Coombs, 1873). At Home Park, Windsor, Whitaker (1860) recorded bones of an elephant (?*M. primigenius*), *B. priscus* and a deer. Abundant fossils mostly of mammoth have been found beneath the centre of Windsor (Dewey and Bromehead, 1915). At the Infantry Barracks large numbers of *R. tarandus*, *B. priscus*, *E. ferus*, *C. lupus* and *Ursus* sp. were all recorded (Dawkins, 1880; Whitaker, 1889). According to Whitaker (1889), the Hurley site yielded an extensive fauna in the mid nineteenth century of 'ox' (?*B. priscus*), *R. tarandus*, elephant (?*M. primigenius*), rhinoceros (?*C. antiquitatis*), *E. ferus* and notably *Hippopotamus amphibius*.

The uniformity of the faunal assemblages from this unit is striking and although much of the material is no longer available for study, it is clear that the fauna is predominantly similar to that from the preceding gravel member. The remains of hippopotamus at the Hurley site must have been reworked from interglacial deposits, as mentioned above.

The Shepperton Gravel is of Late Devensian age 15000–10000 BP (see Section 2.18) and if these faunal assemblages are contemporary with the accumulation, they provide some important Late Devensian mammalian records. The Late Devensian vertebrate faunas are generally poorly known (Stuart, 1982); however *M. primigenius* and *C. antiquitatis* appear to be absent from assemblages post-dating 18000 BP in Britain, and were certainly not present by *c.* 12000 BP. If the dating of the gravel aggradation is correct then the date for the final disappearance of these two species from this country must be reassessed.

Table 4. (*continued*)

	Aggradational unit											
Species	Swanscombe Lower Gravel	Swanscombe Middle Gravel	Boyn Hill	Lynch Hill	Taplow	Trafalgar Square	Brentford	Endsleigh Gardens	Redlands Sands	Reading Town	Kempton Park	Shepperton
Hippopotamidae (Hippopotamuses)												
Hippopotamus amphibius						++	++			+	+	+
Cervidae (Deer)												
Cervus elaphus	++	+		+		+	+	+			+	+
Capreolus capreolus	+			+								
Dama dama						+					+	
Dama dama clactoniana	++	+										
Megaceros giganteus	+	+					+					
Rangifer tarandus											+	++
Bovidae (Bovids)												
Bison priscus							+				++	+
Bos primigenius	++	++					+					
Bos/Bison sp.				++			+		+		++	+
Ovibos moschatus					+							
Saiga tartarica											+	

In conclusion, it appears therefore that the gravel members contain faunal remains of broadly similar assemblages. No one species has definitely been found from all gravel members, but *E. ferus* is present in every stage except possibly the Taplow Gravel. King and Oakley (1936) did list horse from the Taplow Gravel, but did not mention from which site it was found. For this reason its record from the Taplow member must be regarded with caution. Since the Taplow is the only stage to have yielded *O. moschatus*, it is tempting to infer that conditions might have been too severe for horse during this stage. There is some support for extreme cold from the abundant periglacial structures and the paucity of Palaeolithic artefacts in this gravel unit. With the exception of the Boyn Hill Gravel, all gravel aggradations have yielded *M. primigenius*. Overall the faunas support the sedimentary evidence for gravel and sand aggradation under cold climatic conditions. They suggest an herbaceous steppe or at times tundra-like vegetation cover on surrounding land. Biostratigraphically, the faunas are of little value with the exception of the occurrence of hippopotamus in the gravel members. The remains of *H. amphibius* have been recorded from the Cromerian and Ipswichian Stages, but not from the Hoxnian or any cold stage in Britain (Stuart, 1984). No deposits of Cromerian age have yet been identified in the Middle or Lower Thames and therefore it is reasonable to assume that all the hippopotamus fossil material was probably of Ipswichian age. The presence of this species in cold stage gravel aggradations suggests that the remains have been reworked from local interglacial sediments where it is well represented. The gravel aggradations in which hippopotamus was found must therefore post-date the Ipswichian stage and therefore must be assigned to the Devensian. Remains of this animal have been found in every Devensian gravel member yet defined in the study area.

Lastly, mention should be made of the two additional isolated localities that have yielded mammalian fossils. Apart from the complete skeleton of a young *M. primigenius*, the Endsleigh Gardens organic channel filling has also yielded several other remains of this species together with an antler of *C. elaphus* and two bones of *E. ferus* (Hicks, 1892). At Redlands, Reading, Poulton (1880) recovered a variety of often fragmented vertebrate remains that included *M. primigenius*, *Bison/Bos* sp., *E. ferus* and possibly rhinoceros (?*C. antiquitatis*). In neither case do the faunas nor the associated plant remains reliably indicate the environment or possible age of the deposits. The Redlands deposits are probably older than the Devensian Reading Town Gravel (see above), while the Endsleigh Gardens deposits post-date the Lynch Hill Gravel. The faunas are both similar to those of late Ipswichian interglacial age, but are equally consistent with cold stage assemblages (cf. Stuart, 1982). The ages of the deposits and the faunas must await further study.

7

Palaeobotany and biostratigraphy

Palaeobotanical analyses have been undertaken to provide both a biostratigraphical framework for the inorganic deposits and a basis for palaeoecological interpretation of the fossiliferous sediments. The studies of almost all the new sites discussed below have been undertaken with co-workers and the full descriptions of their palaeoecology will be published separately elsewhere. However, a synopsis of the palaeobotany is presented here together with summary pollen diagrams for the unpublished sites. The localities are discussed in stratigraphical order.

Pollen preparation by the author has followed standard chemical methods used in the Subdepartment of Quaternary Research, Cambridge University (West, 1968), including the use of sodium pyrophosphate (Bates, Coxon and Gibbard, 1978). The pollen sum in all cases is total land pollen and spores excluding aquatic taxa. Pollen type conventions follow Andrew (1970) together with some types listed in Birks (1973). Macroscopic plant remains have been extracted using West's (1968) method. Plant nomenclature follows Clapham, Tutin and Warburg (1962) for British species and Tutin *et al.* (1964–80) for non-British taxa.

Fig. 55. Pollen diagram from the organic silts at Nettlebed (reproduced by permission of C. Turner).

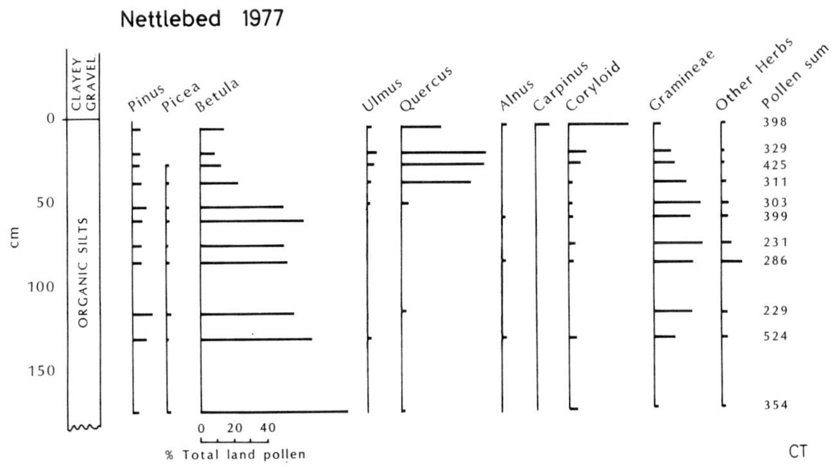

7.1 Nettlebed

The palynological investigations at this site were by C. Turner from whose description (Turner, 1983) the following is taken:

The organic silts here (Section 2.2) have been shown to contain abundant well preserved pollen, but no plant macrofossils or insect remains. The pollen diagram (Figure 55) demonstrates a characteristic Pleistocene interglacial vegetational succession, beginning with a Pre-temperate substage (NeI), continuing through an Early-temperate substage (NeII) and just recording the onset of a Late-temperate substage (NeIII). The Pre-temperate zone is characterised by the dominance of *Betula* with subsidiary *Pinus* and small amounts of *Picea*. *Quercus* dominated the Early temperate forest, though small amounts of *Ulmus* were also present. The onset of Late temperate conditions was heralded by the immigration of *Carpinus* and a marked expansion of *Corylus*. *Alnus* was present in low frequencies during the temperate substages.

The pollen diagram contains no 'exotic' pollen taxa, the only non-native tree being *Picea*. Nevertheless, it is very unlike other Middle and Late Pleistocene pollen diagrams in several respects. One of the most striking features is the absence of temperate elements such as *Hedera* and *Ilex*. Likewise *Tilia* seems to be absent. *Corylus* expands relatively late, while *Picea* occurs early in the interglacial. No trace of *Tsuga* or *Pterocarya* has been found, although a further sequence has still to be investigated and may provide additional information.

From its position in the local sequence of deposits the Nettlebed sediments must be of considerable antiquity. The early appearance of *Picea* and the late expansion of *Corylus* are both features that suggest an early Middle Pleistocene age. Likewise the pollen diagram differs distinctly from those of the Hoxnian or Cromerian (*s.s.*) interglacials, and indeed the Pastonian also. Comparisons with more fragmentary diagrams from the Bramertonian marine sediments are difficult, and the probability is that the Nettlebed deposit represents an interglacial period hitherto undefined in the British Isles. Correlation with sequences in the Netherlands or beyond is fraught with difficulty because of the great distances. However, the sequence appears to fall broadly into the 'Cromerian Complex' of Zagwijn (1974, 1975). The lack of Tertiary-relict species such as *Eucommia* or *Tsuga* preclude a 'Cromerian I' or earlier age (cf. Zagwijn, van Montfrans and Zandstra, 1961). Since 'Cromerian IV' was thought by Zagwijn (1974, 1975) to be the probable equivalent of the Cromerian (*s.s.*), Nettlebed should be equivalent to either 'Cromerian II' (Westerhoven) or 'Cromerian III' (Rosmalen). However, the Nettlebed pollen diagram does not closely resemble those from either Cromerian II or III. Correlation must therefore await further investigation.

In view of these considerations it is tentatively proposed that the sequence should represent the type section of a new biostratigraphical stage provisionally to be termed 'Nettlebedian'.

Fig. 56. Pollen diagram from the organic sediments in borehole Q at Slade Oak Lane, Denham.

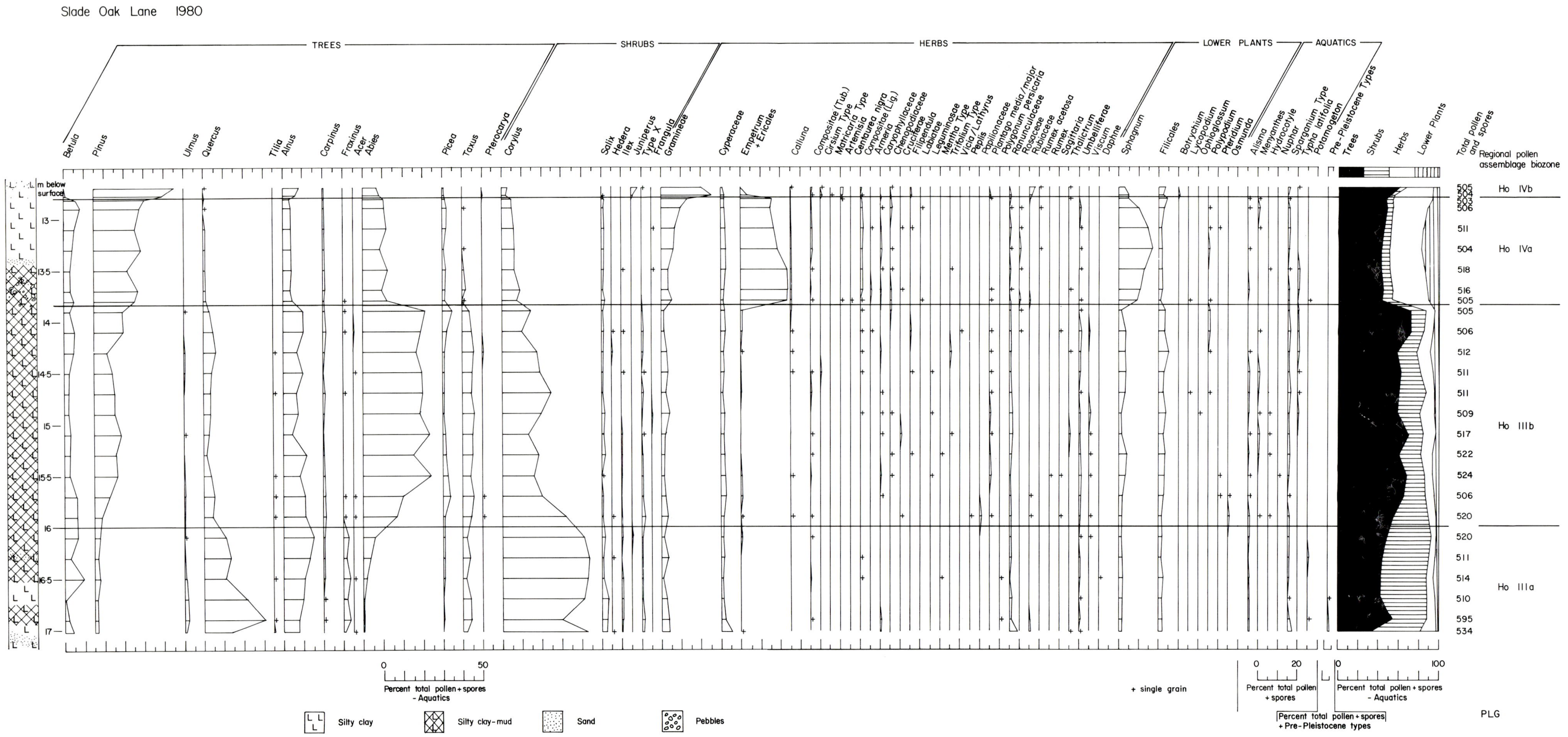

7.2 Slade Oak Lane

Palaeobotany

Detailed investigation of the fossil pollen and macroscopic plant remains have been carried out from borehole Q (Section 2.10) and will be reported fully by Gibbard *et al.* (in prep.). The pollen diagram is shown in Figure 56. The sequence obtained has been subdivided into four pollen assemblage biozones (p.a.b.) which have been correlated with the regional p.a.b.s of the Hoxnian Stage (West, 1956; Turner, 1970).

The biozones recognised are as follows:

HoIIIa: 16.00–17.05 m *Corylus–Quercus–Fraxinus* p.a.b.

Throughout the zone the spectra indicate the total dominance of temperate deciduous mixed oak forest, including *Corylus*, *Quercus*, *Ulmus*, *Betula*, *Fraxinus*, *Alnus* and small amounts of *Tilia*. The occurrence of fully temperate conditions is confirmed by the pollen of frost-susceptible *Hedera* and *Ilex*. The pollen of *Carpinus* and coniferous trees particularly *Abies* and *Picea* become increasingly important throughout the zone. *Taxus* and *Pinus* are also well represented.

The occurrence of the unidentified pollen Type X in this and the subsequent zone further reinforces the Hoxnian age of the deposit since Type X is at present known exclusively from the Hoxnian Stage in this country (Turner, 1970).

Macroscopic remains (A. Hall, personal communication) become progressively more abundant through the zone and include *Betula* fruits, megasporangia of the pond fern *Azolla filiculoides* and achenes of *Urtica dioica*. Over 140 fruits of the extinct sedge *Dulichium arundinaceum* were present in the upper part of the zone. This plant was previously only known from Quinton, Birmingham (Bell, 1970), and Wing, Rutland (Hall, 1978, 1980).

HoIIIb 13.84–16.00 m *Abies–Picea–Corylus–Pinus* p.a.b.

This is by far the best-represented zone in the sequence. It is marked by the dominance of forest, the decline of the pollen of mixed oak forest taxa and the expansion to its acme of conifer pollen, in particular *Abies*. The closed forest seems to have prevented the extensive growth of herbs. The pollen and macroscopic remains of aquatic and waterside plants is abundant and diverse. *Dulichium* occurs in almost all levels and is accompanied by another 'relict' species, *Brasenia schreberi*, previously only found in this country at Wing (Hall, 1978, 1980).

HoIVa 12.80–13.84 m *Pinus*–Ericales–*Sphagnum* p.a.b.

The abrupt decline in the pollen of both deciduous and Late-temperate forest trees (*sensu* Turner and West, 1968), is accompanied by an expansion in *Pinus* pollen. The temperate taxa are replaced by an expansion of *Empetrum* and Ericales pollen, an increase of herb taxa characteristic of open ground, and an increased frequency of spores of the moss *Sphagnum*. Macroscopic plant remains are very poorly represented in this zone. These features together with the increasing inorganic inwashed component in the sediments record a deterioration of soil and climate accompanying opening of the vegetation typical of the end of the Hoxnian interglacial.

HoIVb 12.70–12.80 m *Pinus*–Gramineae p.a.b.

The remaining sediments which continue into the overlying silty clays without a break fall into this zone. The contemporary pollen spectra are characterised by a further abrupt increase in *Pinus* pollen and the continued decrease in temperate trees. Pollen from the latter may have been washed into the pool from the steep sides of the hollow to judge from the poor preservation at this stage. Of note is the occurrence of *Pterocarya* pollen. This tree has been characteristically found in Hoxnian substages HoIIIb and IVa in the British Middle Pleistocene (Turner, 1970). The pollen of *Juniperus* expands at this time and is accompanied by Gramineae. Diverse herbs of unshaded habitats appear in this zone and confirm that large openings existed in the woodland cover by this phase.

Taking the biozones together, there can be no doubt that this sequence represents the second half of the Hoxnian Stage (Late and Post-temperate substages). It is possible not only to match the broad assemblages with those at Hoxne (West, 1956) and Marks Tey (Turner, 1970), but individual curves from Slade Oak Lane are so similar that they could be superimposed on the Marks Tey spectra.

Water-level fluctuation

The occurrence at Slade Oak Lane of a continuous sequence of deposits representing only the second half of the Hoxnian Stage is significant. Lack of weathering in the sediments beneath those yielding temperate pollen indicates that they must have been moist but not submerged for much of the first half of the interglacial. The rise in water level early in substage HoIIIa, which must have taken place to permit deposition of the fossiliferous sediments, has important implications. This rise is contemporary with water-level rises at sites in Hertfordshire (Gibbard and Aalto, 1977) and at Marks Tey (Turner, 1970). That the rise in water level also occurred at this site underlines that it was brought about by a regional water-table rise. This phenomenon may record increased precipitation during the second half of the Hoxnian Stage.

The timing of the formation of the Slade Oak Lane doline and the age of its unfossiliferous infilling sediments will be discussed in Chapter 9.

7.3 Trafalgar Square

Palaeobotany

The sedimentary sequence beneath Canadian Pacific House was sampled in 1980. The upper part, 5–1.5 m OD, was exposed in a section and the remaining deposits beneath were sampled by boring. The pollen samples were taken from blocks used for determination of plant macrofossils, Mollusca and Ostracoda. A full description of the site and its palaeontology is being prepared by Coope *et al.* The pollen analyses are shown in Figure 57. The sequence has been subdivided into the following local biozones:

Ts a −2.20 to −1.90 m *Quercus–Corylus* p.a.b.

This zone occurs within grey silty sand sediment with some brown sand deposited under conditions of variable energy water flow. The dominance of interglacial temperate forests is indicated by the abundance of *Quercus* pollen with subsidiary amounts of *Acer, Fraxinus* and increasing *Corylus* through the zone. Likewise the presence of frost-susceptible *Hedera* and *Ilex* confirm that winters were mild. Herb taxa are abundant and diverse, many of them decreasing in frequency towards the top of the zone. The assemblage includes the pollen of a variety of plants of dry grassland including Gramineae, *Plantago major/media* Type, Chenopodiaceae, Compositae (Liguliflorae) and Umbelliferae which presumably grew on the drier areas of the neighbouring floodplain. There is evidence of considerable inwash of material during this zone indicated by the high frequency of pre-Pleistocene palynomorphs.

Ts b −1.90 to −0.60 m *Corylus Quercus Acer* p.a.b.

The dark grey silty clays of this zone grade upward from the underlying sands and record a fall in current velocities and a stabilisation of the channel. *Corylus* values rise rapidly at the base of the zone to become the dominant tree above. A rich assemblage of temperate forest trees included *Acer, Fraxinus* and *Ulmus* as well as *Quercus*. The conifer *Taxus* was also present at the time. The occurrence of local wet woodland is suggested by the association of *Alnus* pollen together with Filicales spores which certainly include those of the fern *Polypodium*. The continued prevalence of mild winters is demonstrated by the finds of *Hedera, Ilex* and *Viscum* (cf. Iversen, 1944). Throughout the zone the pollen of Gramineae and herb taxa are a very important element of the spectra and reflect the existence of rich dry grassland communities in the vicinity. Plants of disturbed and trampled ground such as *Plantago major/media* Type, *Trifolium* and Compositae (Liguliflorae) are quite well represented and may be related to the activities of large vertebrates known to have inhabited the area (cf. Franks *et al.*, 1958). Damp ground and marsh communities flourished near the channel, while the channel itself contained a lush aquatic vegetation in the slowly flowing water, particularly during the lower half of the zone. Evidence of a slight brackish influence is indicated by the records of *Armeria maritima* and *Plantago maritima*.

Ts c −0.60 to −3.40 m *Quercus–Corylus*–Gramineae–herb p.a.b.

The sediments of this zone show a progressive coarsening becoming more sandy upwards, indicating that the channel was being invaded by increasingly faster flowing water. The pollen spectra show a sharp decline in *Corylus*, accompanied by an expansion of *Quercus* together with the continued presence of *Acer, Fraxinus* and *Pinus* suggesting that deciduous woodland continued to flourish at this time. *Taxus, Alnus* and *Ulmus* all disappear during this zone, but *Picea* is continuously represented for the first time. The short-lived rise in *Salix* suggests a period of wet woodland development at the beginning of the zone. The increased pollen frequency of herbs, particularly those of dry, calcareous grassland throughout the period is very marked and reflects a retreat of the forest vegetation from the immediate area. Plants of marshy ground continue to be represented but at a lower frequency than in the previous zone. Increased water flow is marked by the increased inwash of pre-Pleistocene palynomorphs. A brackish influence is suggested by the presence of *Armeria maritima, Plantago maritima* and members of the Chenopodiaceae.

Fig. 57. Pollen diagram from the fossiliferous sediments at Canadian Pacific House, Trafalgar Square.

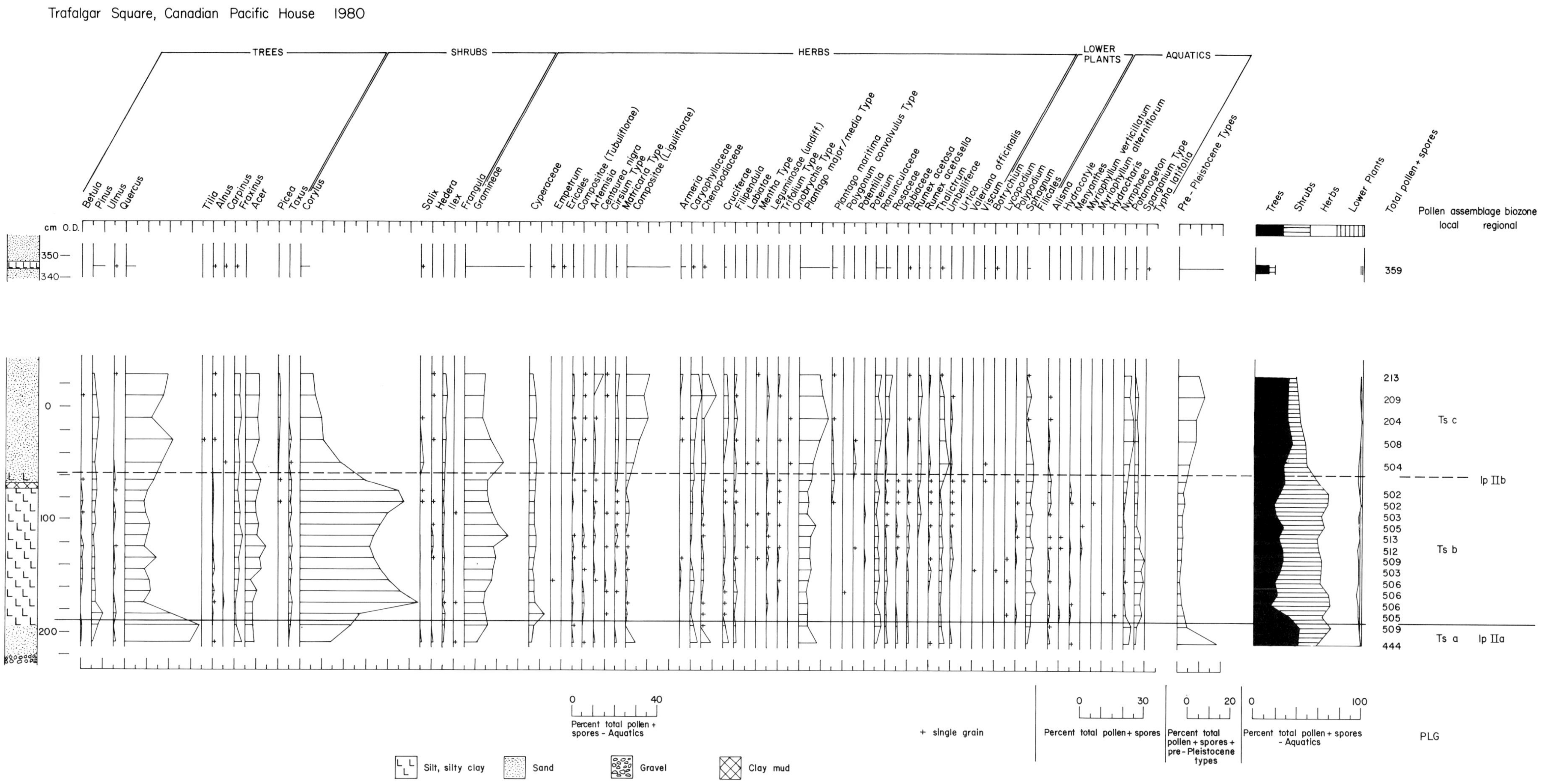

Sedimentary sequence

The pollen sequence at the Canadian Pacific House site is clearly strongly controlled by sediment type. The coarser sediments, i.e. the basal and upper sediments, contain sand bands and yield relatively high herb pollen frequencies. Many of the taxa present at the base disappear as the sediment becomes finer, but reappear as the sediment coarsens towards the top of the sequence. This phenomenon is parallelled by the pre-Pleistocene palynomorph frequencies, which are high in the coarser sediment, but very low during the central part of the sequence. It appears likely that the diverse herb pollen may result from inwash of floodplain sediment. Therefore the pollen of these taxa may be over-represented in the spectra from this site during the initial and later phases (Ts a and Ts c). Furthermore, the changes in proportion of all taxa from local biozones Ts a to Ts c may partially reflect sedimentary phenomena rather than any regional vegetational change. The sedimentary change resulting in the gradation upwards into the current-bedded sand sequence overlying the silts (Section 2.14) is probably associated with deepening and increased flow velocity of water at the site. The very mild brackish influence implied by the pollen is also found in the Mollusca and Ostracoda (e.g. *Cyprideus torosa*) from these sediments (R. C. Preece & J. E. Robinson, personal communication) and strongly suggests that the deepening of the water may have been associated with marine transgression during this phase.

Age of the sequence

The organic deposits at Uganda House, Trafalgar Square, were correlated by Franks (1960) with zone f (IpIIb) of the Ipswichian Stage as developed at the type section (West, 1957) on the basis of pollen assemblages. More recently Coope (1974) supported this correlation on the basis of the coleopteran faunas. Since the deposits at Canadian Pacific House are the lateral extension of those at Uganda House there can be little doubt that they are also of Ipswichian age. The sequence described here is much longer and more detailed than Franks' (1960) analyses, which covered only 96 cm of sediment. It is also more closely related to samples for other fossil groups. Nevertheless, the assemblages he described closely match those of local biozone Ts b and include the same taxa. The entire Canadian Pacific House sequence seems to represent the Ipswichian substage IpIIb by comparison with Bobbitshole, Ipswich (West, 1957), with the possible exception of local biozone Ts a which may be equivalent to IpIIa. The boundary between substages IpIIa and IpIIb is placed conventionally at the expansion of *Corylus* pollen. The correlation is based on the dominance of *Quercus* and *Acer* pollen together with the abundance of *Corylus* pollen.

The Trafalgar Square sequence shows considerable similarity to that from Aveley, Essex in the Lower Thames Valley (West, 1969). Both represent much the same period of time and both show a considerable upward expansion in herb pollen following the decline in *Corylus*.

7.4 Tennessee Pancake House, Whitehall

Palaeobotany

The pollen sequence from the grey silts at Whitehall was sampled from a section by A. Currant (British Museum (Natural History)). Samples were also taken for fossil Mollusca and Ostracoda and the results of the combined analyses are being prepared for publication. Since the silts fill a channel excavated into the sands exposed in the Uganda and Canadian Pacific House sections, they must be younger. The results of the pollen analysis are shown in Figure 58. Many of the comments made for the Trafalgar Square sequence are equally applicable to this, since it appears to be a continuation of the former.

The sequence apparently shows no change in total assemblage throughout, except for a slight change in proportion resulting from progressive silting-up and therefore decreased inwash into the channel. This decreased inwash is illustrated by the upward decline in pre-Pleistocene palynomorphs. Both deciduous and coniferous trees are very poorly represented throughout with an upward decrease in all taxa. This is accompanied by a rise in the frequency of Gramineae, Compositae (Liguliflorae) and Chenopodiaceae. Other grassland herbs are very strongly represented throughout and presumably flourished on the floodplain near the site. Aquatic taxa are limited in number and marsh plants show an upward decline which presumably reflects infilling of the channel. A slight saline influence is implied by the occurrence of *Armeria maritima* and *Plantago maritima*.

Age of the sequence

There is no conclusive evidence of the age of these deposits from the pollen spectra. In isolation they suggest a cold stage environment because of the lack of tree pollen. Conversely, the fossil Mollusca (M. P. Kerney, personal communication) indicate that

Fig. 58. Pollen diagram from the sediments at the Tennessee Pancake House, Whitehall.

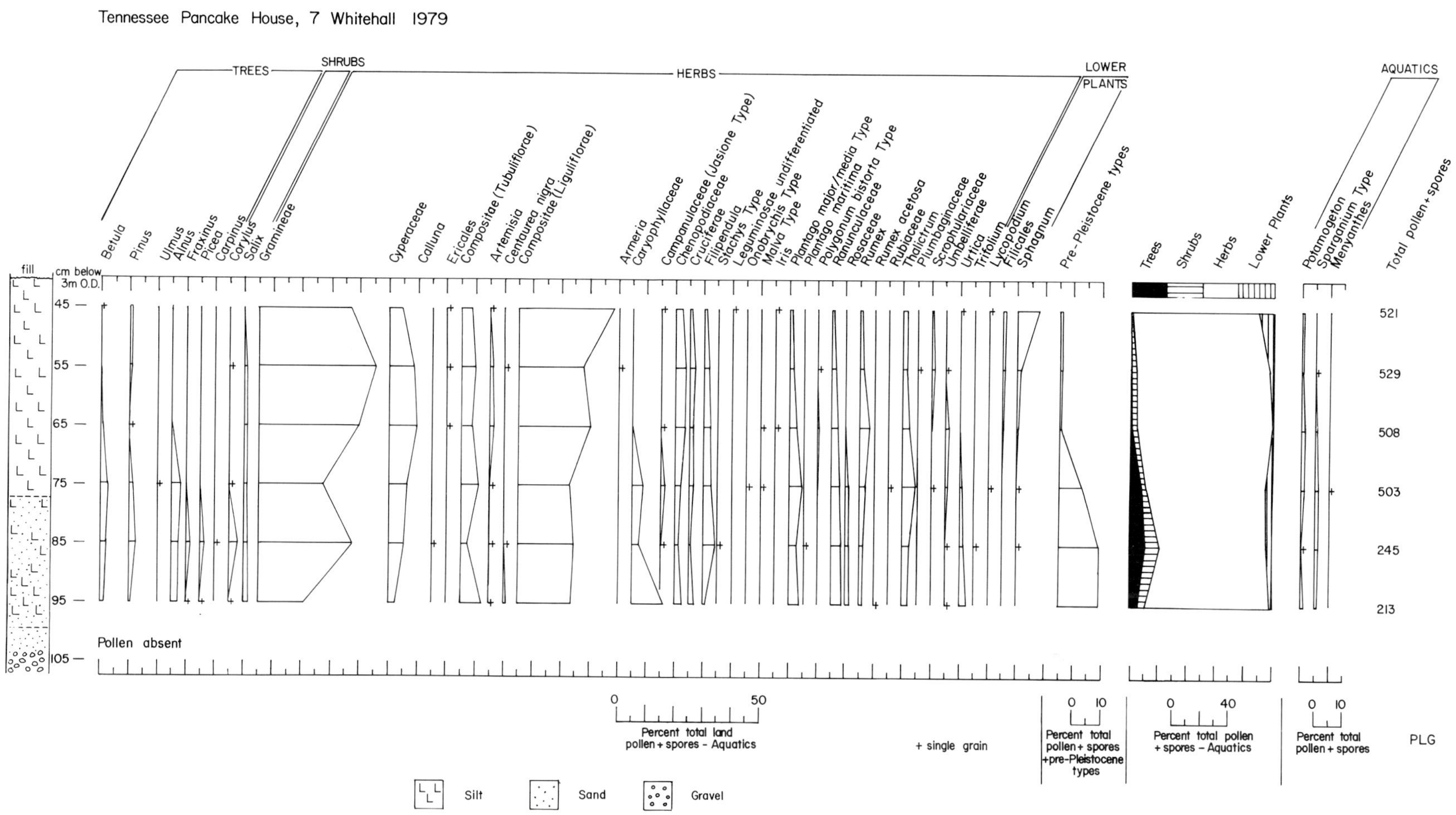

the deposit accumulated under temperate conditions, although reworking of earlier fossils from the underlying sands could have taken place. Since the channel is excavated into the shelly sands of Ipswichian IpIIb age, the channel might also be assumed to represent a later phase of the Ipswichian if the Mollusca are *in situ*. However, if this is correct the lack of tree pollen is remarkable. Increased non-tree pollen (n.a.p.) frequencies have been described from several Ipswichian sites (Phillips, 1974), including Aveley (West, 1969) and Ilford (West, Lambert and Sparks, 1964) in the Thames Valley. This rise in n.a.p. frequencies has been attributed to various factors, but most commonly to the activities of large vertebrates clearing the floodplain. Nevertheless, this does not seem to be the case at Whitehall, because only one vertebrate bone (*Bos* sp.) was found (A. Currant, personal communication). The shelly sands at Trafalgar Square are thought to have accumulated in deep water following development of estuarine conditions. The dominant n.a.p. assemblages may thus have been derived by inwash of pollen from extensive grass and swamp vegetation developed on the surrounding ground and from upstream, as noted above. The slight saline conditions suggest that the same general environmental conditions continued to prevail at this time.

The age of this deposit therefore presents a dilemma which is unresolved at present. Its resolution must await further information.

7.5 Endsleigh Gardens

Plant remains from the mammaliferous loam at this site were examined by C. Reid (in Hicks, 1892). The assemblage included a number of marsh and aquatic plants but none of particular climatic affinities. Reid commented 'that the climate . . . was then cold there cannot . . . be any doubt'. However, Godwin (1975) thought the site interglacial but did not state the reason. It is clear therefore that the age of the deposit cannot at present be determined from its palaeobotany.

7.6 Middle Devensian localities

Within the area studied, fossiliferous plant-bearing deposits of Middle Devensian age are known from six localities within the Kempton Park Gravel: Ismaili Centre (South Kensington), Isleworth,

Fig. 59. Pollen diagram from the fossiliferous sediments at the Ismaili Centre, South Kensington.

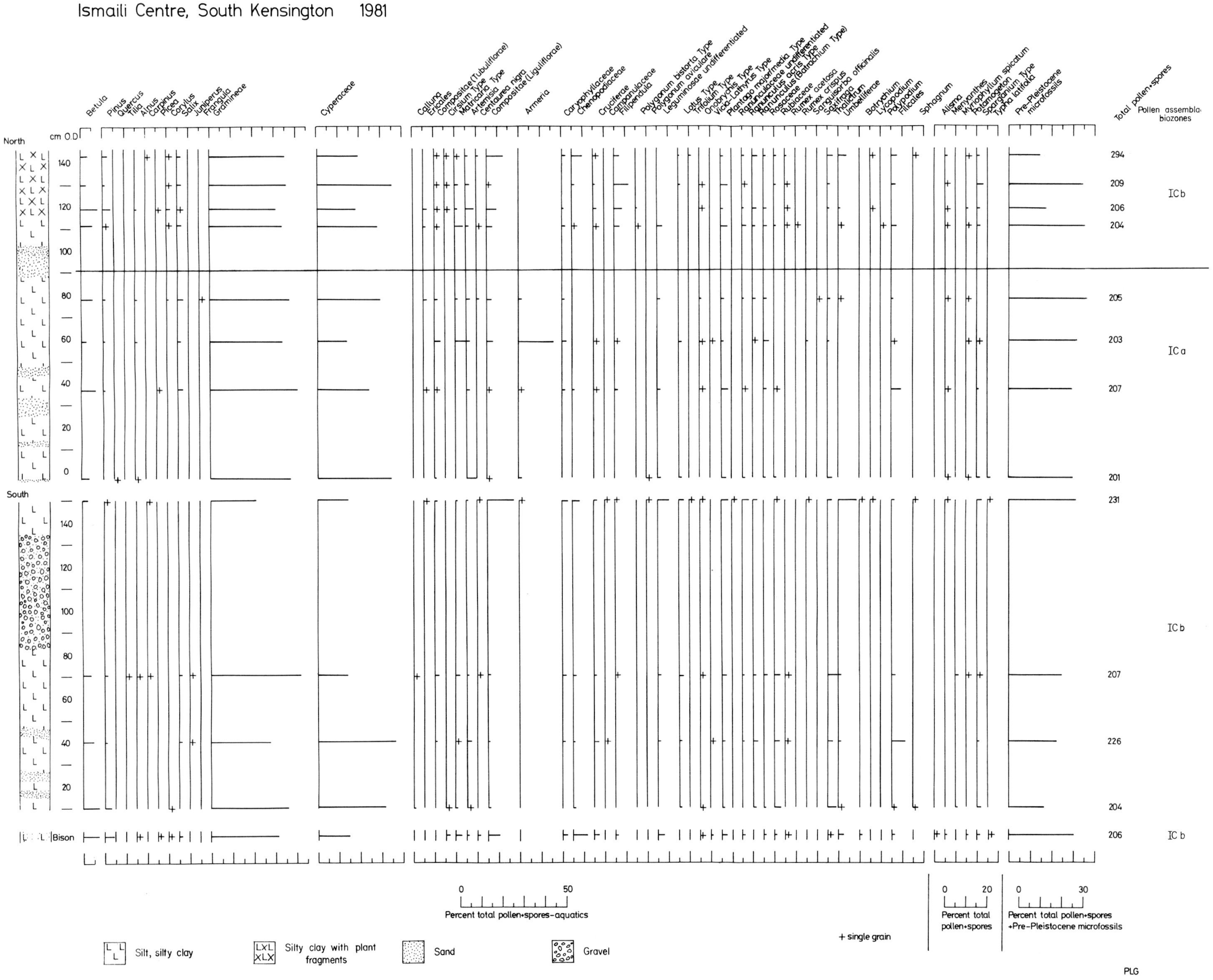

Kempton Park, Marlow, Twickenham and Kew Bridge. Of these the first three are the most important since they have been studied in detail from several aspects. At Marlow (Bell, 1968, 1969) the sediments were not seen *in situ*. The deposits at Twickenham were investigated by Leeson and Laffan (1894), and with the exception of some pollen analyses by the author, have not been examined since. The last site was originally described by Morris (1850) and attempts to relocate his sequence met with only limited success.

The following description will be based on the interpretation of the sequence at the first three sites which represent various parts of the Upton Warren Interstadial Complex (*sensu* Coope and Sands, 1966). This has been determined both by radiocarbon dating and by the fossil coleopteran faunas. The remaining localities will be mentioned after the sequences from the main sites have been described.

Ismaili Centre

Within the gravels at the Ismaili Centre site, a bed of interstratified silt, silty clay and sand up to 2.3 m thick was found. The fine sediments in this bed have yielded remains of various fossil groups which will be published in detail elsewhere by Coope *et al.* Both pollen and plant macrofossil analyses were undertaken using the same samples. Plant macrofossils were generally in poor condition and present in limited numbers (A. Hall, personal communication). Throughout the sequence pollen was poorly preserved and included many degraded grains; for practical reasons these were not counted. In addition reworked pre-Pleistocene microfossils were very common at all levels. The occasional finds of Pleistocene temperate tree pollen grains, often poorly preserved, were probably reworked from local interglacial sediments or else derived by long distance transport. Support for the former hypothesis is found from the other fossil assemblages at the site, all of which contain reworked interglacial elements.

Two vertical profiles were examined, the first and most important sequence from the north face and the second from the south face (Figure 59). Since the north face sequence is the most complete it will be discussed first. Two pollen assemblage biozones can be recognised in the north face section (to which the measurements relate) and are shown in the pollen diagram.

IC a −4 to 90 cm OD Gramineae–Cypraceae–*Armeria* p.a.b.

The pollen assemblages of this zone are dominated by Gramineae and Cyperaceae. The low percentages of arboreal pollen indicate that trees were absent except for local stands of dwarf birch (*B. nana*) also represented by macroscopic remains. The pollen records for *Salix* are not supported by macrofossils, however they were probably derived from dwarf willow nearby. High frequencies of herb pollen are found in all the spectra, but represent only a limited flora. Plants of dry grassland habitats are common and include *Artemisia*, Compositae (Liguliflorae), Caryophyllaceae and *Sanguisorba officinalis*. Local stony ground nearby was colonised by *Saxifraga oppositifolia*. Marshy sedge communities are represented, while aquatic taxa are not abundant. The very high values of *Armeria maritima* pollen are supported by macrofossil remains at the same level. Today this plant occurs on mountain tops, coastal cliffs and salt marshes. *Silene maritima* also found in this zone has a similar distribution. The presence of these plants here may suggest that soils of high salinity possibly induced by high evaporation rates might have occurred locally, perhaps associated with local permafrost. The macrofossil assemblage is also very limited but includes several noteworthy species, such as the 'southern' crucifer *Diplotaxis tenuifolia* (*sensu* Bell, 1969). Overall the palaeobotanical evidence suggests that a cold climate prevailed during this zone.

IC b 90 to 156 cm OD Gramineae–Cyperaceae–Umbelliferae–Compositae (Liguliflorae)–herb p.a.b.

The base of this zone is marked by the appearance of a much more diverse herb pollen assemblage. There is a particularly clear increase in the pollen of dry grassland plants many of which are also recorded as macrofossils. The most striking increase is that of the pollen of Umbelliferae which was almost entirely absent from the previous zone. This corresponds closely with plant macrofossil records for *Anthricus sylvestris*, *Heracleum spondylium* and cf. *Oenanthe* sp. from this zone. The diversity of grassland plants appears to result from the development of a tall herb community. A rich aquatic and waterside flora was also present and includes several taxa absent from zone IC a such as *Eleocharis palustris*, *Schoenoplectus* sp. and *Groenlandica densa*. Tree pollen remains at very low levels, although there is a slight increase in frequencies of *Pinus* suggesting a possible rise in the long distance transported component.

On the basis of the local pollen assemblage biozones defined for the north face, the south face sequence appears to be equivalent only to the upper zone IC b. This is based particularly on the high frequencies of Umbelliferae, Compositae (Liguliflorae) and Chenopodiaceae pollen. Indeed the south face spectra show only slightly differing proportions of individual taxa.

Examination of the insect assemblages from the Ismaili Centre sequence demonstrates that it represents two phases of the Middle Devensian substage. The sediments of zone IC a accumulated under a cold continental climate while those of zone IC b were apparently laid down under temperate conditions thought to be the Upton Warren Interstadial(*s.s.*) on the basis of the Coleoptera and radiocarbon dates (G. R. Coope, personal communication). The increased diversity of the floral assemblages therefore appears to reflect the increased temperatures during the interstadial.

Isleworth

The fossil assemblages from the dark grey laminated silt at Isleworth have been examined by several workers. Analysis of the abundant insect fauna by Coope and Angus (1975) indicated a temperate climate. A radiocarbon date of $43\,140 \pm {}^{1520}_{1280}$ BP (Birm-319) demonstrated the Middle Devensian age of the deposits. More recently, the palaeobotany, Mollusca, Ostracoda and trichopteran larvae were described by Kerney *et al.* (1982). The organic silt is thought to have accumulated in slowly flowing or periodically still water in an abandoned channel or floodplain depression of the river.

The evidence from the various fossil groups analysed provides a consistent environmental picture for the deposits. The pollen analyses indicate treeless conditions. The spectra are dominated by Gramineae, Cyperaceae and herb pollen. A tall herb community inhabited the drier areas in which Umbelliferae was particularly well represented. Patches of bare ground were also present in the vicinity. On damp, possibly periodically inundated ground near the pool, a variety of marsh plants grew. Aquatic vegetation was quite abundant as also indicated by the insect, mollusc and ostracod faunas. The abundance of remains of plants associated with disturbed ground, e.g. *Plantago* spp. and *Coronopus squamatus*, suggests that vegetation near the pool may have been influenced by large mammals grazing and trampling.

The striking similarity of the palaeobotanical assemblages at Isleworth with those at the Ismaili Centre (IC b) strongly supports the insect evidence for correlation of the two sequences.

Table 5. *Pollen counts from isolated samples from Kew Bridge and Twickenham. The counts are percentages of total land pollen and spores excluding aquatic taxa. The counts of pre-Pleistocene types are percentages of total palynomorphs. The two Kew Bridge samples are from levels 440 cm and 465 cm; the two Twickenham samples are listed under the Geological Museum registration number of the bones* (*Bison* sp.).

	Kew Bridge		Twickenham	
	440 cm	465 cm	GS5034	GS5079
Betula	2.01	1.94	0.92	2.08
Pinus	1.07	1.29	—	2.08
Quercus	—	—	0.96	—
Ulmus	0.53	—	—	—
Alnus	—	0.43	0.46	—
Carpinus	—	0.21	0.46	0.42
Corylus	—	0.86	—	—
Salix	3.12	2.16	1.85	1.66
Gramineae	24.19	18.69	38.88	37.99
Cyperaceae	18.27	23.80	16.66	14.58
Artemisia	3.22	0.43	0.46	1.25
Centaurea nigra	—	0.63	0.46	—
Cirsium Type	—	0.27	0.46	0.83
Matricaria Type	—	0.43	3.24	0.83
Solidago Type	—	—	0.92	—
Compositae (Liguliflorae)	12.36	14.06	18.05	10.51
Armeria Type	—	0.21	0.92	0.42
Caryophyllaceae	1.07	1.08	2.77	1.25
Chenopodiaceae	0.53	1.08	0.96	1.25
Campanulaceae	—	—	—	0.42
Cruciferae	1.07	2.16	—	—
Leguminosae (undifferentiated)	1.08	1.68	1.38	1.66
Lotus Type	1.61	1.08	—	0.83
Lathyrus/Vicia Type	1.07	0.43	—	0.42
Plantago major/media Type	4.83	8.44	4.16	10.83
Plantago maritima	—	0.43	—	—
Plantago lanceolata	—	1.08	—	—
Polygonum persicaria Type	—	1.51	0.46	—
Potentilla Type	—	0.21	—	—
Ranunculaceae	3.76	2.96	2.31	0.83
Rosaceae	0.53	0.27	1.85	0.42
Rubiaceae	1.07	1.08	1.38	1.66
Rumex acetosella	3.22	0.21	0.46	1.66
Saxifraga oppositifolia	1.61	—	0.46	—
Thalictrum	—	1.29	1.38	2.08
Umbelliferae	12.36	8.65	5.09	4.16
Heracleum	—	—	—	1.66
Geranium	—	0.21	—	—
Lycopodium	0.53	0.43	—	—
Sphagnum	—	0.21	—	—
Filicales	2.01	0.21	0.46	0.83
Alisma	—	—	0.43	0.42
Menyanthes	0.49	0.21	0.43	—
Myriophyllum spicatum	—	—	—	0.80
Nuphar	—	—	0.86	—
Potomogeton	0.99	0.21	—	0.80
Sparganium Type	6.46	7.69	5.95	1.61

Twickenham

The sequence exposed at Twickenham was described by Leeson and Laffan (1894). Samples of the 'dark loam' underlying gravel at the site were examined by C. Reid and included plants of dry ground, marsh and aquatic environments. Two pollen analyses from sediment adhering to specimens of *Bison* sp. collected by Leeson and Laffan and stored in the Geological Museum, London were examined (Table 5). Both spectra are remarkably similar to those from both Isleworth and the Ismaili Centre (IC b). They are characterised by containing almost no tree pollen, but are dominated by Gramineae and Cyperaceae. They again include a diverse herb assemblage and notably a tall herb community with abundant Umbelliferae is well represented. The latter is also represented in the plant macrofossil collection. It seems in view of their similarity of occurrence, pollen and plant macrofossil assemblages as well as other palaeontological evidence there is a strong likelihood that the Twickenham organic deposit accumulated during the Upton Warren Interstadial (*s.s.*).

Kew Bridge

In an attempt to relocate the fossiliferous sediments described by Morris (1850) a borehole was sunk adjacent to Kew Bridge station. Although the deposits were not relocated, two narrow silt bands in the gravel at 440 cm and 465 cm were encountered. Pollen preserved in the silt bands comprised the same Gramineae, Cyperaceae and diverse herb assemblages (Table 5). As at Twickenham notably high frequencies of Compositae (Liguliflorae) and Umbelliferae were present. The spectra are so similar to those at Isleworth and the Ismaili Centre (IC b) both in their diversity and composition, that it would seem reasonable to tentatively correlate them.

Table 5. (continued)

	Kew Bridge		Twickenham	
	440 cm	465 cm	GS5034	GS5079
Typha latifolia	—	—	0.43	—
Pre-Pleistocene Types	51.04	66.46	70.18	64.48
Total numbers of pollen and spores counted	201	472	233	248

Marlow

Unstratified samples of calcareous organic silt from the upstream outlier of the Kempton Park Gravel at Marlow were described by Bell (1968, 1969). Both pollen and plant macrofossils indicating a treeless, herb-dominated environment were obtained from the silt. A thermophilous insect fauna (Coope, unpublished data) suggested an early Upton Warren Interstadial age, although too little material was available to give a finite radiocarbon date (Birm-184: >31 000 BP: Shotton and Williams, 1971). The flora included a varied assemblage of chalk grassland plants such as *Arenaria ciliata*, *Leontodon autumnalis*, *Diplotaxis tenuifolia* and *Allium scheonoprasum*. Damp ground, fen and aquatic plants were also well represented, many of which were recorded at the previously mentioned sites. Halophytes were common in the assemblages and include *Glaux maritima*, *Silene vulgaris* agg., *Plantago maritima* and possibly *Zannichellia palustris*. The occurrence of highly saline soils may suggest cold, possibly permafrost conditions during part of the time represented by analogy with Earith (Bell, 1968, 1970) and Ismaili Centre (IC a).

Kempton Park

The sequence and fossil assemblages at Kempton Park quarry were described by Gibbard *et al.* (1982). The deposit comprised a bed of grey clayey silt within gravel, that yielded plant remains, Mollusca, Ostracoda and Insecta. Sedimentological and palaeontological results indicated that the silt accumulated in a slowly flowing river channel. A radiocarbon date of 35 230 ± 185 BP (Q-2019) was obtained from plant remains washed from the silt.

The sediments did not contain countable pollen, but samples produced a limited macroscopic plant flora. The assemblage of taxa obtained represented aquatic, marsh and better drained habitats. Apart from leaves of *Salix herbacea* and two fruits determined respectively as *Betula* cf. *nana* and cf. *Salix*, there were no woody taxa. The flora, thus suggested a treeless landscape with vegetation dominated by grasses and sedges with heliophyte herbs such as *Armeria*, *Campanula rotundifolia*, *Linum perenne* agg., *Potentilla anserina*, *Rhinanthus* sp. and *Silene maritima/vulgaris*. A variety of water and marsh plants such as *Groenlandica densa*, *Potamogeton* sp., *Ranunculus* (subgenus *Batrachium*) and *Zannichellia* indicated slowly flowing water with some static water in places.

Several taxa were recorded that are tolerant of 'Arctic' conditions and have been described from 'Arctic' floras by Bell (1970) and Godwin (1975).

The Mollusca and Ostracoda present suggested a fairly clean, well vegetated stream. A variety of environments were indicated by the insect fauna. Evidence from the insects suggested that at the start of sedimentation, the climate was by no means cool, but rapidly deteriorated becoming cold and continental. The date of 35 230 ± 185 BP for the top of the silts and the presence of temperate insect species at the bottom, suggested that the silts accumulated during the period of climatic cooling that immediately followed the temperate Upton Warren Interstadial episode during the Middle Devensian.

Conclusions

The time span of the Middle Devensian sites considered here is shown in Figure 35. From this it would appear that the Upton Warren event is spanned by the three major sites: Ismaili Centre, Isleworth and Kempton Park within the West London area. From the information presented it is possible to assess the palaeobotanical assemblages characterising the three phases. All three phases are dominated by Gramineae and Cyperaceae and lack trees:

Phase 1: Pre-Upton Warren stadial, represented at Ismaili Centre (IC a) and possibly Marlow. This phase is characterised by restricted herb floras of cold aspect. Similar assemblages were described by Bell (1970) from Earith (level E9). The presence of halophytes such as *Armeria maritima*, *Plantago maritima* and *Silene maritima* suggest the possibility of higher than normal salinity in soils during both this and phase 3.

Phase 2: Upton Warren Interstadial (acme) *s.s.*, represented by Ismaili Centre (IC b), Isleworth, Twickenham, Kew Bridge and possibly Marlow. In all the sites examined and indeed at others outside the area, e.g. Earith (E7: Bell, 1970), the interstadial seems to be characterised by a floral diversity much greater than during preceding and subsequent phases. This diversity is expressed throughout the assemblage, for example in the development of a rich aquatic and marsh plant flora. However, the most marked characteristic appears to be the establishment of a tall herb community dominated by the Umbelliferae. Members of the Umbelliferae, particularly *Heracleum spondylium* and *Anthriscus sylvestris* appear to be restricted to this phase. Both have previous interglacial records, but none from cold stages (cf. Godwin, 1975) other than the Upton Warren Interstadial. The presence of these species in particular and the Umbelliferae-dominated tall herb flora in general may provide some evidence for dating. Halophyte plants are relatively rare at this time.

Phase 3: Post-Upton Warren stadial represented by Kempton Park. This phase is marked by a return to more restricted floral assemblages although it appears to include a more diverse flora than that present before the interstadial. Heliophyte herbs are well represented and plants of cold aspect such as *Salix herbacea* reappear.

7.7 Late Devensian sites in the Colne Valley

Two beds of organic sediment were discovered in gravels underlying the present floodplain of the River Colne at West Drayton, Middlesex, and Colnbrook, Buckinghamshire. The deposits and the plant remains they contained were discussed by Gibbard and Hall (1982) and subsequently Coope (1982) described the insect faunas. At neither site was contemporary pollen preserved. The form and stratigraphy of the deposits suggested that the organic sediment accumulated in a shallow channel of the river in slowly flowing conditions. Radiocarbon dates of 13405 ± 170 BP (Q-2021) for Colnbrook and 11230 ± 120 BP (Q-2020) for West Drayton were determined using washed plant remains from each bed.

At both sites, floras indicating cold stage environments with the predominance of open ground taxa were recovered. The only woody plant represented was *Salix* at West Drayton. Notable among the plant records were *Armeria*, *Arenaria ciliata* group, *Betula* cf. *nana*, *Diplotaxis tenuifolia*, *Glaux maritima*, *Linum* cf. *perenne* agg., *Salix herbacea* and *Silene maritima/vulgaris*. All these taxa have been previously recorded from cold stage deposits (Godwin, 1975) and several from nearby Middle Devensian sites (discussed above). The assemblages as a whole suggested a mosaic of vegetation associated with shallow water and streams, but also well drained soils in the vicinity.

The insect faunal assemblages suggest that the Colnbrook deposit formed under a climate of extreme severity. The West Drayton insect remains suggest that they accumulated under the cold climate apparently of Loch Lomond (Younger Dryas) Stadial age.

8

Palaeolithic artefact assemblages

The abundance of stone implements associated with the gravels and sands of the Thames Valley has been known for nearly 300 years. The first hand-axe ever recorded was found in Thames gravels beneath Granville Square, Grays Inn, in 1690 by J. Conyers (Evans, 1860). In the latter half of the nineteenth and the first half of the twentieth century a great era of collection and recording of stone tools flourished. This was the period of handworking and sorting of gravel, and for the same reason as with the vertebrate finds, artefact discoveries have now been reduced by machine working and by exhausting of the implement-rich gravel resources. The Middle Thames Valley is arguably the most prolific region for Palaeolithic artefacts in this country which probably reflects its importance to Palaeolithic man. This is presumably because of the abundance of raw material, i.e. flint, and because the Thames was a substantial river in the past as now. Its valley always offered a natural route into the heart of Britain.

The artefact assemblages of the Thames Valley have been the subject of innumerable studies and many important publications. The most famous local collectors included Evans, Brown and Smith in the nineteenth century and Treacher, Lacaille and, more recently, Wymer in this century. In spite of this there are only two recent synthetic works available on the Thames industries, these being the Gazeteer of Lower Palaeolithic sites (Roe, 1968a), and the important, detailed study by Wymer (1968), already extensively referred to above. Wymer's work has been used as the basic reference and is the source of data assembled in Table 6 to show the artefact assemblages present at individual localities. It has been supplemented by reference to Callow (1976), Collins (1978) and by personal communications with D. Roe, G. de G. Sieveking and J. J. Wymer. The artefacts themselves have not been studied by the author. Detailed technological and typological analyses of artefact assemblages from a range of important localities in the Thames Valley and their relation to sites elsewhere were discussed by Roe (1981).

Attempts to reconstruct the stratigraphy and history of the terrrace deposits using artefact assemblages and hand-axe typology have met with failure. The purpose of the work reported here is not to provide a detailed typological or cultural study, but to relate the vast assemblages that have been collected to the stratigraphy established above and to discuss some of the implications of the distributions.

The classification of pieces is based solely on the typological scheme of Wymer (1968) in which the following types have been recognised (Figure 60):

A Pebble chopper-core (Clactonian)
B Bi-conical chopper-core (Clactonian)
C Proto hand-axe (Clactonian)
D Pointed hand-axe (Acheulian)
E Pointed hand-axe (Acheulian)
F Pointed hand-axe (Acheulian)
G Sub-cordate hand-axe (Acheulian)
H Cleaver (Acheulian)
J Cordate hand-axe (Acheulian)
K Ovate hand-axe (Acheulian)
L Segmental chopping tool (Acheulian)
M Ficron hand-axe (Acheulian)
N Flat butted cordate hand-axe (Acheulian)
Blade (Levalloisian)
Flake (Levalloisian)
Tortoise core (Levalloisian)

The use of alphabet letters for each of the hand-axe forms has been modified, purely for convenience, so that transitional types classified as JK for example have been reclassified as J, or GH as G, etc. in Table 6. This was necessary for the use of computer comparative techniques. This method does not unfortunately allow for the freshness of individual artefacts which could not be included in the techniques used. Almost all the artefacts are composed of flint, but a few are of quartzite or Greensand chert.

Fig. 60. Typological classification of Palaeolithic implements used in this work. See text for explanation of types. (Reproduced by permission of J. J. Wymer). Scale bars represent 5 cm.

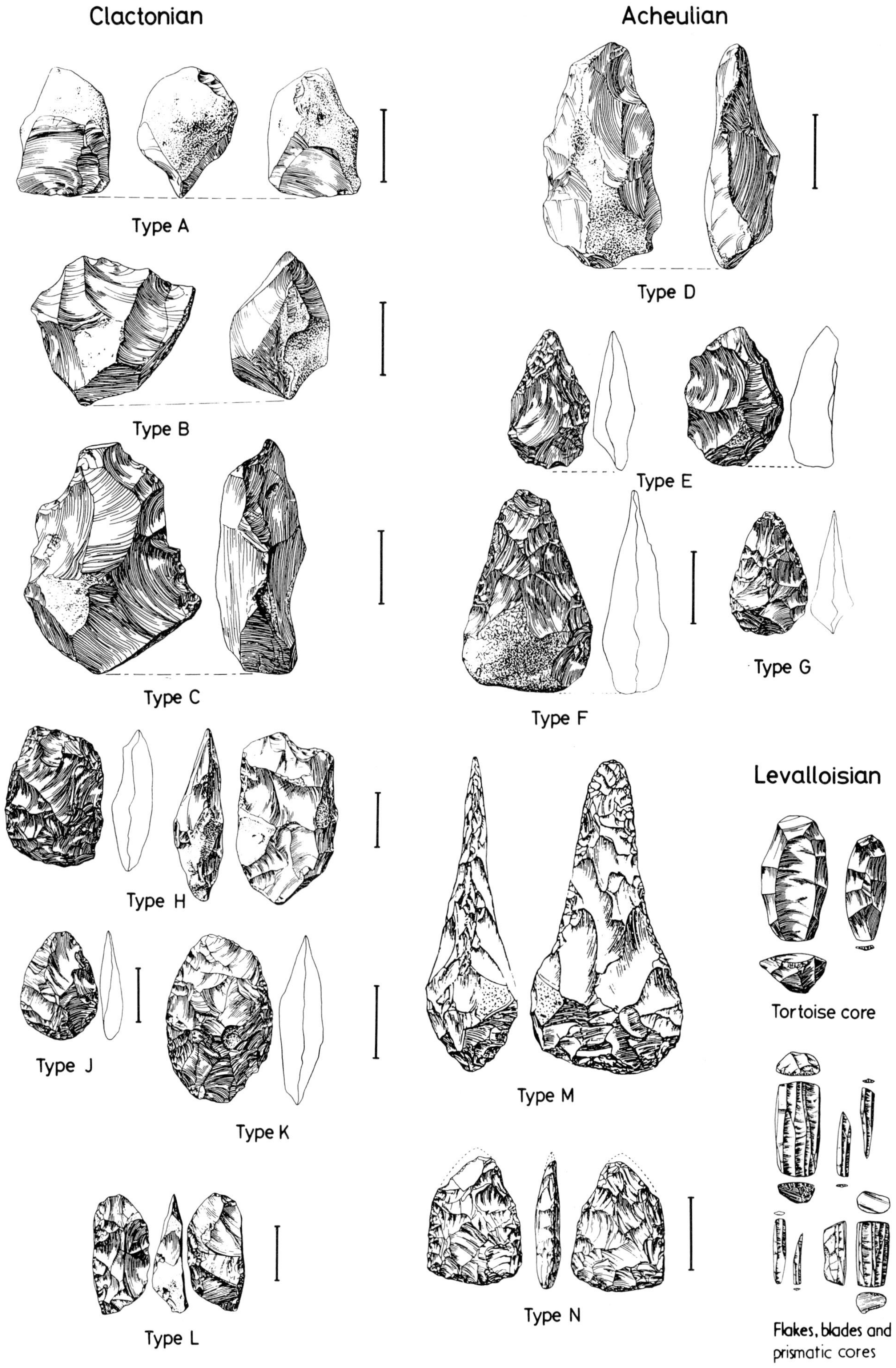

Table 6. *Absolute numbers of Palaeolithic artefacts recorded from the Middle Thames aggradations (modified after Wymer, 1968). The artefact typology is as illustrated in Figure 60 and listed in the text. Additional sites for comparison include unpublished records kindly supplied by J. J. Wymer.*

		Clactonian			Acheulian										Levalloisian		
Site	Grid reference	A	B	C	D	E	F	G	H	J	K	L	M	N	Blade	Flake	Tortoise core
Black Park Gravel																	
Farthingworth Green	SU701766				13	6	7	13		20	11						
Balmore Drive	SU716757						1	1			2						
Emmer Green	SU719761				6	3	7	1		2		1					
Emmer Green, Hathaways	SU718761						1				1						
Kennylands	SU716790				26	10	3	5	1	25	13						
Highland's Farm	SU742813	16	66	8	64	56	3	5	2	70	45						
Herne's Farm	SU743817	1	3		4	2		3		1	2						
Hillingdon (Town)	TQ072824					2	3	4									
Richmond Park	TQ210732				1												
Dartford Heath–Bowman's Lodge	TQ518738	13	55			1		2		25	3	1					
Boyn Hill Gravel																	
Christ Church	SU721721					4	2	2		4							
Toot's Farm	SU707753				20	60	131	23	1	7	6						
North Pit	SU707754				2	2	1	4		1	4						
Highmoor Road	SU707754				3		10	3		3	1						
Harrogate Road	SU705757						4	2									
Kidmore Road	SU706761				15	46	48	8	1	7	3						
Earley	SU744726								1								
Sonning Hill	SU754741				1	2	5	5		4							
Shiplake	SU764783					2	2	1		1							
Henley (Hamilton's pit)	SU760817				2	21	18	1	1	4	6						
Boyn Hill	SU875810				2	4	6	4	1	5	2						
Upper Furze Platt	SU878826				1	5	4			1							
Court House Road (Cooper's pit)	SU876819					2		1		1							
Poyle Farm	SU925838						4	2									
Deverill's (& Cooper's) Burnham	SU947840				13	48	85	22	1	52	18						
Lynch Hill Gravel																	
Roebuck Pit, Tilehurst	SU672752					1	9	12		46	41						
McIlroys Pit, Tilehurst	SU690733				1	1	13										
Denton's Pit, Southcote	SU693723		6		1		1	1	1	2						1	1
Grovelands Pit, Tilehurst	SU693732	1	33		1	17	19	7	2	13	20						
Balmore Drive	SU716755						1	1									
Maiden Erleigh, Earley	SU754713					1	1										
Sonning Cutting	SU762748						4	1			1					?1	
Crockham Well Road	SU764726						1			1	1						
Butts Hill	SU767742		1		5	1	8	2	1								
Twyford (Stanlake Hill)	SU795755						1	1									
Priors Pit, Ruscombe	SU795762				13	19	78	25	2	1	1						
Cookham, Danesfield pit	SU878852				15	23	60	19	11	3	1						
Cannoncourt Farm	SU878831				103	141	272	43	17	2	3						
Lent Rise	SU925818				29	34	131	21	7	2	3						
Farnham Royal	SU957825		4			1											
Lynch Hill	SU949821				1	1											
Baker's Farm	SU958822				38	31	55	8	10	4	1					1	2
Lavender's pit	TQ025802				2	2	6	1	2	3		1					
Iver (GWR)	TQ032800				4	7	10	3	1	3						1	
Yiewsley	TQ071805		2	1	29	66	85	32	5	11	1						
Yiewsley (Collins, 1978)	TQ083803				33	9	6	35	44	9							
Dawley (Hillingdon)	TQ085797				15	34	45	9	1	1	2						
Hayes	TQ095795					2	3										
Southall Gas Works	TQ119797					5	6	1									

Table 6. (*continued*)

		Clactonian			Acheulian										Levalloisian		
Site	Grid reference	A	B	C	D	E	F	G	H	J	K	L	M	N	Blade	Flake	Tortoise core
Southall Norwood Lane	TQ139793					2	1									2	
Heston	TQ120785				1	1	1										
Hanwell + Osterley	—				19	80	56	9		9							
Acton	TQ192808				6	23	24	7		3	1						
Wandsworth	—				3	10	11	7	1	1	1						
Clapham Common	—					3	1				1						
Jermyn Street	TQ295805					3											
Piccadilly Circus	TQ296806				1		7	1									
Oxford Street	TQ286812				1		8	2									
Kings Cross Road	TQ308827						1										
Gray's Inn Road	TQ306825					3	1	2			1						
Pentonville	TQ310835					4	3	2		4							
Holborn	TQ306812						5										
Russell Street	TQ299815						1	1									
Taplow Gravel																	
Henley Road, Donkin Hill	SU724752					1	3	1									
Sonning Town (Charvil Hill)	SU766753					3	4	4		2							
Twyford, Franklin's pit	—						3	3									
Taplow	SU919816				3	4	7	2		3	2						
Taplow, Gibbon's pit	SU917816						5										
Hitcham, Burnham	SU920813				1		3	1									
Wexham	TQ012801				2	1	11	2		1							
Chalvey	SU964798						1										
Salt Hill	SU966803				2	1	2	1									
Braywick	SU895791						2										
Reading Town Gravel																	
Grovelands Church	SU695736				1			1			1						
Station Road, Reading	SU715735					1		1									
Kensington Road	SU698734							1									
(Redlands	SU723727					2	1					1)					
Langley Silt Complex–above Lynch Hill Gravel																	
Iver, Purser's pit	TQ032797														2	3	
Baker's Farm	SU958822															1	2
Lavender's Pit, Iver	TQ025802														1	5	
Iver (GWR)	TQ032800															1	1
Yiewsley + W Drayton	TQ071805														2	4	30
Yiewsley (Collins, 1978)	TQ083803					4				1						4	
Botwell	TQ106795															2	1
Hanwell	TQ150810															4	
Acton, Creffield Road	TQ192808													3	302	155	1
Langley Silt Complex–above Taplow Gravel																	
Taplow	SU919816														1	5	
Langley	TQ012801															2	
Marlow (?) brickpit	SU861873							1						2		1	
Taplow, Wood's Nursery	SU918813														1		
Unknowns																	
Stoke Newington (floor)	TQ338869		1		2	35	24	13	2	1							
Stoke Newington (gravel)					16	98	34	2	2	2	3						
Swanscombe (Lower Gravel)	TQ595745	55	51	4		1											
Swanscombe (Lower Middle Gravel)	TQ595745	1			5	58	17	3									
Swanscombe (Upper Middle Gravel)	TQ595745	2	1		7	51	26	6									
Swanscombe (Upper Gravel)	TQ595745				2	40	45	4	5	32	11	3					
Dorney Wood, Burnham	TQ937857					3	10				2						

8.1 Incorporation of Palaeolithic artefacts in sediments

By far the most Palaeolithic artefacts are from the gravel member sediments which accumulated as we have seen in the braided river environment predominantly under a cold climate. Implements are also found to a lesser extent, in the Silt Complex sediments ('brickearth'). Archaeologists have long acknowledged that these implements must be in a derived state since some are heavily rolled. A quantitative scheme for assessing the degree of abrasion was proposed by Shackley (1974). The amount of transport necessary to produce a rolled appearance from a fresh piece is not known, but may be quite considerable in view of the hardness of flint (for experimental results see Harding *et al.*, 1985). However, there can be little doubt that where flakes, cores or broken hand-axes from sediments can be rejoined or where fresh material of various sizes is concentrated and includes flakes of small size, transport must have been minimal or else of a rather low-energy type that prevented abrasion of individual pieces.

It cannot necessarily be assumed that artefacts were produced immediately prior to being buried in the sediments found today. Artefacts left or dropped onto the ground may become incorporated into sediment by channel migration, by overbank deposition on a floodplain, or by burial in aeolian or slope sediments under cold climates. Again as with vertebrate remains, implements within deposits may be recycled during later phases and redeposited in younger accumulations. Particularly common in this respect is the formation of an implement-bearing gravel 'lag' deposit that arises by removal of fines during channel migration and basal erosion or inflow of soliflucted detritus into a river channel.

It is possible that erosion of pre-existing implement-rich sediment could result in an apparently inverted sequence by the incorporation of stone artefact assemblages into later sediments. This could occur because the upper levels of a sedimentary succession may be first removed exposing progressively lower and therefore earlier sediment as erosion proceeds. Incorporation of indestructible components from eroded sediments might therefore begin with the youngest and finish with the oldest material.

The more restricted occurrence of implements in finer sediment, particularly the Langley Silt Complex ('brickearth') requires a different explanation. It is possible, as suggested by Brown (1886) for the Creffield Road, Acton site, that the industry represents a working floor, i.e. the implements are undisturbed. This origin is equally possible in floodplain overbank depositional situations. However, the concentration of these flake–blades could equally have arisen by sedimentary processes. The sediment in which these artefacts occur is laminated and is interpreted above as a slope wash deposit. In this environment large clasts such as pebbles (including artefacts) can be rolled over smooth silt surfaces particularly on slopes by currents of low velocity. Clasts may then become concentrated in zones where the current falls below the threshold velocity necessary to move the pebble. Such clast concentrations are frequently observed in Silt Complex sediments. This rolling process would not produce discernible wear or abrasion of pieces, comparable to that arising from collision of gravel-sized clasts in a river channel. Moreover, solifluction or remobilisation of the artefact-bearing sediments could lead to their secondary redeposition at lower elevations.

The frequency of artefacts in river valley sediment and particularly in those of the Thames implies that human activity was considerable and it may be that man preferred to live in the valley environment. There are several reasons for thinking this is correct. River valleys would have provided an ideal source of water for drinking, flat land with relatively lush vegetation even during cold periods in summer, and an ideal source of material for stone tool manufacture. In addition, rivers were almost certainly gathering places for large mammals which could have been captured for food. During temperate stages man may well have avoided closed forest because of the dangers from predatory animals and the difficulty of movement, in favour of the generally more open floodplain beside rivers. There are in fact numerous well documented associations of Acheulian industries with stream channels (Roe, 1981). However, the effect of concentration of material by river processes cannot be determined.

8.2 Relationship of the artefacts to the stratigraphical units

The assemblages of artefacts from all sites known to the author that could be related to the stratigraphical units defined above were tabulated (Table 6). Artefacts known to come from within a unit were assigned to that unit. Those from a depositional boundary, for example the interface between gravel and an overlying silt bed may belong to one or other deposit or the intervening period during which the surface was exposed. The data obtained show great variations in sample size, some sites only having produced one artefact compared with occasional sites that have yielded several hundred. In order to compare and contrast assemblages objectively the data were analysed using multivariate methods as for the gravel pebble lithological counts.

Although a number of methods were tested with the data the most useful and interpretable was the ordination procedure Principal Components Analysis (PCA). This method helps the recognition of groups of samples containing typologically similar artefact assemblages regardless of whether or not they originate from the same lithological unit. It also aids the identification of dissimilar assemblages whether from the same or different lithostratigraphical units: i.e. it helps to demonstrate either changes through time or contemporary variation.

Other techniques including Cluster Analysis and Canonical Variates Analysis were applied to the artefact data, but only the PCA gave worthwhile results, and those obtained by the other methods will not be set out here.

8.3 Principal Components Analysis

For these data, a conventional R-mode PCA using a co-variance matrix with square root transformation (see Chapter 3) was found to be the most satisfactory for comparison of all the typologies present. The analytical results are shown in Figure 61. As with the pebble count data, the positions of individual assemblages, represented by their component scores, are plotted on the first and second principal components. For the sake of convenience, assemblages have been separated into their lithostratigraphical members, rather than positioned on a single plot. Nine additional isolated sample assemblages were included in this PCA as unknowns (Figure 61) for comparative purposes. These included four from Swanscombe (Lower Gravel, Lower Middle Gravel, Upper Middle Gravel and Upper Loam: after Wymer, 1968), two from Stoke Newington (basal gravel and beneath the 'brickearth': after Smith, 1894; Wymer, 1968), two from Hoxne (lower industry and upper industry: J. J. Wymer, personal communication) and an assemblage from the hillwash deposit at Dorney Wood, Burnham (Lacaille, 1959; Wymer, 1968).

In this ordination, the first principal component accounts for 29.61% of the total variance. It effectively distinguishes between assemblages containing representative numbers of Acheulian hand-axes, particularly groups E and F of Wymer's system (high

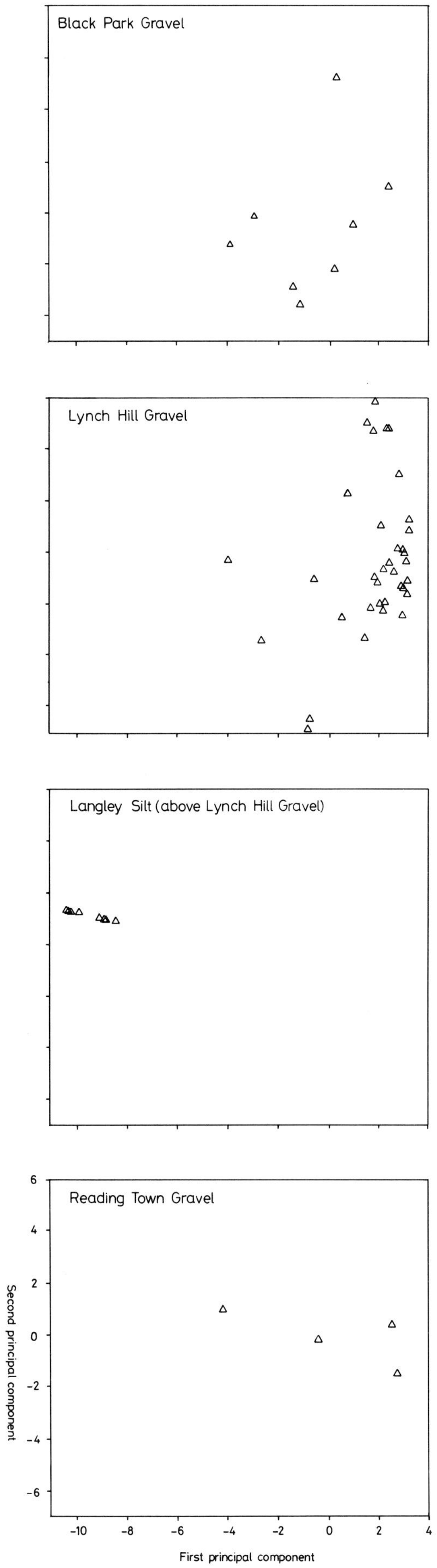

Fig. 61. Principal Components Analysis of Middle Thames Palaeolithic artefact assemblages. The 'Unknown' category includes sites of unknown affinity, as well as sites beyond the study area included for comparison. See text for explanation.

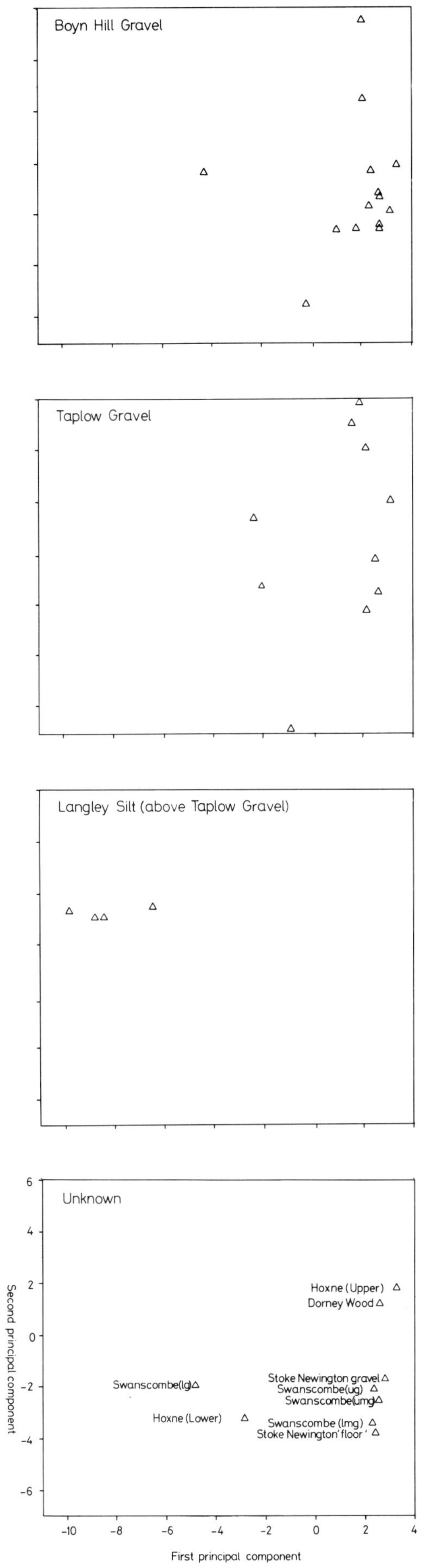

Abbreviations:
lg: lower gravel
lmg: lower middle gravel
ug: upper gravel
umg: upper middle gravel

positive scores >0.64) and those containing Levalloisian (high negative scores <-0.42) and to a lesser extent Clactonian artefacts. The second principal component represents a further 17.21% of the variance and has high positive loadings for Acheulian group F hand-axes (>0.57) and high negative loadings for Acheulian groups E, D and J (<-0.69). The third and remaining principal components show patterns that are too vague to interpret.

From the numerical results obtained by this technique the low percentages of variability expressed by the principal component axes are immediately striking. This clearly underlines the close broad typological similarity of the majority of the assemblages. Secondly, the great variability in sample size creates the problem that the technique tends to discriminate samples comprising one or two artefacts of the same typology (i.e. 100% of that type) and those that include the same type, but with numbers of other types also present. This problem has been reduced as far as possible using the square root transformation.

The main conclusion arising from this analysis is that there is a separation into three broad groupings:

1. The most well defined are the Levalloisian Technique assemblages that occur in the Langley Silt Complex overlying both the Lynch Hill and Taplow Gravels. There are twice as many sites from above the former as from above the latter. The silt overlying the Taplow Gravel has yielded only very few artefacts and might possibly therefore contain an element of sediment and contained artefacts reworked from the higher level spread. This group of sites includes the prolific Creffield Road locality mentioned above.

2. All other Palaeolithic artefact assemblages in the Middle Thames region fall into a second broad grouping comprising those containing Acheulian and Clactonian material. Clactonian assemblages are relatively rare and at no site in the Middle Thames does Clactonian material occur in a pure state. Sites that include Clactonian material are Highland's and Herne's Farms in the Caversham Channel, both of which are included in the Black Park Gravel (Figure 61). The Clactonian industry from the Swanscombe Lower Gravel plots close to those from the Black Park member which is rich in these tools. Clactonian implements were also found in significant numbers in Lynch Hill Gravel at Grovelands, Reading. Their occurrence here and in small numbers at four other Lynch Hill localities may suggest that they were reworked from higher, older deposits. However, since Clactonian artefacts often occur in small numbers in Acheulian industries (Roe, 1981) it is difficult to be sure of the presence of true Clactonian in any mixture of types. Nevertheless, in the Middle Thames some of the mixed assemblages contain far larger components than would reasonably be expected in Acheulian industries and common in the type area. It is for this reason that true Clactonian is thought to have been present at sites like those in the Caversham Channel

3. Assemblages of Acheulian material occur in all five gravel members included in the study. In broad terms there appears to be no reason to separate these gravel deposits on the basis of their artefact content other than possibly by distinguishing those including or excluding Clactonian implements. Both the Lynch Hill and Boyn Hill Gravels contain artefacts of groups D–K, the frequencies of the various forms being approximately the same throughout. These industries appear generally to fall into the Middle Acheulian Culture (Late Middle Acheulian industries) according to Wymer (1968). Levalloisian flakes have been claimed as coming from the gravels (e.g. Lacaille, 1940), but they may come from overlying 'brickearth'.

The assemblages from the sequences at Dorney Wood, Stoke Newington, Hoxne and the Middle and Upper Gravels at Swanscombe all plot with those of the Boyn Hill, Lynch Hill and Taplow Gravels. They therefore contain broadly similar artefact assemblages. The material from the Dorney Wood slope wash deposits (Treacher and White, 1909; Lacaille, 1959) is almost certainly washed-in from higher ground but is thought to post-date the Boyn Hill Gravel on typological grounds (Roe, 1981). The Pleistocene deposits at Stoke Newington, originally described by Smith (1894); and recently re-evaluated by Gibbard (unpublished), have yielded considerable numbers of artefacts and flakes. The geological situation of the finds suggests that they may have been derived from deposits capping Stamford Hill, immediately north of the sites, that include the Lea Valley equivalent of the Lynch Hill Gravel.

8.4 Conclusions and relative ages

It is possible to some extent to ascertain the relative ages of artefact assemblages by comparison with sequences where the industries have been unequivocally related to independent dating evidence. This is based on the assumption that typological similarity indicates similarity of age. However, it must be remembered that the same tasks may have required the manufacture of typologically similar tools over widely separated points in time and therefore attempts to determine the time spread of an individual artefact type may represent an over-simplification.

Black Park Gravel

This member contains local concentrations of artefacts, but for long stretches is apparently devoid of archaeology. In the Caversham Channel the intermixture of Clactonian material and later Acheulian material seems clearly established (Wymer, 1961, 1968) and needs to be explained. At Swanscombe, the Lower Gravel and Lower Loam have yielded exclusively a Clactonian industry. The overlying gravels and sands contain an Acheulian industry the appearance of which therefore post-dates the Lower Gravel and Lower Loam. The Lower Loam and upper part of the Lower Gravel are correlated with the Hoxnian interglacial Stage (Wymer, 1974). This correlation is reinforced by the late Anglian – early Hoxnian age of the Clactonian industry at Clacton itself (Singer *et al.*, 1973; Wymer, 1977b). The evidence from Hoxne demonstrates that the Acheulian industry first appeared in the Early-temperate zone (West and McBurney, 1954), i.e. later than the industry at Clacton. Wymer (1977b) therefore concluded that 'there is reasonable evidence to place the Clactonian industries in the time range spanning the later stages of the Anglian to the early-temperate zone of the Hoxnian.'

The Black Park Gravel is known to be of late Anglian age (Gibbard, 1977, 1979 and above). In view of the dating evidence stated above the Caversham Channel filling would appear to include both late Anglian and Hoxnian artefact elements. Wymer (1961) suggested that the gravels having been deposited late in the Anglian may have been reworked after the Hoxnian, to account for the presence at all levels of the Late Middle Acheulian industry. The latter closely resembles much of the Acheulian from the local Boyn Hill Gravel. A small remnant stream on the floor of the Caversham Channel may have been responsible for reworking the gravel after the division of the Thames. This explanation seems extremely plausible although J. J. Wymer (personal communication) now thinks this unlikely in view of the finds of Acheulian material from the same member at Hillingdon. Later collapse of the gravels into solution hollows makes it almost impossible to clarify the situation further.

The possibility that Early Acheulian material may be present in the Black Park Gravel of the Caversham Channel was suggested by Treacher *et al.* (1948) and Roe (1968b, 1981). However, these typological arguments are not accepted by all archaeologists (e.g. Wymer, 1961, 1968, 1977b) because there is no stratigraphical evidence for separating the disputed pieces from Middle Acheulian material.

Boyn Hill Gravel

This member apparently contains only Acheulian stone tools, many in poor or rolled condition. The industries are much richer in the flat, broad ovates (type K) than the younger Lynch Hill Gravel assemblages (Roe, 1981). Several authors have suggested that the Boyn Hill Gravel aggradation took place under cold conditions on the basis of the coarseness of its gravel (e.g. Lacaille, 1961b). This facies is markedly different from that of the artefact-bearing Swanscombe Middle Gravels which in view of their stratigraphical position and freshness of artefact content presumably pre-date the Boyn Hill-equivalent Fairlop (Orsett Heath) Gravel in the Lower Thames (Gibbard, unpublished; Bridgland and Gibbard, in prep.). The Boyn Hill Acheulian industries themselves were therefore probably partially reworked from previously accumulated sediments during downcutting and gravel and sand aggradation.

Lynch Hill Gravel

The rich artefact assemblages of the Lynch Hill Gravel can be separated from those of the Boyn Hill simply by their greater frequency and the generally fresher character of the material. The Middle Acheulian industries from the Lynch Hill Gravel at important sites particularly Cannoncourt Farm and Baker's Farm have yielded assemblages including characteristic types. These include the fine large ficron hand-axes (type F) and big transverse cleavers (type H), specific forms which are not found in the Swanscombe sequence or the Boyn Hill Gravel (Roe, 1981; Cranshaw, 1983). Roe (1981) included the Cannoncourt Farm and Baker's Farm industries in his Pointed Tradition Group I. The abundance of implements at this stage may arise either from an as yet unidentified contemporary or intervening period of human occupation, during which the manufacture of the same types of Middle Acheulian implements continued as previously, or it may result from increased incision and reworking of remaining pre-existing (?interglacial) sediments during and preceding the Lynch Hill phase.

Taplow Gravel

This member contains very few palaeoliths; all are in a rolled condition and have probably been derived from the higher, Lynch Hill Gravel (Wymer, 1977a). This author noted the presence of a few rolled Levalloisian flakes which he suspected might come from this gravel. One interesting find is the proto-tortoise core from Chalvey Hill, near Sonning Cutting, which was found in this unit. Other than this piece there is virtually no evidence of human occupation either during or immediately preceding the period represented by the Taplow Gravel. Indeed this may have been a phase of very cold conditions.

Langley Silt Complex ('brickearth')

This group of deposits almost certainly includes those of several different ages (see Chapter 9).

The Levalloisian industries associated with the 'brickearths' have already been discussed above. On the basis of a series of sites, notably including Baker's Hole, Northfleet and Crayford, the industries are generally regarded as late Wolstonian to early Devensian (Wymer, 1977b), although an earlier Wolstonian age for the former has been suggested (Roe, 1981). The association of Levalloisian flakes within the so-called 'red brickearth' of Lacaille and Oakley (1936) at Lavender's Pit, Iver may be important. Descriptions of this reddened horizon suggest that it might be a palaeosol. If the dates suggested by Wymer (1977b) are correct then the soil might be of Ipswichian age. The overlying 'grey brickearth' would then be of early Devensian age.

Apart from the Levalloisian material, it is probable that the Langley Silts are also the source of the few 'bout coupé' hand-axes (Mousterian) of the kind discussed by Roe (1981, Chapter VI). Most of these were unstratified finds, but one at least comes from the 'red brickearth' at Iver. By analogy with north-west Europe, this distinctive hand-axe type might be expected to be of early Devensian age, with some later persistence (Roe, 1981).

Reading Town Gravel

A few heavily rolled hand-axes have been found in this unit. Like those in the Taplow they have presumably been derived from the rich assemblages in the adjacent Lynch Hill Gravel.

9

Palaeogeographical evolution of the Middle Thames Valley

From the evidence presented in the preceding chapters the sequence of events in the Middle Thames Valley during the Pleistocene can be reconstructed. The stratigraphical scheme based on all the evidence available is summarised in Table 7 and will be discussed below in chronological succession. Palaeogeographical reconstructions for each of the members are shown in Figure 62.

Unless otherwise stated each aggradational member unit is separated by a phase of valley downcutting from which, by definition, no deposits are preserved. Abandoned upper surfaces of aggradational units have, in almost all cases, been modified either by later sedimentation or degradation. The most common of these modifications resulted from periglacial processes and soil development.

9.1 Nettlebed Gravel (pre-Nettlebedian)

The highest and therefore probably the oldest fluviatile aggradation recognised in this study is the Nettlebed Gravel. It was laid down by an ancestral Thames flowing from Oxfordshire northeastwards across the Middle Thames region. The low content of vein quartz, quartzite and other far-travelled lithologies, in comparison with later units suggests that the river had only limited access to Midland rocks at this early stage. The apparently steeper gradient (90 cm km^{-1} compared to 40–60 cm km^{-1} for later aggradations) implies that the river may have been at an early stage of development in terms of catchment size and discharge. Extension of the gradient downstream strongly suggests that the Nettlebed Gravel is the equivalent of the lithologically similar '400 foot' (110 m) Pebble Gravels of the south Hertfordshire Plateau (Wooldridge, 1927, 1960). By analogy with later units these gravels may have accumulated under a cold climate.

9.2 Nettlebed interglacial deposits (Nettlebedian)

After deposition of the Nettlebed Gravel a period of downcutting occurred. This produced the NNE–SSW-trending channel at Priest's Hill. Reduction of the initially high-velocity flow that deposited pebbly sand led to the accumulation of organic sediments under a temperate climate. These sediments were probably laid down in the slack water of the channel either partially or completely cut-off from the main river. The sediments record the migration of boreal and later temperate forest vegetation into the region. The vegetational succession indicates that the interglacial period is of early Middle Pleistocene age, possibly equivalent to a stage within the Netherlands 'Cromerian Complex' (?Interglacials II or III). Since the basal organic sediments at this site date from the Pre-temperate substage (NeI), the underlying sands must be earlier and may even represent the preceding late glacial phase. The organic infilling of the channel continued into the Late-temperate substage (NeIII) and is truncated by soliflucted debris derived from intermixing of local Nettlebed Gravel and Reading Beds deposited during a subsequent cold period (post-Nettlebedian).

9.3 Post-Nettlebedian to pre-Anglian

After an indeterminate period, downcutting by the river of the bedrock floor followed. Subsequent aggradation of the Stoke Row Gravel marked the first influx of quartz and quartzite-rich gravel into the Middle Thames region. The extraordinarily high frequencies of these lithologies suggests that a major capture, drainage diversion, or possibly even glaciation had occurred during the intervening downcutting phase to bring in material from the south Midlands and even beyond. Throughout subsequent members there is an underlying trend towards dilution

Table 7. *Sequence of events in the Middle Thames Valley during the Pleistocene. *– in Lower Thames; C14 – radiocarbon date; v – vertebrate remains; p – pollen; m – Mollusca; t – temperate; c – cold; a – aeolian activity; gl – glacial.*

Event	Dating evidence	Climate	Stage	
Aggradation of Staines Alluvial Deposits	C14	t	Flandrian	
Deposition of Shepperton Gravel and tributary valley floodplain gravel (*c.* 15 000–10 000 BP)	C14	c	Late	
Downcutting (*c.* 30 000–15 000 BP)				
Deposition of Kempton Park Gravel and tributary equivalents (*c.* 45 000–30 000 BP)	C14	c–t–c	Middle	Devensian
Downcutting				
?Deposition of main mass of Langley Silts		a,c		
Deposition of Reading Town Gravel	v	c	?Early	
?Downcutting				
Deposition of Trafalgar Square and Brentford deposits	p,v	t	Ipswichian	
Deposition of Spring Gardens Gravel		?c		
Downcutting				
Deposition of Taplow Gravel		c		
Downcutting				
Deposition of Lynch Hill Gravel		c	Wolstonian	
Downcutting				
Deposition of Boyn Hill Gravel		c		
Downcutting				
(Deposition of Swanscombe Middle Gravel)*	m	?c, t–c		
Slade Oak Lane organic deposits	p	t	Hoxnian	
?Downcutting				
Deposition of Black Park Gravel		c/gl	late	
Downcutting – following drainage of Moor Mill Lake		gl		
Deposition of Winter Hill Gravel		gl		Anglian
Downcutting				
Deposition of Gerrards Cross Gravel		c/?gl	?early	
Downcutting				
Deposition of Beaconsfield Gravel		c		
Downcutting				
Deposition of Satwell Gravel		c	pre-Anglian	
Downcutting				
Deposition of Westland Green Gravel		c		
Downcutting				
Deposition of Stoke Row Gravel				
Downcutting				
?Deposition of Nettlebed interglacial deposits	p	t	'Nettlebedian'	
?Downcutting				
Deposition of Nettlebed Gravel			pre-Nettlebedian	

Fig. 62. Palaeogeographical evolution of the River Thames during the Pleistocene based on the work presented here. a. Nettlebed Gravel to Winter Hill Gravel, b. Black Park Gravel to Shepperton Gravel.

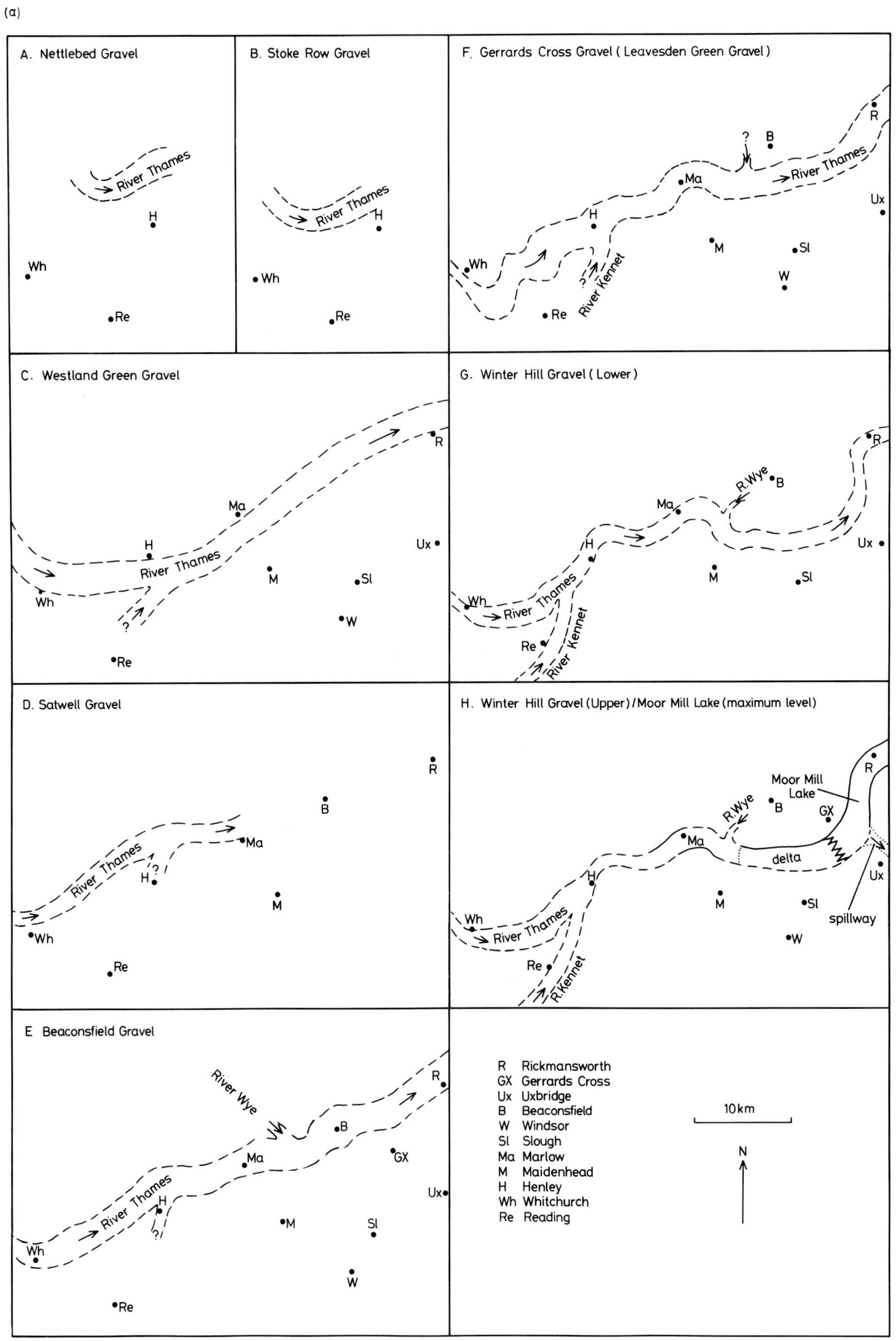

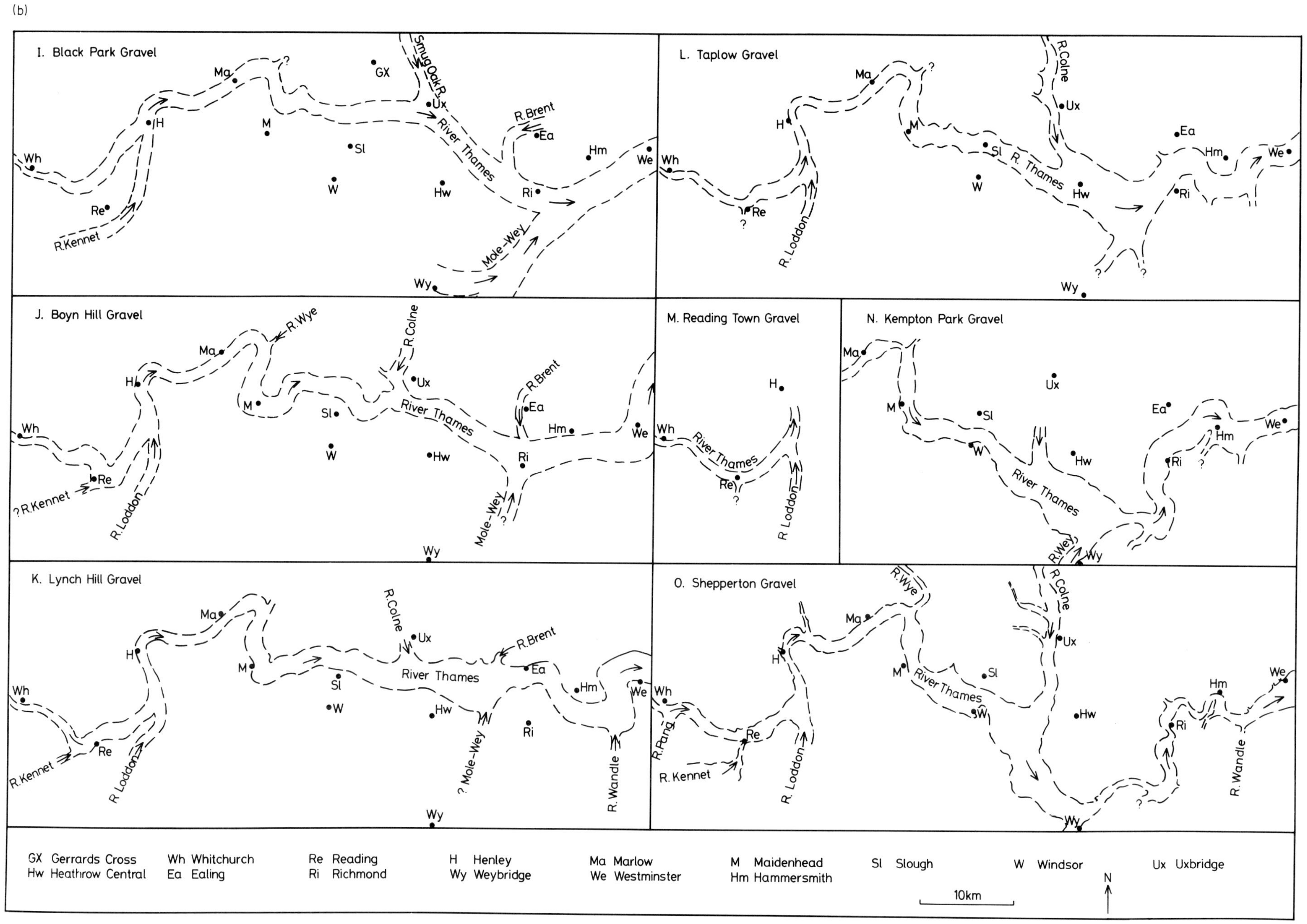
(b)
I. Black Park Gravel
L. Taplow Gravel
J. Boyn Hill Gravel
M. Reading Town Gravel
N. Kempton Park Gravel
K. Lynch Hill Gravel
O. Shepperton Gravel
River Thames
R. Kennet
R. Loddon
R. Colne
R. Brent
R. Wye
Mole-Wey
R. Wandle
R. Pang
GX Gerrards Cross
Hw Heathrow Central
Wh Whitchurch
Ea Ealing
Re Reading
Ri Richmond
H Henley
Wy Weybridge
Ma Marlow
We Westminster
M Maidenhead
Hm Hammersmith
Sl Slough
W Windsor
Ux Uxbridge
10km
N

of the quartz and erratic rocks by the incorporation of additional angular local flint. Rounded flint derived from local Tertiary rocks shows a parallel decline. The lack of downstream equivalents of the Stoke Row Gravel beyond the Middle Thames region suggests that it may have been cut-out by the following Westland Green Gravel or that thus far it has not been recognised.

By Westland Green Gravel times the Thames's course was well established from Goring through the Caversham Channel route ENE across the region into Hertfordshire (Hey, 1965, figure 4). The gravels of this member include erratic lithologies derived from North Wales and the Carboniferous rocks of the Midlands, a fact which led Hey (1965) to suggest that these materials may have been transported into Thames headwaters by glaciers. There is, however, no direct evidence of glaciation at this time in the Thames system although the sediments were deposited under a braided river regime and at some localities include periglacial structures. Downstream extension of this unit into Essex, Suffolk and ultimately Norfolk led Hey (1980) to propose an Early Pleistocene pre-Pastonian age (cf. West, 1980) for this aggradation. While the unit must be of considerable antiquity there must be some doubt that the Westland Green Gravel member of the Middle Thames is so old since it post-dates the Nettlebed interglacial deposits. Resolution of this problem must await further study of the Nettlebed site.

Incision after the Westland Green phase was followed by aggradation of the Satwell Gravel during which time the river seems to have followed a broadly similar course and transported comparable lithologies to those previously.

During the subsequent Beaconsfield Gravel phase, the river for the first time had extensively exposed Chalk bedrock beneath. The gravel therefore includes a marked increase in locally derived angular flint. Down-valley incorporation of flint is accompanied by a proportional dilution of other lithologies. Cross-valley variations are also found particularly resulting from soliflucted regolith introduced from the valley sides and frequently preserved as diamicton beds within the stratified gravels. The gravel and sand deposits of this and later members are arranged in facies typical of the coarse bed-load braided river environment which implies accumulation under a regime of high sediment supply and great variation in discharge. The solution of newly exposed Chalk bedrock seems to have become a major factor at this stage. The sediments of this and subsequent members frequently show evidence of both syndepositional and post-depositional collapse into hollows formed by this process. The former must have resulted from the presence of a talik beneath the river bed. The river continued to flow from Goring via the Caversham Channel across the Middle Thames and into the Vale of St Albans at this time. The Beaconsfield Terrace (Higher gravel train) surface is developed on this member.

The next accumulation represented in the Thames sequence is the Gerrards Cross Gravel. This unit is very well developed throughout the region and is again aligned from Goring through the Middle Thames region and into the Vale of St Albans. The gravels are once more characterised by high frequencies of quartz and erratic lithologies notably including the highest frequencies of north Welsh acidic igneous rocks. McGregor and Green (1978) followed Hey (1965) and Hey and Brenchley (1977) in suggesting possible glaciation into the Upper Thames catchment to account for this. However, SEM studies of sand grains from this unit have failed to support input of quantities of glacial flour, although distant mountain glaciation could have occurred and might explain the lack of this matrix this far downstream. As with the Beaconsfield member, the gravels show excellent suites of sedimentary structures that indicate deposition in a braided river environment. Interbedded lenses of pebbly clay and 'rubble chalk' of periglacial solifluction origin are common within the member. These factors together imply that a periglacial climate occurred during deposition of the unit. Evidence for solutional collapse of Chalk bedrock is commonly associated with this unit. The Harefield Terrace surface is largely developed on these gravels or overlying solifluction deposits.

Because the Gerrards Cross Gravel pre-dates the Anglian Winter Hill Gravel, it has been thought to be of pre-Anglian age (Gibbard, 1974, 1977; Green and McGregor, 1978), the latter authors actually suggesting a Beestonian age. Since no Cromerian deposits or palaeosols have yet been identified in the Middle Thames region it is not possible unequivocally to demonstrate the age of this unit at present. Moreover, there is no reason why the Gerrards Cross Gravel should not be of early Anglian age (e.g. Gunton Stadial: *sensu* Mitchell *et al.*, 1973), since the Winter Hill Gravel is presumably of Lowestoft Stadial age, based on its association with the Chalky Till of the Vale of St Albans (Gibbard, 1977). If the Gerrards Cross Gravel is early Anglian it would explain the possible glacial component without having to resort to unfounded pre-Anglian glaciation to explain the erratic rocks present. Resolution of this problem may be achieved by downstream correlation especially in central Essex as attempted recently by Green, McGregor and Evans (1982).

Fragments of a surface referred to a Rassler Terrace intermediate between the Harefield and Winter Hill Terraces by the Sealys (1956) have been shown to represent either degraded remnants of the former or a surface developed on alluvial or solifluction fans analogous to those in the modern valley near Hambleden. No evidence for a separate Rassler stage has been found and therefore the name has been discontinued.

9.4 Anglian

The Winter Hill Gravel was the next fluvial aggradational unit to be deposited throughout the area following an erosional interval of unknown duration. Lithologically the gravels are very characteristic and therefore easily identified. They are transitional between the quartz-rich older gravels and the flint-rich younger accumulations. This termination of quartz and erratic input suggests a major truncation of Thames headwaters during this stage probably associated with glacial drainage division in the Midlands. It also suggests that if glaciation occurred during previous stages, it had little effect on the lowland catchment of the River Thames. Upstream of Burnham the sediments comprise 'normal' fluvial deposits showing sedimentary structures exactly similar to those of older and younger gravels of braided river origin. However, east of Burnham the gravels are divisible into two subunits which diverge altitudinally downstream. These lithologically indistinguishable subunits can be separated by their differing gradients. The lower subunit, termed the Winter Hill Lower Gravel upon which Hare's (1947) 'lower' Winter Hill Terrace is developed, continues with the same gradient as upstream and can be followed as far east as Fulmer and Rush Green (Figure 62a). Downstream, it turns north-east towards Denham and Rickmansworth where it becomes the Westmill Gravel underlying the Eastend Green Till of the Vale of St Albans (Gibbard, 1974, 1977). However, the Winter Hill Upper Gravel diverges from the lower and remains almost horizontal at 80–82 m OD. In his discussion of the Winter Hill Terrace surface Hare (1947, p. 329) mentioned 'it has little slope, and this fact may have been due to prolonged aggradation in a ponded-up valley.' Gibbard (1974, 1977) showed that the ponding episode responsible for flattening the gradient was equivalent to the deposition of the Moor Mill Laminated Clays in an ice-dammed lake formed when the Anglian Eastend Green Till ice blocked the Thames in the Vale of St Albans.

In view of the evidence from the Middle Thames and the Vale of St Albans, it seems reasonable to interpret the Winter Hill Gravel as in part a fluvial aggradation similar to those of the other major gravel members, as shown by its sedimentary structures, facies and relation to other units. However, downstream the subdivision results from initial fluvial deposition that formed the lower subunit being halted by ponding of the river. This would have caused a delta to form at the upvalley end of the lake, possibly near Burnham. This is confirmed by the lack of gradient, and sedimentary structures present in the upper subunit at Stoke Common. The delta seems to have aggraded eastwards to form this subunit, its upper surface presumably records the actual lake level (80–82 m OD). This explanation accounts for the absence of the upper subunit upstream of Burnham since the river will only have aggraded a delta within the lake, while upstream it would have continued to flow as it did before the lake was formed (Figure 62a).

Readvance of the Anglian ice towards the south-west in the Vale of St Albans as far as the M1 and Aldenham carried the ice into the Moor Mill Lake. This apparently caused the lake water to spill over the London Clay interfluve near Uxbridge into the neighbouring Mole–Wey Valley (Gibbard, 1979). Here an analogous ice-dammed lake, represented by the Coldfall Wood Laminated Beds was ponded in front of the Finchley Till ice lobe (Gibbard, 1979). Initial overspill of the Moor Mill Lake appears to have triggered a chain of events during which spillways were formed from the Thames to the Mole–Wey Valley and from there into the neighbouring valley (?Wandle) and so on until the river reached the unglaciated Medway system. The Thames became confluent with the Medway near Southend-on-Sea from where it flowed north-east along the pre-existing valley to rejoin its original course near Clacton (Bridgland, 1980 and Bridgland and Gibbard, in prep.).

Following drainage of the Moor Mill Lake the Thames adopted and incised its course through the spillway system. The first member aligned along this newly formed course was the Black Park Gravel which can be recognised throughout the Middle Thames region and extends through the present valley to Richmond, Wimbledon and Dartford Heath. At Uxbridge, the Thames was joined by the late Anglian Smug Oak Gravel stream that carried meltwater from the retreating Anglian ice in the Vale of St Albans. This river flowed in a southerly direction from Rickmansworth over the newly drained valley incised into the Moor Mill Lake floor (Figure 62a). Confluence of these streams demonstrates the late Anglian age of this member. An analogous stream seems to have drained the Finchley Valley towards the south-west, depositing the Hanger Hill Gravel and probably joining the Thames near Richmond.

Lithologically, the Black Park Gravel may be distinguished from earlier Thames gravel aggradations by its high content of angular flint and paucity of all other constituents. This first truly flint-rich gravel records the final severing of the headwaters that previously supplied quartz and erratic pebbles at earlier stages. The Black Park Terrace surface occurs on these gravels.

The Black Park Gravel has been shown to be that flooring the Caversham Channel between Caversham and Henley. The presence of Clactonian and both ?Early and Middle Acheulian artefacts in these gravels implies that Palaeolithic man may have been present in the Thames Valley for the first time during this phase. Subsequent downcutting resulted in the river abandoning the Caversham Channel route. This was attributed by Gibbard (1982) to the capture of the Blackwater–Loddon stream system by the Kennet which was confluent with the Thames near Henley up to this stage. These southern tributary streams were draining over London Clay rather than the Chalk of the Caversham Channel. Thus differential erodability may have been a crucial factor in the capture of the Thames from its valley. A remnant stream may have remained on the floor of the Channel after diversion of the Thames and reworked the gravel, at the same time possibly incorporating later flint artefacts, although this is now thought unlikely (J. J. Wymer, personal communication).

The sharp fall in water level brought about by drainage of the Moor Mill Lake (*c.* 15 m) must have had important implications for the groundwater table within the immediate area. Even though the area may have been subject to permafrost, adjacent to the lake a talik almost certainly existed. Such a drop in groundwater level could have triggered the collapse of solution hollows such as that at Slade Oak Lane, Denham. This doline was initially infilled by sorted sand possibly washed in by a stream. The overlying clayey gravel seems to represent detritus that slipped or soliflucted directly into the hollow from the sides during subsidence. The material was probably derived from a mixture of Reading Beds and the overlying Gerrards Cross Gravel. The interbedded clay and sand laminae present at some levels may represent periodic inwash or accumulation in standing water, while the faulting present indicates that the basin was subsiding throughout this time. The lack of both fossils and chemical weathering in these sediments suggests that they accumulated during a cold period, almost certainly the late Anglian, since they are overlain by Hoxnian organic deposits.

9.5 Hoxnian

Interglacial sediments of Hoxnian age are not known from the Middle Thames region with the exception of those at Slade Oak Lane. Here, the rise of water table level in the second half of the Hoxnian Stage led to the accumulation of organic sediments in the shallow, well vegetated pool. The palaeobotanical sequence indicates that dense Late-temperate and later Post-temperate forests colonised the surrounding ground and overhung the pool. Periodic inwash down the steep basin sides possibly following storms brought inorganic sediment into the pool.

During the latter half of Hoxnian substage HoIV the deteriorating climate decreased the vegetation cover and soil erosion began. Slope wash of sediment into the pool brought about the transition into inorganic sediment. Continued climatic deterioration led to a return of periglacial processes. The latter caused solifluction of soil and regolith into the basin presumably early in the Wolstonian. There is no evidence of further basinal subsidence after organic sedimentation began.

9.6 Wolstonian

In the Middle Thames region, the next aggradation is the Boyn Hill Gravel. However, before this can be discussed mention must be made of the relationship of this member to the sequence exposed at Swanscombe in the Lower Thames Valley. Although partially altitudinally equivalent to the upper part of the Swanscombe deposits, the Middle and Upper Gravels, the Lower Thames Boyn Hill equivalent, the Fairlop (Orsett Heath) Gravel (Bridgland and Gibbard, in prep.; Gibbard, unpublished) is significantly different in facies being much coarser and much more variable sedimentologically. Recent investigations (Bridgland and Gibbard, in prep.; Gibbard, unpublished) indicate that the Swanscombe Lower and Upper Middle Gravels accumulated as a channel-fill sequence under a deteriorating, predominantly cold climate, as indicated by ice wedge casts in the sediment (Waechter and Conway, 1977). This conclusion is supported by molluscan faunas (Kerney, 1971) and is contrary to the sequence proposed by Hubbard (1982). Since the Clactonian artefact-bearing Lower Loam and the upper part of the Lower Gravel are conventionally correlated with the Hoxnian Stage (Wymer, 1974), the overlying gravel and sand must be of Wolstonian age (*sensu* Mitchell *et al.*, 1973). Current investigations demonstrate that the Fairlop Gravel post-dates the Lower and Upper Middle Gravels on the basis of stratigraphic position and freshness of artefact content of the latter. If this relationship is correct then the Fairlop/Boyn Hill Gravel must therefore have been deposited during a later phase of the early Wolstonian following an intervening period of downcutting between this and the Swanscombe Middle Gravels. This series of events closely parallels the relationship of the Trafalgar Square sequence to the Kempton Park Gravel.

The Boyn Hill Gravel is the first unit aligned along the present course through Reading. It contains ample evidence for accumulation under a cold climate regime such as braided river facies assemblages and association with periglacial sediments, e.g. soliflucted pebbly gravels. There is no unequivocal evidence for contemporary colonisation of the area by man since the Middle Acheulian artefacts from this unit are mainly in a rolled condition and therefore are probably reworked from older sediments. The only fossil record is of horse.

A Colne Valley equivalent of the Boyn Hill member has been identified near Iver Heath and termed the Broom Warren Gravel. This deposit marks the first evidence of the Colne Valley stream, apart from the Smug Oak Gravel meltwater river.

In spite of suggestions to the contrary, based primarily on artefact assemblages, the Lynch Hill Gravel demonstrably post-dates the Boyn Hill aggradation. The former contains Acheulian artefacts in greater frequency and in fresher condition than the latter and this has led some archaeologists to invert the two units to explain the phenomenon (e.g. Wymer, 1968). Geologically, there are no grounds for this inversion. As stated above, the abundance of implements at this stage may arise either from a yet unidentified intervening period of human occupation, or from incision and reworking of pre-existing, artefact-bearing (?interglacial) sediments during and preceding the Lynch Hill phase. Nevertheless, the possibility of contemporary hand-axe manufacture must be borne in mind. The altitudinal relation of the Lynch Hill Lower Thames equivalent, the Corbets Tey Gravel, to the Swanscombe sequence (Bridgland and Gibbard, in prep.; Gibbard, unpublished) indicates that by this stage the river had deepened its valley below the basal level of the entire Swanscombe accumulation. Therefore, erosion of the interglacial sediments containing both Clactonian and Acheulian material could have taken place. Again if this were repeated throughout the Middle Thames region, it might account for the abundance of artefacts in the Lynch Hill compared to the Boyn Hill member. It would also explain the presence of some of the Clactonian artefacts at this stage, although some are undoubtedly contemporary with the Acheulian material. The numerous vertebrate records from the Lynch Hill Gravel are all of species typical of treeless, generally cold environments such as mammoth and woolly rhinoceros and support the sedimentary evidence for the prevalence of such conditions.

Equivalent aggradations of the Lynch Hill Gravel have been identified from the Loddon, Colne and also the Lea (Gibbard, unpublished) valleys and indicate that the modern drainage system was fully developed at this time. The Lynch Hill Terrace surface overlies the gravels upstream of Marlow, however downstream the surface is developed on the Langley Silt Complex member.

As with previous gravel members sedimentary evidence from the Taplow Gravel indicates that it accumulated under a braided channel regime. However, the presence of Arctic soil structures and ice wedge casts demonstrate the severity of the periglacial climate. The gravels which are extensively developed throughout the Middle Thames region consist almost entirely of flint with only very low frequencies of erratic pebbles reworked from older units particularly the Lynch Hill Gravel. The gravels have yielded a limited assemblage of Palaeolithic artefacts all in poor condition and undoubtedly derived from older gravels. The vertebrate fauna is very limited and again includes species typical of treeless environments such as mammoth, woolly rhinoceros and notably musk ox. The last-mentioned may be indicative of severe winter conditions. A well developed tributary equivalent of the Taplow has been recognised in the Colne Valley and termed the Denham Village Gravel.

As regards age the Taplow Gravel is clearly younger than the Lynch Hill on the basis of its lower altitude and its artefact content. It also pre-dates the downcutting that preceded deposition of the Trafalgar Square sequence. It therefore appears to represent a late Wolstonian phase, as suggested by Wymer (1977b).

The Taplow Terrace surface is developed on the Taplow Gravel upstream of Marlow; downstream the surface occurs on the Langley Silt Complex that overlies the gravels.

The Spring Gardens Gravel, that underlies the interglacial Trafalgar Square Sands and Silts, must have been deposited during an interval before interglacial sedimentation began and after downcutting had ceased. By analogy with other gravel aggradations and on the basis of similarity of gradient this deposit may have been laid down under a cold climate. Since it immediately pre-dates early Ipswichian deposits and is lower than the Taplow Gravel, it must be presumed to date from the latest part of the Wolstonian. That there was no prolonged exposure of the deposits to later modification is suggested by the preservation of channel-like surface features beneath the interglacial sediments, analogous to those on the surface of the Shepperton Gravel. The Spring Gardens Gravel has a limited known distribution although it may be preserved as eroded remnants beneath younger gravel members, in places.

9.7 Ipswichian

The transition to the Ipswichian interglacial is marked by the appearance, following inward migration, of temperate forest vegetation and fauna. Initial sedimentation at Trafalgar Square took place in shallow floodplain channels or depressions with aquatic vegetation and surrounded by grass and herb-dominated low ground nearby. Increasing water depth and flow caused a transition into more widespread sedimentation in the Trafalgar Square area, the change being expressed by different facies at different places. There was a clear separation of distributaries between those where the main water flow was apparently concentrated and quieter areas lacking strong currents. A southward migration of one of these channels seems to have occurred in the Trafalgar Square area during the time represented. The rise in water level recorded by these deposits may be related to sea-level rise during the interglacial. All the deposits examined in the Trafalgar Square area fall into Ipswichian substage IpIIb as probably also do the Brentford Sands. These deposits presumably formed a continuous spread throughout the area during the interglacial, but have since been removed. The possible exception is the Tennessee Pancake House channel filling, which accumulated during a period of treeless vegetation yet it contains temperate Mollusca. The age of this filling must await further research, but it must date from later in the Ipswichian or the early Devensian.

9.8 Devensian

The gravels in the Reading area originally mapped as Taplow, but clearly separable from this and from the Shepperton Gravel, form a separate aggradation unrepresented downstream of the Loddon confluence. This unit, termed the Reading Town Gravel and its Loddon equivalent the Woodley Gravel are higher than the Middle Devensian Kempton Park Gravel and therefore probably older. The presence in the gravels of reworked hippopotamus remains strongly suggests that they accumulated after the Ipswichian interglacial. The sedimentary facies are comparable to those of a braided river regime and therefore probably represent deposition under a cold climate. It seems reasonable therefore to assign this member broadly to a period between the Ipswichian and the Middle Devensian, i.e. early Devensian. This is strongly supported by upstream correlation (Chapter 10). The fossiliferous Redlands Sands may pre-date the Reading Town Gravel if the sequence descriptions are correct. However, the vertebrate assemblage of the former does not suggest a particular age, but is comparable either to late interglacial or cold stage faunas. Nevertheless, in view of their altitudinal position, the Redlands Sands should fall between the late Wolstonian and early Devensian.

As already mentioned, the relationship of the Reading Town Gravel to the Langley Silt Complex cannot be determined because their distributions do not overlap. The latter is widespread below Maidenhead, which coincides with the point where the river crosses onto London Clay bedrock, and reaches considerable thicknesses. It is clear from the descriptions of sections, mode of occurrence and analytical results that these sediments are polygenetic. The bulk of the sediments comprise loess and waterlain or redeposited loess largely derived from local Tertiary rocks, but with a far-travelled component. However, their frequent bimodality indicates remobilisation by slope wash, possibly snow melt, and introduction of additional sand. The presence of isolated pebble bands in the sediments again suggests a similar origin. Recycling of earlier deposited material is extremely difficult to determine in the field because of the nature of the sediment. However, the distribution of archaeological artefact assemblages within the sediment suggests recycling may have taken place. The Langley Silt Complex contains localities that have yielded abundant Levalloisian industries as well as occasional Mousterian hand-axes. Twice as many sites occur above the Lynch Hill Gravel as on the Taplow Gravel. This could result from downslope redistribution of artefact-bearing sediment by slope processes, although it need not.

Because of the apparent polygenetic origin it follows that a single hypothesis cannot explain the formation of the sediment. Nevertheless, there are some points in common. The silts are generally unrelated to the gravel spreads, but simply mantle surfaces of both bedrock and Pleistocene sediments alike. In general the sediments lack fossils mostly as a result of decalcification. The abundance of loess and indications of slope wash suggest a lack of vegetation cover which implies accumulation under a cold stage, possibly periglacial climate. This implication is supported by occasional finds of ice wedge casts. The sediment appears to be largely reworked fine regolith, the loess being formed by local deflation, and redeposition following snow melt by slope wash. Such large amounts of regolith may be available early in a cold stage following loss of vegetation cover during climatic deterioration after an interglacial period. At such a time large volumes of inorganic detritus would be expected to be washed into the river by surface water, causing fine sediment deposition in valleys and on slopes. The sediment supply may therefore be the controlling factor which governs whether loess-dominated fine sediments or gravel and sand-dominated sediments are laid down under the periglacial regime. Indeed it may also control downcutting by the river. Whether these events have a climatic significance and how they are mutually related in time has yet to be demonstrated.

In view of the wide range of sediment types it is probable that the Langley Silt Complex does not simply represent one depositional event, but several. At Iver, for example, the 'red brickearth' described by Lacaille and Oakley (1936) may represent a pre-Ipswichian loess-rich deposit upon which a palaeosol developed. The overlying 'grey brickearth' presumably represents a later Devensian accumulation. The youngest deposit overlain by the Langley Silt Complex is the Kempton Park Gravel, but this cover may be mainly soliflucted from higher level spreads. The silts always overlie Ipswichian interglacial sediments, e.g. at Brentford and Trafalgar Square, and are therefore thought to be largely of Devensian age. However, thermoluminescence dates currently available (Wintle and Gibbard, unpublished) indicate a range of ages from late Wolstonian to Late Devensian. It is obvious that much more research is necessary to clarify the origin and depositional history of the sediments collectively termed the Langley Silt Complex.

According to previous work, by about 45 000 BP the Kempton Park Gravel had begun accumulating throughout the Middle Thames Valley under a cold, continental climate. Climatic amelioration at the Upton Warren Interstadial phase (*c.* 44 000 BP) seems to have resulted in widespread accumulation of organic sediments within floodplain hollows and channels. The floodplain surfaces were seemingly colonised by both plants and animals and diverse, often dense, but treeless plant communities developed on the land beyond the river. A variety of vertebrate and invertebrate animals were present at the time. To allow such an accumulation of sediment and colonisation of the floodplain, the intensity and magnitude of the rivers' nival flood must have been markedly reduced although it was certainly not eliminated. This no doubt resulted from increased vegetation of the land, degradation of permafrost permitting increased groundwater flow, and less sediment supply because of slope stabilisation. The climatic deterioration following the Upton Warren Interstadial (*s.s.*) seems to have reactivated the river and it resumed its active braided habit as more sediment became available and the flood events once more increased in magnitude following degradation of the vegetation cover. Erosion of the previously deposited organic beds, associated with the deposition of the gravel and sand, left them as isolated bodies within the unit. Abandoned channel fills such as that at Kempton Park record the return to cold, more continental climates at about 36 000 BP. The contemporary development of permafrost is evidenced by ice wedge casts preserved at Isleworth. After completion of deposition of the Kempton Park Gravels at about 30 000 BP, ice wedges continued to grow in the gravels and these extremely large structures penetrated the entire thickness of the deposits. A tributary equivalent aggradation in the Wye Valley is termed the Byfleet Gravel.

Later, solifluction of London Clay and Langley Silt Complex sediments from valley sides onto the Kempton Park Gravel occurred. It is also possible that contemporary loess deposition took place during this interval although it has yet to be proved conclusively (Wintle and Gibbard, unpublished). The Isleworth ('Upper Floodplain') Terrace is developed on the Kempton Park Gravel or these overlying silts.

Downcutting and incision to a lower level by the river was followed by aggradation of the Shepperton Gravel. These events were paralleled by equivalent events in the tributary valleys. The Shepperton Gravel almost totally comprises flint accompanied by very minor quantities of other lithologies, the elimination of which progressed from the Black Park Gravel to this member. The Shepperton member and indeed tributary equivalents display sedimentary structures and facies exactly comparable with those of earlier units that indicate deposition under a braided channel environment. That this deposition took place under a cold, probably periglacial climate is evidenced by the presence of a vertebrate fauna of steppe and tundra elements and organic channel fill deposits preserved in the Colne Valley. These indicate contemporary treeless shrub and herb-dominated local vegetation in adjacent areas and on stable parts of the floodplain. Radiocarbon dates from these deposits suggest that aggradation of this member began at about 15 000 BP and continued until approximately 10 000 BP. This was therefore the last major phase of gravel and sand aggradation in the Thames as well as its tributary valleys. The intervening downcutting phase between the Shepperton and Kempton Park members seems to have lasted approximately 15 000 years (30 000–15 000 BP). Where the river is deeply incised into its valley, for example between Henley and Marlow, alluvial fans of chalk-rich flint gravel formed and interdigitated with the fluvial deposits. The complex dry valley network present throughout much of the area appears to be graded into this aggradation, a factor which suggests that they last functioned regularly during this interval. The widespread existence of permafrost within the underlying bedrock must be the explanation for this feature.

The presence of collapse features in the gravels and sands of this unit closely resembles those associated with older members. The occurrence of these features suggests that syndepositional Chalk bedrock solution has been active under a periglacial climate during the accumulation of the sediments. Larger collapse structures beneath Central London may have also been produced by Chalk bedrock solution resulting from upwelling of artesian springs and possibly enlarged by fluvial scour during gravel emplacement. The contemporary collapse of gravel and sand into these hollows has in some places continued into the Flandrian and underlines that periglacial conditions need not be invoked to explain their operation.

9.9 Flandrian

Following climatic amelioration at the end of the Devensian the river seems to have stopped aggrading sand and gravel, and became stabilised within the flow channels it was occupying at that time. This stabilisation was undoubtedly brought about by a marked reduction in spring flooding arising from mild winters, a drastically reduced sediment supply brought about by colonisation of the land by lush temperate forest vegetation on the valley sides as well as the more favourable areas of the floodplain surface, and lastly as a result of groundwater rather than surface water drainage following the decay of permafrost. This stabilisation of ground surface led to the river receiving almost negligible amounts of sediment. This factor enabled marl and tufa to accumulate in floodplain pools. With the development of woodland on the floodplain, peat began forming and as this sediment thickened it progressively buried the relict braided river surface. Mesolithic people colonised the emergent gravel bars early in the Flandrian and used the flint to manufacture implements which were later buried by accretion of floodplain sediments. During the Flandrian channels became slowly filled mostly with fine often highly fossiliferous sediments with occasional coarser flood-derived deposits. This process has continued until in some places only one flowing channel remains. The clearance of woodland by man for agriculture has resulted in the increased inwash of silt and clay into the river. Such clearances have increased sediment input into rivers by as much as 100-fold in the USA (Ursic and Dendy, 1965; for review see Gregory and Walling, 1973). This clearance has had a two-fold effect; it has caused increased flooding because of more rapid through-put of precipitation and it has caused deposition of fine inorganic sediments on the floodplain. These overbank deposits are of considerable thickness in places and would still accumulate at the present day if the river flow was not regulated. It appears therefore from this information that the present course of the River Thames is inherited from the latest Devensian time and that there has been little or no erosion since then.

10

Correlation with neighbouring areas

The continuity of the fluvial aggradations recognised throughout the Middle Thames Valley and in some of the tributary valleys strongly implies that equivalent deposits might exist in neighbouring up and downstream areas. Although some mention of lateral equivalents has already been made in earlier chapters, this section presents a summary of possible stratigraphical correlates in adjacent areas (Table 8). The correlations are made on the assumption that differential crustal warping has not been demonstrated between the areas.

10.1 Upper Thames Valley

The Upper Thames Valley is usually considered to be that area of the Thames drainage system upstream of the Goring Gap. This region includes the valleys of three main streams: the Thames and its tributaries the Evenlode and the Cherwell. The Pleistocene deposits of the Oxford area were originally mapped by Pocock (1908) and subsequently the sequence throughout the area was thoroughly evaluated by Sandford (1924, 1926, 1954). Detailed investigations of the relationship of the Midlands glacial sequence to the Thames system by Bishop (1958) and Tomlinson (1963) led them to propose revised correlations for the terrace deposits. Reviews of the sequence by Shotton (1973), Goudie and Hart (1975) and most recently Briggs and Gilbertson (1980) have led to further revision. Few attempts have been made to equate the sequence in the Upper Thames Valley with that downstream. Goudie (1976, p. 3) wrote 'It has not yet proved possible to link this sequence with that of the Middle . . . and Lower Thames'.

Correlation of the Middle and Upper Thames region in this study is presented in Table 8 using upstream altitudinal extension of the gradient of each gravel member. This method has been shown to be acceptable for the Floodplain gravels (Gibbard, 1981 and below). It is the only practical technique since the gravels upstream of the Goring Gap lack appreciable quantities of flint derived from the Chalk, but frequently include large quantities of Jurassic limestone.

The highest and therefore probably the oldest deposits in the Upper Thames region are grouped into the so-called Plateau or Northern Drift. These deposits consist of limestone-free, non-local quartz and quartzite-rich red sandy to clayey gravels that have most frequently been interpreted as marine or glacial sediments (Sandford, 1924; Arkell, 1947; Goudie and Hart, 1975; Shotton *et al.*, 1980). The considerable height range (135 m), facies variation, lack of stratification and lack of local lithologies have been cited as evidence of a glacial origin. The absence of local rocks is, however, surprising if glacial deposition is involved since tills are generally characterised by containing abundant local bedrock clasts (Dreimanis, 1976).

It seems more probable that the Northern Drift has been completely decalcified, a process that would remove almost all local lithologies and destroy any sedimentary structures. Post-depositional downslope movement and periglacial processes would further modify the sediments. It is this mechanism which R. W. Hey (personal communication) invokes to interpret the Northern Drift as a series of predominantly fluvial deposits. It was previously also suggested by Shotton *et al.* (1980). Hey (1965) had already suggested that it is possible to identify an upstream equivalent of the Stoke Row Gravel (Westland Green Gravel) in Wychwood Forest. If this interpretation is correct then the Northern Drift may include upstream equivalents of the pre-Winter Hill Gravel members of the Middle Thames Valley. By extending the member gradients, upstream equivalents should occur at Oxford as follows:

Middle Thames gravel member	Approximate predicted surface–base heights at Oxford (m OD)
Westland Green Gravel	182–*c.*179
Satwell Gravel	150–144
Beaconsfield Gravel	140–133
Gerrards Cross Gravel	123–119

In the Evenlode Valley, Arkell (1947) recognised two terrace gravel aggradations that he termed: (1) the Combe Terrace gravel,

Table 8. *Correlation of the Middle Thames sequence with those in neighbouring areas. See text for explanations.* [1]*Devoy (1979),* [2]*Mitchell et al.* (1973).

Upper Thames	Middle Thames	Vale of St Albans (west)	Lower Thames provisional nomenclature	Stage[2]
Modern alluvium	Staines Alluvial Deposits	Floodplain organic deposits	Tilbury alluvium[1]	Flandrian
Northmoor/ Floodplain gravels Summertown–Radley upper gravel	Shepperton Gravel Kempton Park Gravel ?Langley Silt Complex Reading Town Gravel	Colney Street Gravel — — —	unnamed East Tilbury Marshes Gravel ?'brickearth' —	Late Middle Devensian Early
Summertown–Radley shelly sands	Trafalgar Square/Brentford deposits	—	Aveley/ Northfleet deposits	Ipswichian
Summertown–Radley lower gravel Wolvercote Terrace gravel Hanborough Terrace gravel	?Spring Gardens Gravel Taplow Gravel Lynch Hill Gravel Boyn Hill Gravel	— Denham Village Gravel — —	?Sandy Lane sand Mucking Gravel Corbets Tey Gravel Fairlop Gravel Swanscombe Middle Gravel	?late Wolstonian ?early
Sugworth channel deposits(?)	(Slade Oak Lane deposits)	(Hatfield Organic Deposits)	Swanscombe Lower Loam	Hoxnian
Freeland Terrace gravel(?) Combe Terrace gravel Northern Drift	Black Park Gravel Winter Hill Gravel (Upper) (Lower) Gerrards Cross Gravel pre-Gerrards Cross aggradations	Smug Oak Gravel Moor Mill Laminated Clays Westmill Gravel Leavesden Green Gravel pre-Leavesden Green aggradations	Swanscombe Lower Gravel Dartford Heath Gravel initiation of Lower Thames	late Anglian early or pre-Anglian pre-Anglian

with a surface elevation of 116–120 m OD (i.e. 53.5 m above the river) and (2) the Freeland Terrace gravel, with the surface at 107–112 m OD (i.e. 44.3 m above the river) at the type localities. The gravels comprise typical Northern Drift lithologies and seldom exceed 1.5 m in thickness. This should be compared with thicknesses of at least three times as much for limestone-bearing gravels (Sandford, 1924). Although rejected by Arkell (1947) there can be little doubt that this results from decalcification, since similar erratic-rich pockets are present in lower level aggradations, e.g. Hanborough Terrace gravel, where they have been decalcified (Sandford, 1924).

Upstream projection of the Winter Hill Gravel gradient from Mapledurham, where it is 97 m (top)–92.5 m OD (base), to Oxford (northing SP 05), a distance of 31 km, gives a height range of 112 m (top)–107.5 m OD (base). The Combe Terrace gravel would lie at 108.5–107 m OD at Oxford which, allowing for the reduced gravel thickness mentioned already, would equate with this unit. Similarly, extension of the Black Park Gravel gradient from Chazey Wood near Mapledurham where it is 81.5–75.8 m OD to Oxford gives a height range of 101–96.5 m OD. The Freeland Terrace gravel would lie at 99–97.5 m OD at Oxford and therefore appears to be equivalent to Black Park Gravel. Assuming that the correlations are correct then the Combe Terrace gravel should be of Anglian (Lowestoft Stadial) age and the Freeland Terrace gravel should be late Anglian.

The Hanborough Terrace is the next lowest in the Upper Thames and occurs at approximately 80.5 m OD in the Oxford area. It has been conventionally correlated with the Boyn Hill Terrace of the Middle Thames in the past (Shotton *et al.*, 1980). The Hanborough Terrace gravels have yielded rolled vertebrate bones of *Palaeoloxodon antiquus*, (straight tusked elephant), *Bos primigenius* (aurochs) and *Cervas elephus* (red deer) which are normally associated with temperate forest conditions. The deposits were therefore originally thought to represent an interglacial accumulation. However, cold stage molluscan faunas from channel deposits within the gravels led Briggs and Gilbertson (1973) to propose an early Wolstonian age for the unit. Furthermore, the sedimentary facies of the gravels and sands and their association with periglacial structures indicates deposition in a cold climate, braided river environment (Briggs and Gilbertson, 1980). The vertebrate remains must therefore be reworked. If the Boyn Hill Gravel gradient is projected upstream from Reading, where it is at 70–66 m OD, it intersects with the Hanborough Terrace at *c.* 80–77 m OD in the Oxford area. There is no reason to dispute correlation of the Hanborough Terrace gravel and the Boyn Hill Gravel on the basis of altitude, sedimentary structures and relative age.

In contrast, upstream projection of the Lynch Hill Gravel is more problematic. In the Reading area the Lynch Hill appears to be converging with the Boyn Hill Gravel because of the shallower gradient of the latter. Extension of the Lynch Hill gradient upstream from Reading where the unit is at 60–55 m OD would also intersect with the Hanborough Terrace gravel at Oxford. It may therefore be that the Hanborough Terrace gravel is a composite aggradation including an element of Boyn Hill and later Lynch Hill-equivalent deposits analogous to the Floodplain (Northmoor) gravels (see below). These correlations correspond with available archaeological and palaeontological evidence from the Hanborough gravels.

The Taplow Gravel can be shown to be the downstream equivalent of the Wolvercote Terrace gravel using similar criteria. The latter is present at 70–67 m OD at Oxford while the Taplow occurs at *c.* 54–40 m OD at Caversham. Upstream projection of the Taplow member gradient intersects with the Wolvercote unit and they are therefore equated. This correlation seems to be satisfactory on the basis of descriptions of the sedimentary structures, interpreted depositional environment and archaeology (Sandford, 1924, 1926; Bishop, 1958; Tomlinson, 1963; Briggs and Gilbertson, 1980). The Wolvercote Terrace gravel includes greater frequencies of flint than the other limestone-rich gravels and was interpreted by Bishop (1958) as representing discharge of material from the Wolstonian glaciation of the Midlands into the Upper Thames system (Goudie and Hart, 1975).

The lowest terrace surface in the Upper Thames Valley is the Summertown–Radley, the deposits underlying which are divisible into three subunits. These comprise a lower 'Summertown–Radley Terrace' gravel and an upper 'Summertown–Radley Terrace' gravel separated in places, e.g. Magdalen College, Oxford, by shelly sands that have yielded *Hippopotamus amphibius* and *P. antiquus* remains together with a temperate mollusc fauna. At other places temperate vertebrate material was almost certainly present as a 'lag' at the erosional contact of the upper and lower units (Sandford, 1924, 1926). The probable equivalent of the upper gravels has been studied in detail by Bryant (1983b) at Stanton Harcourt and has been shown to occur in facies typical of deposition in a braided river under a periglacial regime. The presence of *Hippopotamus* in the interglacial sediments beneath the upper gravels indicates that they are probably of Ipswichian age (Sutcliffe, 1960; Stuart, 1982). This led Goudie and Hart (1975) to suggest an Early Devensian age for the upper gravels. This has recently been confirmed by thermoluminescence dates of *c.* 90 000 BP from silt beds in this unit by A. Wintle (personal communication). The lower gravels beneath the interglacial levels include *Mammuthus primigenius* and therefore represent an immediately pre-Ipswichian cold stage, i.e. late Wolstonian. These lower gravels occur in an analogous position to the Spring Gardens Gravel of the London area, but the great intervening distance precludes correlation.

Downstream extension of the Summertown–Radley Terrace surface (63 m OD at Oxford) at a gradient of 45–55 cm km^{-1} to Reading intersects precisely with the surface of the Reading Town Gravel at 46 m OD. This evidence for correlation of the Upper Summertown–Radley Terrace gravel with the Reading Town member is reinforced by the records of *Hippopotamus* from the latter (Chapter 6). The presumed Early Devensian age of the Reading Town member is in accord with that of the Upper Thames deposit. The interglacial sediments and underlying gravels are apparently absent in the Reading area.

The Upper Thames Floodplain gravels (Northmoor Terrace of Arkell, 1947) have been shown by Briggs and Gilbertson (1976, 1980) to include both Middle and Late Devensian aggradations. Radiocarbon dates of organic channel fills in the gravels give two groups of 40 000–29 000 BP and 13 500–10 500 BP (Briggs and Gilbertson, 1980; Aalto, Coope and Gibbard, 1984). All these deposits contain fossil assemblages that indicate deposition under treeless predominantly cold stage climates. Correlation of the gravel underlying the modern floodplain of the Upper Thames with those downstream in the Reading area was mentioned by Gibbard (1981) and is shown in Figure 44. This demonstrates that there is complete continuity of slope between the two areas through the Goring Gap. The radiocarbon dates confirm the presence beneath the Upper Thames floodplain of eroded remnants of a time-equivalent aggradation of the Kempton Park Gravel of the Middle Thames, at the same level as the Late Devensian equivalent of the Shepperton Gravel. This is to be expected since the Kempton Park and Shepperton Gravel members converge upstream in the Middle Thames; indeed the former is apparently absent beyond Marlow. In view of the Upper Thames' relationship it is possible that fragments of Kempton Park Gravel may be present beneath the Shepperton Gravel in the western part of the Middle Thames Valley. In the Upper Thames, the Middle and Late Devensian deposits can only be separated where datable organic deposits are preserved because of the nature of the gravel and sand depositional structures.

If the sequence of correlations presented above is correct then one major problem arises: the age of the Sugworth channel fills. These deposits of fluvial sands and organic material fill channels cut into the Kimmeridge Clay and have been assigned to the second half of the Cromerian interglacial Stage on the basis of palaeontological evidence (summarised in Shotton *et al.*, 1980). The correlations shown above place the Sugworth deposits (92–85.5 m OD at Sugworth) intermediate in altitude between the Black Park correlative, the Freeland Terrace gravel (predicted elevation 97–95.5 m OD at Sugworth), and the Hanborough Terrace gravel (elevation 79–76 m OD at Sugworth). These two units are late Anglian and early Wolstonian in age respectively (Table 8). It is therefore very difficult to understand how the Sugworth channel deposits can be of Cromerian age in this position, since there is no suggestion that they have been displaced. Indeed it would seem more likely that they would be of Hoxnian age in this position. It is therefore necessary to consider the palaeontological evidence for a Cromerian age.

The flora present indicated a biozone IIIb age of a Middle Pleistocene interglacial stage, but did not enable discrimination between the Hoxnian and Cromerian. However, the spectra lack crucial taxa thought to be characteristic of the Hoxnian, e.g. pollen type X. The Ostracoda limit the age to Hoxnian or earlier (Robinson, 1980). The Mollusca contain a few specimens of *Nematurella runtoniana* and *Valvata goldfussiana* (Gilbertson, 1980), which in Britain have hitherto been found in deposits of Cromerian age (Kerney, 1977). Very little work has been done on interglacial insect faunas, most records coming from Ipswichian deposits, e.g. Bobbitshole (Coope, 1974). Not surprisingly, in the absence of work on the Hoxnian or Cromerian faunas, the assemblage was found to be most similar to those of the Ipswichian (Osborne, 1980). As correlation with the Ipswichian was unacceptable stratigraphically, no date was suggested. By far the strongest evidence for a Cromerian age derives from the vertebrate assemblage. The records of the Middle Pleistocene rodents *Mimomys savini*, *Sorex* sp. and *Pliomys episcopalis* suggest a pre-Hoxnian age (Stuart, 1980). The evolution of *M. savini* into *Arvicola cantiana* presumably represents an irreversible lineage that occurred only during CrIV (Stuart and West, 1976). Since *M. savini* is present at Sugworth, there is strong support for a pre-CrIV age. *Dicerorhinus etruscus* (a rhinoceros) has also been recorded from the site. This species was previously thought to be extinct after the Cromerian in Britain, but it has now tentatively been identified from Swanscombe (A. Currant, personal communication). Again, present knowledge of the Hoxnian vertebrate fauna is very sparse.

The over-riding problem is that of a comparable contemporary environment. Nowhere else in Britain is there a fluvial channel fill of biozone IIIb age known from either the Hoxnian or Cromerian Stages. Since most Hoxnian sites are kettle hole lakes, the flora and fauna surrounding such sites may well differ significantly from those adjacent to river point bar complexes like that preserved at Sugworth. It is almost certainly this factor which causes the resemblance between the Sugworth insect fauna and that of Ipswichian deposits, most of which are fluvial channel fills. This differing facies may equally explain many of the variations from typical Hoxnian floral and invertebrate faunal assemblages.

There is therefore a dilemma in the interpretation of the results in terms of the geological and palaeontological evidence from Sugworth. It is particularly difficult to explain the vertebrate evidence since in at least one instance it is dependent on an evolutionary lineage rather than on changes of assemblages which can potentially recur at various times. Nevertheless, the present author would favour a Hoxnian HoIIIb rather than a Cromerian CrIIIb age for the reasons stated above.

10.2 Vale of St Albans

The Vale of St Albans represents the north-eastward extension of the Middle Thames area into Hertfordshire. It has been shown by various workers to be the abandoned valley that was occupied by the Thames before the river was diverted by the Anglian ice sheet (see Gibbard, 1974, 1977 for summary). The sequence of correlations of the Middle Thames deposits with those in the Vale has already been discussed several times above, particularly with reference to diversion of the river. The correlations based on lithology of the gravels, altitude, palaeocurrents and stratigraphical relationships are slightly modified from earlier work (Gibbard, 1974, 1977, 1983; Green and McGregor, 1978) and are summarised in Table 8.

10.3 Lower Thames Valley

The Lower Thames is taken here to include that area of the Thames Valley downstream of Central London. While the area has been the subject of intensive study at numerous significant single sites, there have been few attempts to correlate the regional sequence with that in the Middle Thames. Notable studies have been made by Dewey, Bromehead and Dines (1924), Dines and Edmunds (1925), King and Oakley (1936) and later Wooldridge and Linton (1955), but all met with limited success. This is because the Lower Thames Valley contains a highly complex sequence of fragmented interglacial as well as cold stage deposits, the relationships of which have not been rigorously established. At present the area is under reinvestigation by the author and it is therefore rather premature to present any detailed correlations. Nevertheless, a preliminary evaluation of the gravel and sand aggradations has been attempted by Bridgland and Gibbard (in prep.), as already mentioned above. Some preliminary members have been proposed using altitudinal methods. Four aggradational units were recognised and equated with units in the Middle Thames as well as the Southend-on-Sea area. The correlations are summarised in Table 8.

APPENDIX

Pebble counts from the Middle Thames region

The pebble counts are derived from examination of sieve grades 33, 16 and 8 mm, and are given as percentages. Grid references for sample sites are given in the text.

	Flint (angular)	Flint (rounded)	Flint (total)	Vein quartz	Quartzite	Green-sand chert	Green-sand	Chalk	Red siltstone	Ironstone	Igneous & meta-morphic	Calcite	Brown sand-stone	Lime-stone	Palaeo-zoic chert	Black shale	Grit	Red sand-stone	Brown silt	Green silt-stone	Banded mud-stone	Total number of pebbles
Floodplain Gravel																						
Shepperton	91.21	4.22	95.43	2.28	1.23	1.05	0	0	0	0	0	0	0	0	0	0	0	0	0	0	0	569
Bray	92.06	3.43	95.49	2.65	1.09	0.62	0.16	0	0	0	0	0	0	0	0	0	0	0	0	0	0	642
Furze Platt	89.34	3.88	93.22	2.67	0.73	1.21	0	2.18	0	0	0	0	0	0	0	0	0	0	0	0	0	825
Kempton Park	87.77	4.91	92.68	5.13	1.20	0.98	0	0	0	0	0	0	0	0	0	0	0	0	0	0	0	916
Twynersh	89.28	4.99	94.27	3.41	1.10	1.22	0	0	0	0	0	0	0	0	0	0	0	0	0	0	0	821
Lower Feltham	89.67	4.15	93.82	4.03	1.00	0.88	0	0	0	0	0.12	0	0	0	0.12	0	0	0	0	0	0	794
Ismaili Centre B	83.15	7.42	90.57	5.57	1.32	2.51	0	0	0	0	0	0	0	0	0	0	0	0	0.84	0	0	753
Ismaili Centre A	76.53	9.30	85.83	5.28	2.32	5.49	0	0	0	0	0.21	0	0	0	0	0	0	0	0	0	0	473
Kew Bridge	88.87	3.86	92.73	1.70	0.77	4.17	0	0	0	0	0.16	0	0.16	0	0.31	0	0	0	0	0	0	647
Sonning	87.43	4.69	92.12	1.64	2.79	0	0	2.28	0	0	0.50	0	0	0	0.12	0	0	0	0	0	0	788
Taplow Gravel																						
Furze Platt	89.23	5.15	94.38	3.09	0.92	1.60	0	0	0	0	0	0	0	0	0	0	0	0	0	0	0	873
Furze Platt	86.44	5.50	91.94	3.73	1.57	1.96	0	0.79	0	0	0	0	0	0	0	0	0	0	0	0	0	509
Langley	90.38	3.81	94.19	2.99	1.32	1.49	0	0	0	0	0	0	0	0	0	0	0	0	0	0	0	603
Taplow	88.95	2.48	91.43	4.95	1.52	2.10	0	0	0	0	0	0	0	0	0	0	0	0	0	0	0	525
Cranford Park	82.15	6.88	89.03	6.25	2.50	2.03	0	0	0.16	0	0	0	0	0	0	0	0	0	0	0	0	639
Ferns pit	88.15	5.27	93.42	1.83	3.21	1.54	0	0	0	0	0	0	0	0	0	0	0	0	0	0	0	778
Twyford	36.88	5.40	92.28	1.54	1.39	4.78	0	0	0	0	0	0	0	0	0	0	0	0	0	0	0	648
Caversham	91.33	2.64	93.97	3.64	1.88	0.50	0	0	0	0	0	0	0	0	0	0	0	0	0	0	0	796
Harlington	88.73	4.66	93.39	4.01	1.03	1.42	0	0	0	0	0	0	0.12	0	0	0	0	0	0	0	0	772
Sipson	85.19	7.40	92.59	5.01	1.02	1.13	0	0	0	0	0.11	0	0	0	0.11	0	0	0	0	0	0	878
Lynch Hill Gravel																						
Iver	87.24	6.37	93.61	2.51	2.51	1.36	0	0	0	0	0	0	0	0	0	0	0	0	0	0	0	439
Yiewsley (S part of quarry)	86.19	7.13	93.32	2.58	1.37	2.73	0	0	0	0	0	0	0	0	0	0	0	0	0	0	0	659
Yiewsley (N part of quarry)	84.44	9.48	93.92	3.58	1.79	0.72	0	0	0	0	0	0	0	0	0	0	0	0	0	0	0	568
Lynch Hill	82.99	8.34	91.33	4.09	2.67	1.88	0	0	0	0	0	0	0	0	0	0	0	0	0	0	0	635
Switchback Lane	84.40	6.36	90.76	4.39	1.52	2.24	0	1.08	0	0	0	0	0	0	0	0	0	0	0	0	0	1116
Cocksherd Wood	82.47	6.36	88.83	5.01	2.50	3.66	0	0	0	0	0	0	0	0	0	0	0	0	0	0	0	519
Remenham (lower)	80.76	6.77	87.53	3.69	1.08	7.54	0	0	0	0	0.15	0	0	0	0	0	0	0	0	0	0	650
Remenham (upper)	81.31	8.85	90.16	5.41	1.47	2.79	0	0	0	0	0.16	0	0	0	0	0	0	0	0	0	0	610
Twyford	80.08	6.89	86.97	5.65	2.36	5.02	0	0	0	0	0	0	0	0	0	0	0	0	0	0	0	452
Prospect Park	80.40	4.57	84.97	7.15	4.16	2.57	0	0	0	0	0.71	0	0.42	0	0	0	0	0	0	0	0	699
Boyn Hill Gravel																						
Kiln Wood (1 m)	79.06	8.72	87.78	6.14	1.65	4.11	0	0	0	0	0	0	0.29	0	0	0	0	0	0	0	0	535
Kiln Wood (3 m)	80.02	8.15	88.17	6.14	1.65	4.01	0	0	0	0	0	0	0	0	0	0	0	0	0	0	0	846
Crow Piece Lane	77.44	10.88	88.32	5.04	2.36	3.94	0	0	0	0.31	0	0	0	0	0	0	0	0	0	0	0	634
Boyn Hill	82.30	6.58	88.88	6.58	2.05	2.46	0	0	0	0	0	0	0	0	0	0	0	0	0	0	0	243
Farnham Royal Church	67.58	7.73	75.31	13.96	4.73	5.98	0	0	0	0	0	0	0	0	0	0	0	0	0	0	0	401
Shiplake	80.03	7.35	87.38	6.00	3.90	2.70	0	0	0	0	0	0	0	0	0	0	0	0	0	0	0	666
Edgehill Road, Reading	83.68	5.71	89.39	6.03	4.24	0.32	0	0	0	0	0	0	0	0	0	0	0	0	0	0	0	613
Gravel Hill, Chazey Wood	87.98	3.86	91.84	3.86	3.46	0	0	0	0	0	0.20	0	0.40	0	0.20	0	0	0	0	0	0	491
Hitcham	81.82	7.44	89.26	6.20	2.06	2.48	0	0	0	0	0	0	0	0	0	0	0	0	0	0	0	848

Appendix: Pebble counts from the Middle Thames region

	Flint (angular)	Flint (rounded)	Flint (total)	Vein quartz	Quartzite	Green-sand chert	Green-sand	Chalk	Red siltstone	Ironstone	Igneous & meta-morphic	Calcite	Brown sand-stone	Lime-stone	Palaeo-zoic chert	Black shale	Grit	Red sand-stone	Brown silt	Green silt-stone	Banded mud-stone	Total number of pebbles
Black Park Gravel																						
Kiln Wood, Wingroves	72.46	11.51	83.97	9.47	3.93	2.63	0	0	0	0	0	0	0	0	0	0	0	0	0	0	0	570
Black Park A	75.54	9.07	84.61	8.48	3.74	2.95	0	0	0	0	0	0.19	0	0	0	0	0	0	0	0	0	507
Hillingdon	77.73	8.39	86.12	6.16	2.91	4.10	0	0.17	0	0	0.34	0	0.17	0	0	0	0	0	0	0	0	584
Lambourne Wood	77.83	11.17	89.00	8.17	1.33	1.50	0	0	0	0	0	0	0	0	0	0	0	0	0	0	0	600
West Burnham	80.88	8.37	89.25	3.79	1.90	5.06	0	0	0	0	0	0	0	0	0	0	0	0	0	0	0	633
St Nicholas School, Taplow	75.77	9.10	84.87	10.07	2.51	2.51	0	0	0	0	0	0	0	0	0	0	0	0	0	0	0	516
Iver Heath Church	81.86	8.19	90.05	6.19	2.21	1.55	0	0	0	0	0	0	0	0	0	0	0	0	0	0	0	452
Wickenden pipeline	76.73	8.96	85.69	6.44	3.93	0	0	0	0	0	0	0	0	0	0	0	0	0	0	0	3.93	613
Highland's Farm	81.26	5.13	86.39	6.34	5.89	0.75	0	0	0	0	0.15	0	0	0	0.45	0	0	0	0	0	0	662
Morgan's Farm	86.15	3.49	89.64	5.87	4.32	0	0	0	0	0	0	0	0.13	0	0	0	0	0	0	0	0	715
Chazey Wood	85.28	5.94	91.22	5.05	3.71	0	0	0	0	0	0	0	0	0	0	0	0	0	0	0	0	673
Henley Hill	84.37	8.95	93.32	4.55	1.82	0	0	0	0	0.15	0	0	0	0	0.15	0	0	0	0	0	0	659
Winter Hill Gravel																						
Rush Green (M40)	69.59	3.44	78.03	12.83	6.25	2.53	0	0	0	0	0	0	0.17	0.17	0	0	0	0	0	0	0	592
Dorney Wood	73.59	8.06	81.65	10.69	6.25	1.41	0	0	0	0	0	0	0	0	0	0	0	0	0	0	0	496
Wingroves, Burnham	69.86	9.81	79.67	15.06	2.51	2.51	0	0	0	0	0.22	0	0	0	0	0	0	0	0	0	0	438
Burnham Beeches (north)	74.81	7.97	82.78	10.98	3.12	2.91	0	0	0	0	0	0	0.21	0	0	0	0	0	0	0	0	929
Oak End Wood	74.37	7.42	81.79	9.58	3.56	3.95	0	0	0	0	0.84	0	0	0.28	0	0	0	0	0	0	0	355
Winter Hill	74.60	6.40	81.00	9.80	2.40	6.60	0	0	0	0	0	0	0	0	0.20	0	0	0	0	0	0	500
Tatling End	84.10	8.76	92.86	5.06	1.61	1.61	0	0	0	0	0.28	0	0	0	0	0	0	0	0	0	0	434
Bourne End Road	71.92	10.78	82.70	12.33	1.20	3.42	0	0	0.17	0	0.17	0	0	0	0	0	0	0	0	0	0	584
E Burnham Common	77.10	8.18	85.28	9.20	2.02	3.29	0	0	0	0	0	0	0.40	0	0	0	0	0	0	0	0	489
Croxley Green	78.03	7.09	85.12	9.15	4.11	0.91	0	0	0	0	0.68	0	0	0	0	0	0	0	0	0	0	437
Denham Aerodrome	80.76	4.64	85.40	7.21	4.32	3.04	0	0	0	0	0	0	0	0	0	0	0	0	0	0	0	624
Binfield Heath	68.12	4.73	72.85	18.57	8.56	0	0	0	0	0	0	0	0	0	0	0	0	0	0	0	0	549
Rosehill Kiln	62.24	7.21	69.45	16.44	13.25	0	0	0	0	0	0.17	0	0.50	0	0	0	0.17	0	0	0	0	596
Crosscroft	59.87	6.74	66.61	21.50	9.30	0	0	0	0	0	0.32	0	1.44	0	0.32	0	0.32	0	0	0	0	623
Mapledurham	67.94	5.88	72.82	17.05	8.52	0	0	0	0	0	0.58	0	0	0	0	0	0	0	0	0	0	340
Goring	66.50	4.00	70.50	20.25	7.75	0	0	0.25	0	0.50	0.50	0	0	0	0.25	0	0	0	0	0	0	400
Cooks Wood (Hill End)	63.54	13.24	76.78	15.27	1.83	5.70	0	0	0	0	0	0	0.20	0	0.20	0	0	0	0	0	0	491
Henley Park	62.21	9.95	72.16	18.09	9.50	0	0	0	0	0	0	0	0.22	0	0	0	0	0	0	0	0	442
Stoke Common	75.81	7.97	82.78	9.90	4.41	1.41	0	0	0	0.25	0.80	0	0.50	0	0.25	0	0	0	0	0	0	667
Gerrards Cross Gravel																						
Littleworth Common	66.50	11.30	77.80	16.00	3.72	2.21	0	0	0	0	0	0	0.25	0	0	0	0	0	0	0	0	406
Highlands Wood	48.33	7.12	55.45	31.96	12.42	0	0	0	0	0	0.15	0	0	0	0	0	0	0	0	0	0	660
Slade Oak Lane	61.08	9.14	70.22	20.23	3.89	3.89	0	0	0.58	0	0	0	0.19	0.77	0	0	0	0	0	0	0	514
Burtley Wood (M40)	46.61	14.05	60.66	27.10	8.26	3.80	0	0	0	0	0	0	0	0.16	0	0	0	0	0	0	0	605
Jenkins Farm Wood	52.24	12.62	64.86	23.16	9.74	2.07	0	0	0	0.16	0	0	0	0	0	0	0	0	0	0	0	626
Broad Lane Wooburn	83.54	15.66	99.20	0.32	0.16	0.16	0	0	0	0	0	0	0.16	0	0	0	0	0	0	0	0	632
Egypt Woods	59.83	8.76	68.59	18.84	9.92	2.48	0	0	0	0	0	0	0	0	0	0.17	0	0	0	0	0	605
Gerrards Cross	47.83	18.80	66.73	23.62	7.28	1.77	0	0	0	0	0.20	0	0.39	0	0	0	0	0	0	0	0	508
Gerrards Cross	38.07	18.93	57.00	31.72	8.09	1.38	0	0	1.78	0	0	0	0	0	0	0	0	0	0	0	0	507
Hedsor Court	89.02	6.69	95.71	3.26	0.88	0.17	0	0	0	0	0	0	0	0	0	0	0	0	0	0	0	583
Westwood	42.35	15.29	57.64	27.23	10.42	1.35	0	0	0	0	0.34	0	1.18	0	0	0	1.01	0	0	0.34	0	595
Westwood	46.64	13.04	59.68	27.55	9.01	1.48	0	0	0	0	0.54	0	0.13	0	0.67	0	0.94	0	0	0	0	744
Croxley Green	54.65	9.31	63.96	25.26	8.86	1.33	0	0.30	0	0	0.15	0	0	0	0	0	0	0	0	0.15	0	677
Remenham Place	38.88	9.97	48.85	41.31	8.82	0	0	0	0	0	0.58	0	0	0	0.14	0	0	0	0	0	0	702
Rotherfield Peppard	51.37	5.98	57.35	26.17	15.83	0	0	0	0	0	0.65	0	0	0	0	0	0	0	0	0	0	619
Old Copse Sonning	57.88	3.73	61.61	25.25	12.65	0	0	0	0	0	0.48	0	0	0	0	0	0	0	0	0	0	615
Tilehurst	33.62	12.06	45.68	37.75	15.51	0	0	0	0	0	0.68	0	0.34	0	0	0	0	0	0	0	0	580

Appendix: Pebble counts from the Middle Thames region

	Flint (angular)	Flint (rounded)	Flint (total)	Vein quartz	Quartzite	Green-sand chert	Green-sand	Chalk	Red siltstone	Ironstone	Igneous & meta-morphic	Calcite	Brown sand-stone	Lime-stone	Palaeo-zoic chert	Black shale	Grit	Red sand-stone	Brown silt	Green silt-stone	Banded mud-stone	Total number of pebbles
Beaconsfield Gravel																						
Valentine Wood	43.00	5.64	48.64	36.07	14.17	0	0	0	0	0	0.64	0	0.32	0	0	0	0	0	0	0	0.16	621
Beaconsfield Station (Lr.)	37.62	9.81	47.43	42.92	8.68	0.96	0	0	0	0	0	0	0	0	0	0	0	0	0	0	0	622
Beaconsfield Station (Up.)	51.38	11.08	62.46	28.21	7.80	1.51	0	0	0	0	0	0	0	0	0	0	0	0	0	0	0	397
Welders Lane A	42.10	12.74	54.84	26.59	16.06	1.94	0	0	0	0.55	0	0	0	0	0	0	0	0	0	0	0	543
Seer Green	48.07	8.47	56.54	27.99	14.36	0.92	0	0	0	0	0	0	0	0	0.18	0	0	0	0	0	0	361
Juniper Lane, Flackwell	88.65	10.29	98.94	0.79	0	0.26	0	0	0	0	0	0	0	0	0	0	0	0	0	0	0	379
Flackwell Heath	76.04	10.48	86.52	7.48	5.39	0.30	0	0	0	0	0	0	0	0	0.30	0	0	0	0	0	0	334
Deans Wood	42.09	12.85	54.94	35.75	8.01	1.30	0	0	0	0	0	0	0	0	0	0	0	0	0	0	0	537
Warren Wood	47.80	10.46	58.26	29.84	10.60	0	0	0	0	0	0.99	0	0	0	0.28	0	0	0	0	0	0	707
Rowe Wood	52.93	10.58	63.51	24.36	11.47	0	0	0	0	0	0.51	0	0.25	0	0	0	0	0	0	0	0	784
Withy Copse	43.01	3.57	46.58	34.62	17.08	0	0	0	0	0	1.08	0	0.31	0	0.31	0	0	0	0	0	0	644
Holtspur Lane	42.16	11.82	53.98	32.05	10.68	1.42	0	0	0	0	1.42	0	0.28	0	0	0	0	0.14	0	0	0	702
Goring Heath	44.03	3.02	47.05	35.81	14.86	0	0	0	0	0	0.89	0	0.32	0	0.32	0	0	0.32	0	0	0.43	987
Satwell Gravel																						
Satwell	26.29	6.02	32.31	42.87	11.02	0	0	0	0	0	0.98	0	0	0	0	0	0	0	0	0	0	514
Westland Green Gravel																						
Cold Harbour	84.45	4.62	89.07	7.77	2.52	0	0	0	0	0	0.63	0	0	0	0	0	0	0	0	0	0	476
Cray's Pond	75.04	13.15	88.19	6.96	4.64	0	0	0	0	0	0.19	0	0	0	0	0	0	0	0	0	0	517
Stoke Row Gravel																						
Stoke Row	30.62	15.44	46.06	49.86	3.79	0	0	0	0	0	0	0	0	0	0.27	0	0	0	0	0	0	369
Checkenden	62.62	30.63	93.25	3.70	2.35	0	0	0	0	0	0.33	0	0	0	0	0	0	0.33	0	0	0	297
Nettlebed Gravel																						
Nettlebed D	42.24	39.50	81.74	14.00	4.11	0	0	0	0	0	0	0	0	0	0.12	0	0	0	0	0	0	729
Nettlebed B	43.75	37.50	81.25	15.10	3.36	0	0	0	0	0	0	0	0	0	0.28	0	0	0	0	0	0	416
Nettlebed C	52.99	31.98	84.97	10.62	4.17	0	0	0	0	0	0	0	0	0	0.22	0	0	0	0	0	0	719
Nettlebed E	53.35	31.94	85.29	10.19	4.17	0	0	0	0	0	0	0	0	0	0.33	0	0	0	0	0	0	551
Kilridge Wood	59.85	32.81	92.66	3.85	2.97	0	0	0	0	0	0	0	0	0	0.52	0	0	0	0	0	0	281
Loddon gravels																						
Woodley	83.65	7.01	90.66	0.55	0	8.79	0	0	0	0	0	0	0	0	0	0	0	0	0	0	0	728
Beech Lane	73.17	8.18	81.35	0.34	0	18.29	0	0	0	0	0	0	0	0	0	0	0	0	0	0	0	574
Whistley Park	86.88	6.29	93.17	0.25	0.12	6.42	0	0	0	0	0	0	0	0	0	0	0	0	0	0	0	778
Unknown																						
Davenport Wood	69.12	17.32	86.44	9.39	3.22	0.81	0	0	0	0	0	0	0.13	0	0	0	0	0	0	0	0	745
Rassler Wood	67.41	11.11	78.52	10.68	9.57	1.23	0	0	0	0	0	0	0	0	0	0	0	0	0	0	0	324
Chalfont St Peter	83.76	6.26	90.02	5.96	3.43	0.60	0	0	0	0	0	0	0	0	0	0	0	0	0	0	0	671

References

Aalto, M. M., Coope, G. R. and Gibbard, P. L. 1984. Late Devensian river deposits at Thrupp House Farm, Abingdon, Berkshire, England. *Proc. Geol. Ass.*, **95**, 65–79.

Abbott, W. J. L. 1892. The sections exposed in the foundations of the new Admiralty offices. *Proc. Geol. Ass.*, **12**, 346–56.

Allen, J. R. L. 1970. Studies in fluviatile sedimentation: a comparison of fining-upward cyclothems, with special reference to coarse-member composition and interpretation. *J. Sedim. Petrol.*, **40**, 298–323.

Allen, T. J. 1978. Disposition of the terraces of the River Thames in the vicinity of Yiewsley. In: *Early Man in West Middlesex* (Ed. D. Collins). Her Majesty's Stationery Office: London, pp. 5–10.

American Society for Testing and Materials (ASTM) 1964. *Procedures for Testing Soils*. Am. Soc. for Testing and Materials: Philadelphia, pp. 95–101.

Andrew, R. 1970. The Cambridge pollen reference collection. In: *Studies in the Vegetational History of the British Isles* (Ed. D. Walker and R. G. West). University Press: Cambridge, pp. 225–31.

Arkell, W. J. 1947. The geology of the Evenlode Gorge, Oxfordshire. *Proc. Geol. Ass.*, **58**, 87–114.

Avery, B. W., Bullock, P., Catt, J. A., Rayner, J. H. and Weir, A. H. 1982. Composition and origin of some brickearths on the Chiltern Hills, England. *Catena*, **9**, 153–74.

Banfield, A. W. F. 1974. *The Mammals of Canada*. University of Toronto Press: Toronto.

Barrow, G. 1919. Some future work for the Geologists' Association. *Proc. Geol. Ass.*, **30**, 36–48.

Bate, D. M. A. 1937. Note on recent finds of *Dama clactoniana* (*Cervus browni* auctt.) in London and Swanscombe. *Proc. Prehist. Soc.*, **3**, 460–3.

Bates, C. D., Coxon, P. and Gibbard, P. L. 1978. A new method for the preparation of clay-rich sediment samples for palynological investigation. *New Phytol.*, **81**, 459–63.

Bell, F. G. 1968. Weichselian Glacial Floras in Britain. Ph.D. thesis, University of Cambridge.

Bell, F. G. 1969. The occurrence of southern, steppe and halophyte elements in Weichselian (last glacial) floras from southern Britain. *New Phytol.*, **68**, 913–22.

Bell, F. G. 1970. Late Pleistocene floras from Earith, Huntingdonshire. *Phil. Trans. R. Soc. Lond.*, **B 258**, 347–78.

Berry, F. G. 1979. Late Quaternary scour hollows and related features in central London. *Q. J. Engng. Geol.*, **12**, 9–29.

Bird, J. B. 1967. *The Physiography of Arctic Canada*. The Johns Hopkins University Press: Baltimore.

Birks, H. J. B. 1973. *Past and Present Vegetation of the Isle of Skye – A Palaeoecological Study*. University Press: Cambridge.

Bishop, W. W. 1958. The Pleistocene geology and geomorphology of three gaps in the Midlands Jurassic escarpment. *Phil. Trans. R. Soc. Lond.*, **B 241**, 255–306.

Blackith, R. E. and Reyment, R. A. 1971. *Multivariate Morphometrics*. Academic Press: London and New York.

Blake, J. H. 1903. The geology of the country around Reading. *Mem. Geol. Surv. G. B.*

Bluck, B. J. 1967. Sedimentation of beach gravels: an example from south Wales. *J. Sedim. Petrol.*, **37**, 128–56.

Bluck, B. J. 1979. The structure of coarse grained braided stream alluvium. *Trans. R. Soc. Edinburgh*, **70**, 181–221.

Boulton, G. S. 1971. Till genesis and fabric in Svalbard, Spitsbergen. In: *Till: A Symposium* (Ed. R. P. Goldthwait). Ohio State University Press: Columbus.

Breuil, H. 1934. De l'importance de la solifluction dans l'études des terrains quaternaire de la France et des pays voisins. *Rev. Géogr. Phys. Géol. Dynamique*, **7**, 269–331.

Bridgland, D. R. 1980. A reappraisal of Pleistocene stratigraphy in North Kent and East Essex and new evidence concerning the former courses of the Thames and Medway. *Quaternary Newsl.*, **32**, 15–24.

Bridgland, D. R. and Gibbard, P. L. in prep. A preliminary reappraisal of the Pleistocene stratigraphy of the Lower Thames Valley.

Briggs, D. J. and Gilbertson, D. D. 1973. The age of the Hanborough Terrace of the River Evenlode, Oxfordshire. *Proc. Geol. Ass.*, **84**, 155–73.

Briggs, D. J. and Gilbertson, D. D. 1976. River terraces in the Oxford area and non-marine molluscan faunas of terrace gravels in the Upper Thames basin. In: *Field Guide to the Oxford Region* (Ed. D. Roe). Quaternary Research Association: London, pp. 6–19.

Briggs, D. J. and Gilbertson, D. D. 1980. Quaternary processes and environments in the Upper Thames Valley. *Trans. Inst. Brit. Geogr.*, **5**, 53–65.

Bromehead, C. E. N. 1912. On the diversions of the Bourne near Chertsey. 'Summary of progress for 1911'. *Geol. Surv. G.B.*, Appendix II.

Bromehead, C. E. N. 1925. The geology of north London. *Mem. Geol. Surv. G.B.*

Brown, J. A. 1883. On probable glacial deposits or evidence of the action of ice near Ealing. *Proc. Geol. Ass.*, **8**, 172–84.

Brown, J. A. 1886. The Thames Valley surface deposits of the Ealing district and their associated Palaeolithic floors. *Q. J. Geol. Soc. Lond.*, **42**, 192–200.

Brown, J. A. 1887. Excursion to the Mount, Ealing and Horsington Hill. *Proc. Geol. Ass.*, **10**, 172–81.

Brown, J. A. 1889. On the discovery of *Elephas primigenius* associated with flint implements at Southall. *Proc. Geol. Ass.*, **10**, 361–72.

Brown, J. A. 1896. Excursion to Hanwell, Dawley and West Drayton. *Proc. Geol. Ass.*, **14**, 118–20.

Bryant, I. D. 1983a. The utilisation of Arctic analogue studies in the interpretation of periglacial river sediments from southern Britain. In: *Background to Palaeohydrology: A Perspective* (Ed. K. J. Gregory). J. Wiley and Sons: London, pp. 415–31.

Bryant, I. D. 1983b. Facies sequences associated with some braided river deposits of Late Pleistocene age from southern Britain. In: *Modern and Ancient Fluvial Systems* (Ed. J. D. Collinson). Proc. Int. Fluvial Conf. Keele 1982, pp. 267–75.

Bull, A. J. and Wooldridge, S. W. 1925. Geomorphology of the Mole Gap. *Proc. Geol. Ass.*, **34**, 1–10.

Bull, A. J., Gossling, F., Green, J. F. N., Hayward, H. A., Turner, E. A. and Wooldridge, S. W. 1934. The River Mole: its physiography and superficial deposits. *Proc. Geol. Ass.*, **45**, 35–69.
Burchell, J. P. T. 1934. The Middle Mousterian Culture and its relation to the Coombe Rock of post-early Mousterian times. *Antiq. J.*, **14**, 33–9.
Callow, P. 1976. *The Lower and Middle Palaeolithic of Britain and Adjacent Areas of Europe*. Ph.D. thesis, University of Cambridge.
Cant, D. J. 1976. Selective preservation of flood stage deposits in a braided fluvial environment. *Geol. Soc. Can. Progr. Abstr.*, 1, 77.
Castleden, R. 1980. Fluvioglacial pedimentation: a general theory of fluvial valley development in cool temperate lands, illustrated from western and central Europe. *Catena*, **7**, 135–52.
Catt, J. A. 1977. Loess and coversands. In: *British Quaternary Studies* (Ed. F. W. Shotton). University Press, Oxford, pp. 221–9.
Catt, J. A. 1978. The contribution of loess to soils in lowland Britain. In: *The Effect of Man on the Landscape: The Lowland Zone* (Ed. S. Limbrey and J. G. Evans). *Council Brit. Archaeol. Res. Rep.*, **21**, 12–20.
Catt, J. A. and Hodgson, J. M. 1976. Soils and geomorphology of the Chalk in south-east England. *Earth Surf. Processes*, **1**, 181–93.
Cheetham, G. H. 1975. Late Quaternary Palaeohydrology, with Reference to the Kennet Valley. Ph.D. thesis, University of Reading.
Cheetham, G. H. 1980. Late Quaternary palaeohydrology: the Kennet Valley case-study. In: *The Shaping of Southern England* (Ed. D. K. C. Jones). Academic Press: London, pp. 203–23.
Church, M. 1972. Baffin Island sandurs: a study of Arctic fluvial processes. *Geol. Surv. Canada Bull.*, No. 216.
Clapham, A. R., Tutin, T. G. and Warburg, E. F. 1962. *Flora of the British Isles.* University Press: Cambridge.
Clarke, M. R. and Dixon, A. J. 1981. The Pleistocene braided river deposits in the Blackwater area of Berkshire and Hampshire, England. *Proc. Geol. Ass.*, **92**, 139–57.
Clayton, K. M. 1977. River terraces. In: *British Quaternary Studies* (Ed. F. W. Shotton). University Press: Oxford, pp. 153–67.
Collins, D. 1978. *Early Man in West Middlesex*. Her Majesty's Stationery Office: London.
Collinson, J. D. 1970. Bed forms in the Tana River, Norway. *Geogr. Ann.*, **52A**, 31–55.
Coombs, J. A. 1873. On a recently exposed section at Battersea. *Proc. Geol. Ass.*, **3**, 33–8.
Coope, G. R. 1974. Interglacial Coleoptera from Bobbitshole, Ipswich, Suffolk. *J. Geol. Soc. Lond.*, **130**, 333–40.
Coope, G. R. 1977. Fossil coleopteran assemblages as sensitive indicators of climatic changes during the Devensian (Last) cold stage. *Phil. Trans. R. Soc. Lond.*, **B 280**, 313–40.
Coope, G. R. 1982. Coleoptera from two Late Devensian sites in the Lower Colne Valley, West London, England. *Quaternary Newsl.*, **38**, 1–6.
Coope, G. R. and Angus, R. B. 1975. An ecological study of a temperate interlude in the middle of the last glaciation, based on fossil Coleoptera from Isleworth, Middlesex. *J. Anim. Ecol.*, **44**, 365–91.
Coope, G. R. and Sands, C. H. S. 1966. Insect faunas of the last glaciation from the Tame Valley, Warwickshire. *Proc. R. Soc. Lond.*, **B 165**, 389–412.
Cooper, J. E. 1907. Holocene Mollusca from Staines. *Proc. Malac. Soc. Lond.*, **7**, 310–11.
Cooper, J. E. 1913. A Holocene deposit at Boveney. *Proc. Malac. Soc. Lond.*, **10**, 318–19.
Cooper, J. E. 1922. Note on a Holocene deposit at Penton Hook. *Proc. Malac. Soc. Lond.*, **15**, 35–6.
Corner, R. 1975. The Tana Valley terraces. A study of the morphology and sedimentology of the lower Tana Valley terraces between Mansholmen and Maskejokka, Finmark, Norway. *Uppsala Universitet Naturgeografiska Institutionen Rapport*, No. 38.
Cotton, C. A. 1940. Classification and correlation of river terraces. *J. Geomorph.*, **3**, 27–37.
Cranshaw, S. 1983. Handaxes and cleavers: selected English Acheulian industries. *Brit. Archaeol. Rep.*, No. 113.
Dalrymple, J. B. 1960. The Nature, Origin, Age and Correlation of some of the Brickearths and Associated Soils of Southeast England. Ph.D. thesis, University of London.
Davis, J. C. 1973. *Statistics and Data Analysis in Geology*. J. Wiley and Sons: New York and Chichester.
Dawkins, W. B. 1880. *Early Man in Britain.* London.
Devoy, R. J. N. 1979. Flandrian sea level changes and vegetational history of the lower Thames estuary. *Phil. Trans. R. Soc. Lond.*, **B 285**, 355–410.
Dewey, H. 1926. Palaeoliths in Hyde Park. *Antiq. J.*, **6**, 73–5.
Dewey, H. and Bromehead, C. E. N. 1915. The geology of the country around Windsor and Chertsey. *Mem. Geol. Surv. G.B.*
Dewey, H. and Bromehead, C. E. N. 1921. The geology of south London. *Mem. Geol. Surv. G.B.*
Dewey, H., Bromehead, C. E. N. and Dines, H. G. 1924. The geology of the country around Dartford. *Mem. Geol. Surv. G.B.*
Dines, H. G. and Edmunds, F. H. 1925. The geology of the country around Romford. *Mem. Geol. Surv. G.B.*
Dreimanis, A. 1976. Tills: their origin and properties. In: *Glacial Till* (Ed. R. F. Legget). *R. Soc. Canada Spec. Publ.*, **12**, 11–49.
Dunkley, P. N. 1979. The sand and gravel resources of the country around Maidenhead and Marlow. *Inst. Geol. Sci. Min. Assess. Rep.*, No. 42.
Edmunds, F. H. 1931. Some gravel-filled pipes in the London Clay at Battersea, 'Summary of progress for 1930'. *Geol. Surv. G.B.*, Appendix II, 12–30.
Embleton, C. and King, C. A. M. 1975. *Periglacial Geomorphology*. Edward Arnold: London.
English Mechanic, The 1882. The fossil lions of Charing Cross. 20 October, p. 49.
Evans, J. 1860. On the occurrence of flint implements in undisturbed beds of gravel, sand and clay. *Archaeologia*, **38**, 280–307.
Fahnestock, R. K. 1963. Morphology and hydrology of a glacial stream – White River, Mt. Rainier, Washington. *U.S. Geol. Surv. Prof. Paper*, **422-A**, 1–70.
Fisk, H. N. 1944. *Geological Investigation of the Alluvial Valley of the Lower Mississippi River.* Corps of Engineers, U.S. Army.
Franks, J. W. 1960. Interglacial deposits at Trafalgar Square, London. *New Phytol.*, **59**, 145–52.
Franks, J. W., Sutcliffe, A. J., Kerney, M. P. and Coope, G. R. 1958. Haunt of the elephant and rhinoceros: the Trafalgar Square of 100,000 years ago – new discoveries. *Illustrated London News*, 14 June, pp. 1011–13.
French, H. M. 1971. Ice cored mounds and patterned ground, southern Banks Island, Western Canadian Arctic. *Geogr. Ann.*, **53A**, 32–8.
French, H. M. 1975. Pingo investigations and terrain disturbance studies, Banks Island, District of Franklin. *Geol. Surv. Canada*, **75-1A**, 459–64.
French, H. M. 1976. *The Periglacial Environment.* Longman Group: London.
Friend, P. F. and Moody-Stuart, M. 1972. Sedimentation of the Wood Bay Formation (Devonian) of Spitsbergen: regional analysis of a late orogenic basin. *Norsk Polarinst. Skr.*, **157**, 1–77.

Frye, J. C. and Leonard, A. R. 1954. Some problems of alluvial terrace mapping. *Am. J. Sci.*, **252**, 242–51.
Galloway, R. W. 1961. Solifluction in Scotland. *Scott. geogr. Mag.*, **77**, 75–87.
Gentleman's Magazine 1758. Elephant bones in England. p. 166.
Gibbard, P. L. 1974. Pleistocene Stratigraphy and Vegetational History of Hertfordshire. Ph.D. thesis, University of Cambridge.
Gibbard, P. L. 1977. Pleistocene history of the Vale of St. Albans. *Phil. Trans. R. Soc. Lond.*, **B 280**, 445–83.
Gibbard, P. L. 1978. Quaternary geology and landform development of the Vale of St. Albans. In: *Field Guide to the Vale of St. Albans* (Ed. J. Rose and P. L. Gibbard). Quaternary Research Association: London, pp. 9–29.
Gibbard, P. L. 1979. Middle Pleistocene drainage in the Thames Valley. *Geol. Mag.*, **116**, 35–44.
Gibbard, P. L. 1981. Views on the Pleistocene of the Thames Valley. *Quaternary Newsl.*, **34**, 32–35.
Gibbard, P. L. 1982. Terrace stratigraphy and drainage history of the Plateau Gravels of north Surrey, south Berkshire and north Hampshire, England. *Proc. Geol. Ass.*, **93**, 369–84.
Gibbard, P. L. 1983. The diversion of the Thames – a review. In: *Diversion of the Thames Field Guide* (Ed. J. Rose). Quaternary Research Association: Cambridge. pp. 8–23.
Gibbard, P. L. and Aalto, M. M. 1977. A Hoxnian interglacial site at Fishers Green, Stevenage, Hertfordshire. *New Phytol.*, **72**, 505–23.
Gibbard, P. L. and Hall, A. R. 1982. Late Devensian river deposits in the lower Colne Valley, West London, England. *Proc. Geol. Ass.*, **93**, 291–300.
Gibbard, P. L., Bryant, I. D. and Hall, A. R. *in prep.* A Hoxnian interglacial doline filling at Slade Oak Lane, Denham, Buckinghamshire, England.
Gibbard, P. L., Coope, G. R., Hall, A. R., Preece, R. C. and Robinson, J. E. 1982. Middle Devensian river deposits beneath the 'Upper Floodplain' terrace of the River Thames at Kempton Park, Sunbury, Surrey, England. *Proc. Geol. Ass.*, **93**, 275–90.
Gilbert, G. K. 1877. Report on the geology of the Henry Mountains. *U.S. Geographical and Geological Survey of the Rocky Mountain Region.* Department of the Interior: Washington.
Gilbertson, D. D. 1980. The palaeoecology of Middle Pleistocene Mollusca from Sugworth, Oxfordshire. *Phil. Trans. R. Soc. Lond.*, **B 289**, 107–18.
Godwin, H. 1964. Late Weichselian conditions in south-eastern Britain: organic deposits at Colney Heath, Hertfordshire. *Proc. R. Soc. Lond.*, **B 160**, 258–75.
Godwin, H. 1975. *History of the British flora.* University Press: Cambridge.
Gordon, A. D. and Birks, H. J. B. 1974. Numerical methods in palaeoecology: II. Comparison of pollen diagrams. *New Phytol.*, **73**, 221–49.
Goudie, A. R. 1976. The Oxford region. In: *Field Guide to the Oxford Region* (Ed. D. Roe). Quaternary Research Association: London. pp. 1–5.
Goudie, A. R. and Hart, M. G. 1975. Pleistocene events and forms in the Oxford region. In: *Oxford and its Region* (Ed. C. G. Smith and D. I. Scargill). University Press: Oxford. pp. 3–13.
Green, C. P. and McGregor, D. F. M. 1978. Pleistocene gravel trains of the River Thames. *Proc. Geol. Ass.*, **89**, 143–56.
Green, C. P. and McGregor, D. F. M. 1980. The terraces of the River Thames and their palaeohydrological implications. In: *The Shaping of Southern England* (Ed. D. K. C. Jones). Academic Press: London. pp. 172–202.
Green, C. P., Hey, R. W. and McGregor, D. F. M. 1980. Volcanic pebbles in Pleistocene gravels of the Thames in Buckinghamshire and Hertfordshire. *Geol. Mag.*, **117**, 59–64.
Green, C. P., McGregor, D. F. M. and Evans, A. 1982. Development of the Thames drainage system in Buckinghamshire and Hertfordshire. *Geol. Mag.*, **119**, 281–90.
Gregory, K. J. and Walling, D. E. 1973. *Drainage Basin Form and Process.* Edward Arnold: London.
Hall, A. R. 1978. Some new palaeobotanical records for the British Ipswichian interglacial. *New Phytol.*, **81**, 805–12.
Hall, A. R. 1980. Late Pleistocene deposits at Wing, Rutland. *Phil. Trans. R. Soc. Lond.*, **B 289**, 135–64.
Hanson, C. B. 1980. Fluvial taphonomic processes: models and experiments. In: *Fossils in the Making* (Ed. A. K. Behrensmeyer and A. P. Hill). University of Chicago Press: Chicago. pp. 156–81.
Harding, P. and Gibbard, P. L. 1984. Excavations at Northwold Road, Stoke Newington, north east London, 1981. *Trans. Lond. Middx. Archaeol. Soc.*, **34**, (in press).
Harding, P., Gibbard, P. L., Lewin, J., Macklin, M. G. and Moss, E. H. 1985. The transport and abrasion of flint hand axes in a gravel-bed river. In: *The Human Uses of Flint and Chert: Proceedings of the Fourth International Flint Symposium* (Ed. G. Sieveking and M. H. Newcomer). University Press: Cambridge (in press).
Hare, F. K. 1947. The geomorphology of a part of the Middle Thames. *Proc. Geol. Ass.*, **58**, 294–339.
Harms, J. C. and Fahnestock, R. K. 1965. Stratification, bedforms and flow phenomena (with an example from the Rio Grande). In: *Primary Sedimentary Structures and their Hydrodynamic Interpretation* (Ed. G. V. Middleton). *Soc. Econ. Palaeontol. Mineral. Spec. Publ.*, **12**, 84–115.
Harms, J. C., Southard, J. B., Spearing, D. R. and Walker, R. G. 1975. Depositional environments as interpreted from primary sedimentary structures and stratification sequences. *Soc. Econ. Palaeontol. Mineral.*, Short Course 2.
Hawkins, H. L. 1926. On the former course of the Kennet between Theale and Pangbourne. *Proc. Geol. Ass.*, **37**, 442–6.
Hedberg, H. D. 1976. *International Stratigraphic Guide.* J. Wiley and Sons: New York.
Hein, F. J. and Walker, R. G. 1977. Bar evolution and development of stratification in the gravelly, braided, Kicking Horse River, B.C. *Can. J. Earth Sci.*, **14**, 562–70.
Hey, R. W. 1965. Highly quartzose pebble gravels in the London Basin. *Proc. Geol. Ass.*, **76**, 403–20.
Hey, R. W. 1976. The terraces of the Middle and Lower Thames. *Studia Soc. Sci. Torunensis*, **8C**, 115–22.
Hey, R. W. 1980. Equivalents of the Westland Green Gravels in Essex and East Anglia. *Proc. Geol. Ass.*, **91**, 279–90.
Hey, R. W. and Brenchley, P. J. 1977. Volcanic pebbles from Pleistocene gravels in Norfolk and Essex. *Geol. Mag.*, **114**, 219–25.
Hey, R. W., Krinsley, D. H. and Hyde, P. J. W. 1971. Surface textures of sand grains from the Hertfordshire Pebble Gravels. *Geol. Mag.*, **108**, 377–82.
Hicks, H. 1892. On the discovery of mammoth and other remains in Endsleigh Street and on sections exposed in Endsleigh Gardens, Gordon Street and Tavistock Square, London. *Q. J. Geol. Soc. Lond.*, **48**, 453–68.
Higginbottom, I. E. and Fookes, P. G. 1971. Engineering aspects of periglacial features in Britain. *Q. J. Engng. Geol.*, **3**, 85–117.
Hinde, G. J. 1897. Notes on the gravels of Croydon and its neighbourhood. *Proc. Croydon Nat. Hist. Scient. Soc.*, **4**, 219–33.
Hollin, J. 1977. Thames interglacial sites, Ipswichian sea levels and Antarctic ice surges. *Boreas*, **6**, 33–52.
Holmes, T. V. 1892. The new railway from Grays Thurrock to Romford: sections between Upminster and Romford. *Q. J. Geol. Soc. Lond.*, **48**, 365–72.

Holyoak, D. T. 1980. Late Pleistocene Sediments and Biostratigraphy of the Kennet Valley, England. Ph.D. thesis, University of Reading.

Horton, A. 1977. Nettlebed. In: *Southeast England and the Thames Valley* (Ed. E. R. Shephard-Thorn and J. J. Wymer). Guide book for excursion A5 X INQUA Congress, pp. 16–18.

Howard, A. D. 1959. Numerical systems of terrace nomenclature: a critique. *J. Geol.*, **67**, 239–43.

Howard, A. D., Fairbridge, R. W. and Quinn, J. H. 1968. Terraces – fluvial. In: *The Encyclopedia of Geomorphology* (Ed. R. W. Fairbridge). Reinhold Book Corp.: New York, pp. 1117–23.

Howard, M. M. 1952. *Planorbis acronicus* Férussac in a Holocene deposit in Surrey. *J. Conch. Lond.*, **23**, 261–4.

Howe, J. A. and Skeats, E. W. 1903. Excursion to Denham and Gerrards Cross to the new railway cutting on the Great Western Railway. *Proc. Geol. Ass.*, **18**, 188–91.

Hubbard, R. N. L. B. 1982. The environmental evidence from Swanscombe and its implications for Palaeolithic archaeology (Ed. P. E. Leach). *Council Brit. Archaeol. Res. Rep.*, **48**, 3–7.

Hudleston, W. H. and Price, F. G. H. 1873. On excavations on the site of the new Law Courts. *Proc. Geol. Ass.*, **3**, 43–64.

Hutchinson, J. N. 1980. Possible late Quaternary pingo remnants in central London. *Nature*, **284**, 253–5.

Iversen, J. 1944. *Viscum*, *Hedera* and *Ilex* as climate indicators. *Geol. Fören. Stockh. Förh.*, **66**, 463–83.

Jessen, K. and Milthers, V. 1928. Stratigraphical and palaeontological studies of interglacial freshwater deposits in Jutland and northwest Germany. *Danm. Geol. Unders.*, RII, 48.

Johnson, D. 1944. Problems of terrace correlation. *Bull. Geol. Soc. Am.*, **55**, 793–818.

Jones, D. K. C. 1981. *Southeast and southern England*. Methuen: London.

Jukes-Brown, A. J. and White, H. J. O. 1908. The geology of the country around Henley-on-Thames and Wallingford. *Mem. Geol. Surv. G.B.*

Kennard, A. S. 1944. The Crayford brickearths. *Proc. Geol. Ass.*, **55**, 121–69.

Kennard, A. S. and Woodward, B. B. 1906. On sections of the Holocene alluvium of the Thames at Staines and Wargreave. *Proc. Geol. Ass.*, **19**, 252–8.

Kerney, M. P. 1959. Pleistocene Non-marine Mollusca of the English Interglacial Deposits. Ph.D. thesis, University of London.

Kerney, M. P. 1971. Interglacial deposits in Barnfield pit, Swanscombe and their molluscan faunas. *J. Geol. Soc. Lond.*, **127**, 69–93.

Kerney, M. P. 1977. British Quaternary non-marine Mollusca: a brief review. In: *British Quaternary Studies* (Ed. F. W. Shotton). University Press: Oxford, pp. 31–42.

Kerney, M. P., Gibbard, P. L., Hall, A. R. and Robinson, J. E. 1982. Middle Devensian river deposits beneath the 'Upper Floodplain' terrace of the River Thames at Isleworth, West London. *Proc. Geol. Ass.*, **93**, 385–93.

King, W. B. R. and Oakley, K. P. 1936. The Pleistocene succession in the lower part of the Thames Valley. *Proc. Prehist. Soc.*, **2**, 52–76.

Krinsley, D. H. and Doornkamp, J. C. 1973. *Atlas of Quartz Sand Surface Textures*. University Press: Cambridge.

Kurtén, B. 1964. The evolution of the polar bear *Ursus maritimus* Phipps. *Acta Zool. Fenn.*, **108**, 1–26.

Lacaille, A. D. 1936. The Palaeolithic sequence at Iver, Bucks. *Antiq. J.*, **16**, 420–43.

Lacaille, A. D. 1939. The Palaeolithic contents of the gravels at East Burnham, Bucks. *Antiq. J.*, **19**, 166–76.

Lacaille, A. D. 1940. The Palaeoliths from the gravels of the Lower Boyn Hill Terrace around Maidenhead. *Antiq. J.*, **20**, 245–71.

Lacaille, A. D. 1959. Palaeoliths from brickearth in south-east Buckinghamshire. *Records of Bucks.*, **16**, 274–88.

Lacaille, A. D. 1961a. Mesolithic facies in Middlesex and London. *Trans. Lond. Middx. Archaeol. Soc.*, **20**, 101–50.

Lacaille, A. D. 1961b. The Palaeoliths from Boyn Hill, Maidenhead. *Antiq. J.*, **41**, 154–85.

Lacaille, A. D. 1963. Mesolithic industries beside Colne Waters in Iver and Denham, Buckinghamshire. *Records of Bucks.*, **17**, 143–81.

Lacaille, A. D. and Oakley, K. P. 1936. See Lacaille, A. D. 1936.

Lane-Fox, A. 1872. Palaeolithic implements associated with *Elephas primigenius* in Thames Valley gravels at Acton. *Q. J. Geol. Soc. Lond.*, **28**, 449–65.

Lawrence, G. F. 1930. Antiquities from the Middle Thames. *Archaeol. J.*, **86**, 69–98.

Leeson, M. D. 1891. The Saiga antelope in Britain. *Geol. Mag.*, **8**, 94.

Leeson, M. D. and Laffan, G. B. 1894. On the geology of the Pleistocene deposits in the valley of the Thames at Twickenham, with contributions to the fauna and flora of the period. *Q. J. Geol. Soc. Lond.*, **50**, 453–62.

Leopold, L. B., Wolman, M. G. and Miller, J. P. 1964. *Fluvial Processes in Geomorphology*. W. H. Freeman: San Francisco.

Lill, G. O. and Smalley, I. J. 1978. Distribution of loess in Britain. *Proc. Geol. Ass.*, **89**, 57–65.

Lobley, J. L. 1887. Excursion to the Brent Valley, Dollis Hill and Hampstead. *Proc. Geol. Ass.*, **10**, 148–53.

Loveday, J. 1962. Plateau deposits of the southern Chiltern Hills. *Proc. Geol. Ass.*, **73**, 83–102.

Lucas, J. 1887. *Hydrogeological Survey*. Stanford: London.

Lundqvist, G. 1949. The orientation of the block material in certain species of earth flow. *Geogr. Annalr.*, **37**, 335–47.

McGee, W. J. 1891. The Pleistocene history of north eastern Iowa. *U.S. Geol. Surv. Ann. Rep.*, **11**, 189–577.

McGregor, D. F. M. and Green, C. P. 1978. Gravels of the River Thames as a guide to Pleistocene catchment changes. *Boreas*, **7**, 197–203.

McGregor, D. F. M. and Green, C. P. 1983. Post-depositional modification of Pleistocene terraces of the River Thames. *Boreas*, **12**, 23–34.

Miall, A. D. 1977. A review of the braided river depositional environment. *Earth Sci. Rev.*, **13**, 1–62.

Miall, A. D. 1978. Lithofacies types and vertical profile models in braided river deposits: a summary. In: *Fluvial Sedimentology* (Ed. A. D. Miall). *Can. Soc. Petrol. Geol.*, **5**, 597–604.

Mitchell, G. F., Penny, L. F., Shotton, F. W. and West, R. G. 1973. *A Correlation of the Quaternary Deposits in the British Isles. Geol. Soc. Lond. Spec. Rep.*, No. 4.

Monckton, H. W. 1893. On the occurrence of boulders and pebbles from the glacial drift in gravels south of the Thames. *Q. J. Geol. Soc. Lond.*, **49**, 308–24.

Monckton, H. W., Mangles, H. A. and Salis, R. F. D. 1904. Excursions to the Farnham gravel pits on 23rd April and to brickfields and gravel pits at Dawley, between Hayes and West Drayton, on 30th April 1904. *Proc. Geol. Ass.*, **18**, 409–14.

Morris, J. 1838. On the deposits containing Carnivora and other mammalia in the valley of the Thames. *Mag. Nat. Hist.*, **2**, 539–46.

Morris, J. 1850. On the occurrence of mammalian remains at Brentford. *Q. J. Geol. Soc. Lond.*, **6**, 201–4.

Oakley, K. P. 1936. The Pleistocene deposits of the Iver district. *Antiq. J.*, **16**, 432–8.

Oakley, K. P. 1937. Field meeting at Taplow, Burnham and Iver, Bucks. *Proc. Geol. Ass.*, **48**, 276–9.

Oakley, K. P. 1952. Swanscombe Man. *Proc. Geol. Ass.*, **63**, 271–300.

Osborne, P. J. 1980. The insect fauna of the organic deposits at Sugworth and its environmental and stratigraphic implications. *Phil. Trans. R. Soc. Lond.*, **B 289**, 119–33.
Ovey, C. D. (Ed.) 1964. *The Swanscombe Skull.* A survey of research on a Pleistocene site. *R. Anthropol. Inst. Occas. Paper*, No. 20.
Owen, R. 1856. Description of a fossil cranium of the musk-buffalo from the 'lower level drift' at Maidenhead, Bucks. *Q. J. Geol. Soc. Lond.*, **12**, 124–31.
Peake, D. S. 1971. The age of the Wandle gravels in the vicinity of Croydon. *Proc. Croydon Nat. Hist. Scient. Soc.*, **14**, 145–76.
Peake, D. S. 1982. A reappraisal of the Pleistocene history of the River Wandle and its basin. *Proc. Croydon Nat. Hist. Scient. Soc.*, **17**, 89–116.
Penn, J. S. and Rolls, J. D. 1981. Problems in the Quaternary development of the Thames valley around Kingston: a framework for archaeology. *Trans. Lond. Middx. Archaeol. Soc.*, **32**, 1–12.
Phillips, L. M. 1974. Vegetational history of the Ipswichian/Eemian interglacial in Britain and continental Europe. *New Phytol.*, **73**, 589–604.
Pocock, T. I. 1903. On the drifts of the Thames Valley near London. 'Summary of progress for 1902'. *Geol. surv. G.B.*, Appendix VIII.
Pocock, T. I. 1908. The geology of the country around Oxford. *Mem. Geol. Surv. G.B.*
Potter, P. E. and Pettijohn, F. J. 1963. *Paleocurrents and Basin Analysis.* Springer-Verlag: Berlin, Heidelberg.
Poulton, S. B. 1880. On mammalian remains and tree trunks in Quaternary sands at Reading. *Q. J. Geol. Soc. Lond.*, **36**, 296–306.
Preece, R. C. and Robinson, J. E. 1982. Mollusc, Ostracod and plant remains from early postglacial deposits near Staines. *London Nat.*, **61**, 6–15.
Prestwich, J. 1856. Note on the gravel near Maidenhead in which the skull of the musk-buffalo was found. *Q. J. Geol. Soc. Lond.*, **12**, 131–3.
Prestwich, J. 1857. *The Ground Beneath Us; Its Geological Phases and Changes*, etc. J. van Voorst: London.
Prestwich, J. 1864. On the loess of the valleys of the south of England, and of the Somme and the Seine. *Phil. Trans. R. Soc. Lond.*, **154**, 247–309.
Prestwich, J. 1890. On the relation of the Westleton Beds or pebbly sands of Suffolk, to those of Norfolk and their extension inland, etc. *Q. J. Geol. Soc. Lond.*, **46**, 120–54.
Quinn, J. H. 1957. Paired river terraces and Pleistocene glaciation. *J. Geol.*, **65**, 149–66.
Rackham, D. J. 1978. Evidence for changing vertebrate communities in the Middle Devensian. *Quaternary Newsl.*, **25**, 1–3.
Robinson, J. E. 1980. The ostracod fauna of the interglacial deposits at Sugworth, Oxfordshire. *Phil. Trans. R. Soc. Lond.*, **B 289**, 99–106.
Roe, D. A. 1968a. Gazetteer of British Lower and Middle Palaeolithic sites. *Council Brit. Archaeol. Res. Rep.*, No. 8.
Roe, D. A. 1968b. British Lower and Middle Palaeolithic hand-axe groups. *Proc. Prehist. Soc.*, **34**, 1–82.
Roe, D. A. 1981. *The Lower and Middle Palaeolithic Periods in Britain.* Routledge and Kegan Paul: London.
Rose, J. 1979. River terraces and sea level change. *Brighton Polytech. Geogr. Soc. Mag.*, **3**, 13–30.
Ross, B. R. M. 1931. The physiographic evolution of the Kennet–Thames. *Rep. Brit. Assoc.*, London, p. 368.
Rust, B. R. 1972. Structure and process in a braided river. *Sedimentology*, **18**, 221–46.
Salter, A. E. 1896. 'Pebbly gravel' from Goring Gap to the Norfolk Coast. *Proc. Geol. Ass.*, **14**, 389–404.
Salter, A. E. 1905. On the superficial deposits of central and parts of southern England. *Proc. Geol. Ass.*, **19**, 1–56.
Sandford, K. S. 1924. The river gravels of the Oxford district. *Q. J. Geol. Soc. Lond.*, **80**, 113–79.
Sandford, K. S. 1926. Pleistocene deposits. In: *The Geology of the Country around Oxford* (Ed. J. Pringle). *Mem. Geol. Surv. G.B.*, pp. 104–72.
Sandford, K. S. 1954. River development and superficial deposits. In: *The Oxford Region* (Ed. A. F. Martin and R. W. Steel). University Press: Oxford, pp. 21–4.
Saner, B. R. M. and Wooldridge, S. W. 1929. River development in Essex. *Essex Nat.*, **22**, 244–50.
Sealy, K. R. and Sealy, C. E. 1956. The terraces of the Middle Thames. *Proc. Geol. Ass.*, **67**, 369–92.
Shackley, M. 1974. Stream abrasion of flint implements. *Nature*, **248**, 501–2.
Shephard-Thorn, E. R. and Wymer, J. J. 1977. *Southeast England and the Thames Valley.* Guidebook for excursion A5. X INQUA Congress.
Sherlock, R. L. 1924. The superficial deposits of south Buckinghamshire and south Hertfordshire and the old course of the Thames. *Proc. Geol. Ass.*, **35**, 1–28.
Sherlock, R. L. and Noble, A. H. 1922. The geology of the country around Beaconsfield. *Mem. Geol. Surv. G.B.*
Shotton, F. W. 1973. The English Midlands. In: *A Correlation of the Quaternary Deposits in the British Isles* (ed. G. F. Mitchell *et al.*). *Geol. Soc. Lond. Spec. Report*, No. 4.
Shotton, F. W. and Williams, R. E. G. 1971. Birmingham University radiocarbon dates V. *Radiocarbon*, **13**, 141–56.
Shotton, F. W. and Williams, R. E. G. 1973. Birmingham University radiocarbon dates VII. *Radiocarbon*, **15**, 451–68.
Shotton, F. W., Goudie, A. S., Briggs, D. J. and Osmaston, H. A. 1980. Cromerian interglacial deposits at Sugworth near Oxford, England, and their relation to the Plateau Drift of the Cotswolds and the terrace sequence of the Upper and Middle Thames. *Phil. Trans. R. Soc. Lond.*, **B 289**, 55–86.
Shrubsole, O. A. 1890. On the valley gravels about Reading, with especial reference to the Palaeolithic implements. *Q. J. Geol. Soc. Lond.*, **46**, 582–94.
Shrubsole, O. A. and Whitaker, W. 1902. Excursion to Reading. *Proc. Geol. Ass.*, **17**, 381–2.
Singer, R., Wymer, J. J., Gladfelter, B. G. and Wolff, R. 1973. Excavation of the Clactonian Industry at the golf course, Clacton-on-Sea, Essex. *Proc. Prehist. Soc.*, **39**, 6–74.
Smith, H. J. U. 1949. Physical effects of Pleistocene climatic changes in non-glaciated areas: eolian phenomena, frost action and stream terracing. *Bull. Geol. Soc. Am.*, **60**, 1485–1516.
Smith, R. A. 1917. Plateau deposits and implements. *Proc. Prehist. Soc. East Anglia*, **2**, 392–408.
Smith, R. A. and Dewey, H. 1913. Stratification at Swanscombe. *Archaeologia*, **64**, 177–204.
Smith, R. A. and Dewey, H. 1915. Researches at Rickmansworth: report on excavations made in 1914 on behalf of the British Museum. *Archaeologia*, **66**, 195–224.
Smith, W. G. 1894. *Man the Primeval Savage.* London.
Sperling, C. B. H., Goudie, A. S., Stoddart, D. R. and Poole, G. G. 1977. Dolines in the Dorset chalklands and other areas in southern Britain. *Trans. Inst. Brit. Geogr.*, new series **2**, 205–23.
Squirrell, H. C. 1974. The sand and gravel resources of the country around Gerrards Cross, Buckinghamshire. *Rep. Inst. Geol. Sci.*, No. 74/14.
Squirrell, H. C. 1976. The sand and gravel resources of the Thames and Kennet

Valleys, the country around Pangbourne, Berkshire. *Inst. Geol. Sci. Min. Assess. Rep.*, No. 21.
Squirrell, H. C. 1978. The sand and gravel resources of the country around Sonning and Henley. *Inst. Geol. Sci. Min. Assess. Rep.*, No. 32.
Stevens, J. 1882. On the earliest known traces of man in the Thames drift at Reading. *Trans. Berks. Arch. Architect. Soc.*, pp. 1–18.
Stuart, A. J. 1974. Pleistocene history of the British vertebrate fauna. *Biol. Rev.*, **49**, 225–66.
Stuart, A. J. 1976. The history of the mammal fauna during the Ipswichian/Last interglacial in England. *Phil. Trans. R. Soc. Lond.*, **B 276**, 221–50.
Stuart, A. J. 1977. British Quaternary vertebrates. In: *British Quaternary Studies* (Ed. F. W. Shotton). University Press: Oxford, pp. 69–81.
Stuart, A. J. 1980. The vertebrate fauna from the interglacial deposits at Sugworth, near Oxford. *Phil. Trans. R. Soc. Lond.*, **B 289**, 87–97.
Stuart, A. J. 1982. *Pleistocene Vertebrates in the British Isles.* Longman: London.
Stuart, A. J. 1984. Pleistocene occurrences of *Hippopotamus* in Britain. *Quatär-paläeontologie* (in press).
Stuart, A. J. and West, R. G. 1976. Late Cromerian fauna and flora at Ostend, Norfolk. *Geol. Mag.*, **113**, 469–73.
Sutcliffe, A. J. 1960. Joint Mitnor Cave, Buckfastleigh. *Trans. Proc. Torquay Nat. Hist. Soc.*, **13**, 1–26.
Sutcliffe, A. J. 1964. The mammalian fauna. In: *The Swanscombe Skull* (Ed. C. D. Ovey). *R. Anthropol. Inst. Occas. Paper*, **20**, 85–111.
Sutcliffe, A. J. and Kowalski, K. 1976. Pleistocene rodents of the British Isles. *Bull. Br. Mus. Nat. Hist. (Geol.)*, **27**, 33–147.
Sweeting, M. M. 1972. *Karst Landforms.* Macmillan: London.
Thomas, M. F. 1961. River terraces and drainage development in the Reading area. *Proc. Geol. Ass.*, **72**, 415–36.
Tomlinson, M. 1963. The Pleistocene chronology of the Midlands. *Proc. Geol. Ass.*, **74**, 187–202.
Treacher, L. 1896. Palaeolithic man in east Berkshire. *Berks, Bucks and Oxon. Arch. J.*, new series **2**, pp. 16–18, 39–43.
Treacher, L. 1904. On the occurrence of stone implements in the Thames Valley between Reading and Maidenhead. *Man*, **10**, 17–19.
Treacher, L. 1916. Excursion to Bourne End. *Proc. Geol. Ass.*, **27**, 107–9.
Treacher, L. and White, H. J. O. 1901. Excursion to Twyford and the Wargrave outlier. *Proc. Geol. Ass.*, **17**, 176.
Treacher, L. and White, H. J. O. 1909. Excursion to Maidenhead. *Proc. Geol. Ass.*, **21**, 198–201.
Treacher, L., Arkell, W. J. and Oakley, K. P. 1948. On the ancient channel between Caversham and Henley, Oxfordshire and its contained flint implements. *Proc. Prehist. Soc.*, **14**, 126–54.
Tricart, J. 1948. Méthode d'étude des terrasses. *Bull. Soc. Géol. France*, Ser. 5, No. 17, pp. 559–75.
Trimmer, W. K. 1813. An account of some organic remains found near Brentford, Middlesex. *Phil. Trans. R. Soc. Lond.*, **53**, 131–7.
Turner, C. 1970. The Middle Pleistocene deposits at Marks Tey, Essex. *Phil. Trans. R. Soc. Lond.*, **B 257**, 373–440.
Turner, C. 1983. Nettlebed interglacial deposits. In: *Diversion of the Thames Field Guide* (Ed. J. Rose). Quaternary Research Association: Cambridge, pp. 67–8.
Turner, C. and West, R. G. 1968. The subdivision and zonation of interglacial periods. *Eiszeitalter Gegenw.*, **19**, 93–101.
Tutin, T. G., Heywood, V. H., Burges, N. A., Valentine, D. H., Walters, S. M. and Webb, D. A. 1964–80. *Flora Europaea*, Vols. 1–5. University Press: Cambridge.
Ursic, S. J. and Dendy, F. E. 1965. Sediment yields from small watersheds under various land uses and forest covers. *Proc. Federal Inter-Agency Sedim. Conf. U.S.D.A. Misc. Publ.*, **970**, 47–52.
Waechter, J. d'A. and Conway, B. W. 1977. Barnfield pit, Swanscombe. In: *Southeast England and the Thames Valley* (Ed. E. R. Shephard-Thorn and J. J. Wymer). Guide book for excursion A5. X INQUA Congress, pp. 38–44.
Walder, P. 1967. The composition of the Thames gravels near Reading, Berkshire. *Proc. Geol. Ass.*, **107**, 107–19.
Washburn, A. L. 1968. Weathering, frost action and patterned ground in the Mesters Vig district, Northeast Greenland. *Medd. om Grønland*, No. 176.
Washburn, A. L. 1972. *Periglacial Processes and Environments.* Edward Arnold: London.
Watson, E. 1969. The slope deposits in the Nant Iago valley near Cader Idris, Wales. *Bull. Periglacjalny*, **18**, 95–113.
West, R. G. 1956. The Quaternary deposits at Hoxne, Suffolk. *Phil. Trans. R. Soc. Lond.*, **B 239**, 265–356.
West, R. G. 1957. Interglacial deposits at Bobbitshole, Ipswich. *Phil. Trans. R. Soc. Lond.*, **B 241**, 1–31.
West, R. G. 1968. *Pleistocene Geology and Biology.* Longmans Green: London.
West, R. G. 1969. Pollen analysis from interglacial deposits at Aveley and Grays, Essex. *Proc. Geol. Ass.*, **80**, 271–82.
West, R. G. 1980. *Pre-glacial Pleistocene of the Norfolk and Suffolk Coasts.* University Press: Cambridge.
West, R. G. and McBurney, C. M. B. 1954. The Quaternary deposits at Hoxne and their archaeology. *Proc. Prehist. Soc.*, **20**, 131–54.
West, R. G., Lambert, J. M. and Sparks, B. W. 1964. Interglacial deposits at Ilford, Essex. *Phil. Trans. R. Soc. Lond.*, **B 247**, 185–212.
Whitaker, W. 1860. Geology of Reading. *The Geologist*, **3**, 390–1.
Whitaker, W. 1864. The geology of parts of Middlesex, Hertfordshire, Buckinghamshire, Berkshire and Surrey. *Mem. Geol. Surv. G.B.*
Whitaker, W. 1889. The geology of London and parts of the Thames Valley. *Mem. Geol. Surv. G.B.*
White, H. J. O. 1896. On the distribution and relations of the Westleton and glacial gravels in Oxfordshire and Berkshire. *Proc. Geol. Ass.*, **14**, 11–23.
White, H. J. O. 1899. On the origin of the high level gravel with Triassic debris adjoining the valley of the Upper Thames. *Proc. Geol. Ass.*, **15**, 157–74.
White, H. J. O. 1902. On a peculiarity in the course of certain streams in the London and Hampshire Basins. *Proc. Geol. Ass.*, **17**, 399–413.
White, H. J. O. 1904. Excursion to Henley-on-Thames. *Proc. Geol. Ass.*, **18**, 414.
White, H. J. O. 1908. Scenery and superficial deposits. In: The geology of the country around Henley-on-Thames and Wallingford (Ed. A. J. Jukes-Brown and H. J. O. White). *Mem. Geol. Surv. G.B.*, 77–103.
Williams, P. F. and Rust, B. R. 1969. The sedimentology of a braided river. *J. Sedim. Petrol.*, **39**, 649–79.
Wintle, A. 1981. Thermoluminescence dating of Late Devensian loess in southern England. *Nature*, **289**, 479–80.
Wintle, A. and Huntley, D. J. 1982. Thermoluminescence dating of sediments. *Quat. Sci. Rev.*, **1**, 31–54.
Woodward, H. B. 1909. The geology of the London district. *Mem. Geol. Surv. G.B.*, 49–139.
Wooldridge, S. W. 1927. The Pliocene history of the London Basin. *Proc. Geol. Ass.*, **38**, 49–132.
Wooldridge, S. W. 1938. The glaciation of the London Basin and the evolution of the Lower Thames drainage system. *Q. J. Geol. Soc. Lond.*, **94**, 627–67.
Wooldridge, S. W. 1960. The Pleistocene succession in the London Basin. *Proc. Geol. Ass.*, **71**, 113–29.

References

Wooldridge, S. W. and Kirkaldy, J. F. 1936. River profiles and denudation chronology in England. *Geol. Mag.*, **73**, 1–16.

Wooldridge, S. W. and Linton, D. L. 1955. *Structure, Surface and Drainage in Southeast England*. George Philip: London.

Wooldridge, S. W. and Smetham, D. J. 1931. The glacial drifts of Essex and Hertfordshire and their bearing upon the agricultural and historical geography of the region. *Geogr. J.*, **78**, 243–69.

Wright, W. B. 1937. *The Quaternary Ice Age*. Macmillan: London.

Wymer, J. J. 1961. Lower Palaeolithic succession in the Thames Valley and the date of the ancient channel between Caversham and Henley. *Proc. Prehist. Soc.*, **27**, 1–27.

Wymer, J. J. 1968. *Lower Palaeolithic Archaeology in Britain*. John Baker: London.

Wymer, J. J. 1974. Clactonian and Acheulian industries in Britain. Their chronology and significance. *Proc. Geol. Ass.*, **85**, 391–421.

Wymer, J. J. (Ed.) 1977a. Gazetteer of Mesolithic sites in England and Wales. *Council Brit. Archaeol. Res. Rep.*, 20.

Wymer, J. J. 1977b. The archaeology of man in the British Quaternary. In: *British Quaternary Studies* (Ed. F. W. Shotton). University Press: Oxford. pp. 93–106.

Zagwijn, W. H. 1974. Les problèmes de la subdivision du Pleistocene moyen des Pays-Bas. *Bull. Ass. Français Étude Quat.* No. 40–41, 105–7.

Zagwijn, W. H. 1975. Variations in climate as shown by pollen analysis, especially in the Lower Pleistocene of Europe. In: *Ice Ages: Ancient and Modern* (Ed. A. E. Wright and F. Moseley). *Geol. J.*, **6**, 137–52.

Zagwijn, W. H., van Montfrans, H. M. and Zandstra, J. G. 1961. Subdivision of the Cromerian in the Netherlands: pollen analysis, palaeomagnetism and sedimentary petrology. *Geol. en Mijnbouw*, **50**, 41–58.

Zeuner, F. E. 1959. *The Pleistocene Period*. Hutchinson: London.

UNIVERSITY COLLEGE LIBRARY CARDIFF